MW01000275

THE HISTORY OF AMERICAN WEATHER

Early American Hurricanes 1492-1870

David M. Ludlum

American Meteorological Society
45 Beacon Street
Boston 8, Massachusetts

Copyright 1963 by David M. Ludlum
of Princeton, New Jersey
Second printing, 1989
Cover design by Eric Sloane
of Cornwall Bridge, Connecticut

Printed by Lancaster Press, Inc.,
of Lancaster, Pennsylvania

Published and sold by the
American Meteorological Society
45 Beacon Street, Boston 8,
Massachusetts

Permission to use figures, tables, and **brief** excerpts from this publication in scientific and educational works is hereby granted, provided the source is acknowledged. All rights reserved. No part of this publication may be reproduced, stored in a retrieval system, or transmitted in any form or by any means, electronic, mechanical, photocopying, recording, or otherwise, without the prior written permission of the publisher.

ISBN 0-933876-16-5

The design of THE HISTORY OF AMERICAN WEATHER is to relate the facts concerning the development of the science of meteorology in the Americas and to describe the principal weather events in our climatic past by means of a series of historical monographs.

Early American Hurricanes 1492–1870
Early American Tornadoes 1586–1870
Early American Winters 1604–1820
Early American Winters, II 1821–1870

The task of assembling the record of our early weather history is a monumental one due to the multiplicity and diversity of sources. All references to the location of meteorological records or descriptions of weather events, either in manuscript or in printed form, will be most welcome. Address: Historical Monograph Editor, American Meteorological Society, 45 Beacon Street, Boston 8, Massachusetts.

In Honor of

WILLIAM C. REDFIELD (1789–1857)
of New York City,
who initiated the scientific
investigation of the American hurricane.

CONTENTS

FOREWORD

The purpose of this work is to set down in chronological order, as far as available historical sources permit, the meteorological situations attending the occurrence of hurricanes prior to 1870 which have either closely approached or actually crossed the Atlantic and Gulf coastlines of the present United States.

There has been very little written about the historical past of the American climate possessing real scientific significance for the present-day meteorologist. In the field of hurricanes, Sidney Perley's *Historic Storms of New England* (1891) and Marjory Douglas' *Hurricane* (1958) do treat the subject in full-scale from the historical view, but their approaches are from the human interest story. Neither contributes much toward adequate meteorological descriptions of individual hurricanes nor supplies a critical chronology of occurrences.[1]

Several regional studies for a particular area have been compiled, such as the Texas coast, the Florida peninsula, the Charleston area, and the Outer Banks of North Carolina, but these present only a restricted regional view, usually lack the desired historical depth for the early period, and are often scanty as to meteorological details.[2] The standard works of Ivan Ray Tannehill (1938–55) and of Gordon Dunn and Banner Miller (1960) are admirable volumes whose chief emphasis concerns a description of the physical nature of tropical storms. Of necessity, they relegate chronologies to a minor role in an appendix.[3]

This study has been divided into two periods. The first section extends from the voyages of Columbus, through the years of exploration and colonization, to the end of 1814. This year makes a convenient breakpoint, for the first Federal attempts to institute a national weather observing service date from the last full year of the War of 1812 and the Napoleonic upheavals. Thereafter, the task of the meteorological historian is greatly assisted by the availability of a growing volume of nationwide observational data and by a growing awareness among scientific men of the weather problem, especially in regard to safe navigation at sea.

With the cessation of hostilities in 1815, the nations of the world entered on an extended era of peace for the first time in a century. An exciting period followed when the ships of all nations could ply the West Indian and American waters freely with only the weather elements as a major adversary. The second section of this study encompasses these great years of the American merchant marine, one of whose most significant developments brought an increasing recognition of the threat to safe navigation posed by tropical storms. There followed an intensive study of the hurricane. Many a ship and crew were saved from a watery grave by the practical application of this new knowledge. Rules for prudent navigation in hurricane situations were composed, widely circulated among mariners, and ultimately incorporated into all manuals of sailing instructions.

The second section of this study, spanning this period, carries through the Civil War down to 1870, the year the U. S. Signal Corps established its storm warning system and ringed the American coastline with modern weather observing stations. Again, the historian's task is changed, since his search for material can now be concentrated on the great mass of readily available printed meteorological data that has poured from the presses in every year since 1870.

Two deviations have been made from the prescribed geographical limits. The stage has been set with a brief introduction treating the weather experiences of Christopher Columbus in the New World, since he was the first European to encounter and to describe a true West Indian hurricane. A later section deals with the outstanding hurricane season of 1780 in the West Indies when three remarkable October storms, successively sweeping the shipping lanes of the Gulf, Caribbean, and western Atlantic, caught many warring vessels at sea, with important consequences following for the military fortunes of the contending parties in the American War of Independence.

In compiling and arranging the account of individual hurricanes, an effort has been made to present all significant meteorological data available. These have been set down, often at length and with full quotation, with a view of serving as a starting place for the local historian interested in reconstructing the story of the weather past of his own locality or region. Lengthy descriptions of local damage not pertinent to the meteorological description of a storm have generally been omitted, as have many interesting accounts of moving human experiences in these often catastrophic events. In the case of the Great Hurricane of 1752 at Charleston, however, the complete local press coverage has been reproduced to show at firsthand how a colonial editor handled his biggest weather news story of the century.

References to source material have been placed at the

end of each case to assist the historian who wishes to delve deeper into local hurricane situations, and a bibliography of pertinent American publications on the subject will be found at the end. A series of local and regional studies of hurricanes and other meteorological events is greatly needed—only if we know the extremes of past violence of the elements can we prepare to cope with future onslaughts of nature.

PREVIOUS EARLY HURRICANE CHRONOLOGIES

Aside from several novel accounts attempting to describe the nature of a single hurricane as observed locally, serious study of the occurrence, frequency, and areal distribution of tropical storms in American waters dates from 1831. In the July issue of the *American Journal of Science and the Arts,* William C. Redfield published the first of his meteorological papers: "Remarks on the prevailing storms of the Atlantic Coast, of the North American States." [4] He propounded the hypothesis that hurricanes possess a rotary wind system circulating around a definite center and the whole mass of whirling winds exhibit a regular progressive motion. In August of the same year the Great Barbados Hurricane of 1831 occurred, and this event brought Lt. Col. William Reid of the Royal Engineers to the scene. Reid was charged with the task of reconstructing the buildings and facilities of the devastated island; and from his studies of the extreme physical violence and enormous destruction of that famous storm, he went on to a lifetime concern with probing the nature of the hurricane.

For the next quarter of a century Redfield and Reid carried on a friendly, enlightened correspondence, exchanging freely the fruits of their researches in storm dynamics and commenting on many related scientific matters of the day. Redfield's theories on hurricane structure and movement were developed in a series of papers or brief monographs published between 1831 and 1857, mainly in the *American Journal of Science and the Arts* and occasionally in British and American marine journals.

Though he never published a formal chronology of hurricane occurrences, Redfield did maintain a manuscript "Record Book of Storms" in which he jotted down notes and pasted clippings about major storm events both past and contemporary. During his lifetime he generously shared these notes with fellow students of the subject, and after his death the Redfield family deposited his manuscript material, appropriately, in the Yale University Library for the future use of scholars.

At the time Redfield commenced his studies and for several years following, he was unaware of the previous suggestions of the existence of a vortex at the center of a tropical storm circulation which had been made by James Capper in India and by William Dunbar in Louisiana, oddly enough both announced in the same year—1801. Nor was he aware for some years of the contemporary work of Prof. Heinrich W. Dove in Germany who correctly described wind movement around an extra-tropical storm system. Redfield's just fame rests on his laborious assembly of data from ship logs, the press, and interviews with seamen, and on his subsequent orderly reduction of these to proper time and place so that the great mass of data could be first mapped and then analyzed. He was able to ascertain the region of origin, the path of advance, and some of the peculiarities of movement of tropical storms. Throughout his studies, Redfield insisted that one must first determine *what* a hurricane was before the question as to *why* and *how* it had developed could be answered. It was in the description of hurricane movement that Redfield made his significant contribution to meteorology.

Colonel Reid also took up the study of individual hurricanes in order to arrive at certain conclusions concerning their structure and behavior. His first published material appeared in 1838 in *The Law of Storms,* a collection of chapters describing storm occurrences in both the West Indies and the Far East.[5] Reid did not compile a formal listing of hurricanes and typhoons. Rather he presented detailed studies of certain storms, primarily those occurring in the year 1780 and in the decade of the 1830's, by an exhaustive examination of ship logs. From these he drew conclusions as to the rotary circulation and progressive motion of these storms that were in close agreement with the findings of Redfield.

There were others contemporary with Redfield and Reid who as historians or geographers attempted to compile extensive lists of hurricanes. The focus of their view, however, was almost exclusively confined to the West Indies and they gave few references to the United States. M. Moreau de Jonnes in 1822 listed 63 hurricanes in the period 1495 to 1821.[6] A British naval lieutenant, "Stormy Jack" Evans, brought out a tabulation totaling some 70 cases in 1848.[7] A more extensive chronological list for the period 1494–1846 was compiled by the historian of Barbados, Sir Robert H. Schomburgk, and contained some 127 storms which had done damage in the West Indies.[8] Lastly, Prof. Alexander Keith Johnston of Edinburgh also listed 127 hurricanes in the second edition of his famous *Physical Atlas* in 1855.[9]

The basic work of Redfield and Reid probably inspired Andres Poey y Aguirre of Havana to compile his classic list of all hurricanes known to history. In pursuing his researches the young Cuban made a trip to New York City where Redfield opened all his notes and compilations. Poey also visited Europe where he consulted with Dove and Buys Ballot, two leading meteorologists there. Poey's work was first published in *The Journal of the Royal Geographical Society* of London in 1855.[10] It contained a chronological table of 400 "cyclonic hurricanes" plus a listing of some 450 books or articles relating to the storm problem. The work was translated into French and republished at Paris in 1862. An expanded edition of the bibliography listing 1,008 publications on the subject appeared in the *Annales Hydrographiques* in 1865 and was issued as a separate in the following year.

At this time in Washington, Lorin Blodget was carrying on extensive researches which lead to the publication in 1857 of his impressive *Climatology of the United States*.[11] Among its 536 pages was a "list of Hurricanes on the Coast of the South Atlantic States, and on the North Coast of the Gulf of Mexico," wherein were arranged 55 American storm notices with brief data on each. Some of these were major hurricanes.

Another exclusively American mainland list came from the labors of Prof. Increase Lapham of Wisconsin. He was associated closely with the formation of the U. S. Army Signal Corps storm-warning service. As one of his duties he surveyed all available meteorological literature and presented in 1872 a 6-page listing of storms of all types known to have affected the American continent.[12] He uncovered several hurricane occurrences not found in previous lists.

Prof. Elias Loomis of Yale subjected the new weather maps of the Signal Corps to intensive analysis in the 1870's and 1880's. In one his papers published over the years in the *American Journal of Science,* he enumerated 30 hurricanes which had approached the American mainland in the period dating back to 1815.[13]

Aside from Perley's work on New England storm history, students of early American hurricanes seem to have passed from the scene for a quarter of a century. True, there were studies of contemporary storms or those of very recent occurrence, notably by Frank Bigelow (1897), Edward B. Garriott (1900), William H. Alexander (1902), Oliver L. Fassig (1913), Charles L. Mitchell (1924 and 1928), and Isaac Cline (1926).[14]

Interest in the historical background of the hurricane problem did not revive until the great Florida storms of the middle and late 1920's demanded a more exact knowledge of the past frequency of these destructive visitations. At this time F. G. Tingley of the Climatology Section of the U. S. Weather Bureau commenced his monumental task of compiling a file of notes and references to all known hurricane occurrences. He charted their courses on individual annual maps for the years from 1493 to 1930. Though never published, the mass of manuscript material has been preserved in the Marine Section, Office of Climatology, U. S. Weather Bureau, as the greatest single source of references to original hurricane data. Two co-workers in the Weather Bureau, however, have made good use of Tingley's labors by publishing useful American hurricane listings, though neither appears to have given Tingley credit for his basic spadework.

In 1929, Prof. A. J. Henry of the Weather Bureau presented in the *Monthly Weather Review* an article: "The Frequency of Tropical Cyclones that closely approach or enter Continental United States," a list of 79 hurricanes which had affected the American coastline.[15] A decade later Ivan Ray Tannehill brought out the first edition (1938) of his useful and interesting reference work: *Hurricanes*.[16] This contained a "List of Tropical Storms of the North Atlantic, including the Gulf of Mexico and Caribbean Sea, from 1494 to 1900" and was based on Tingley's data plus a few new entries. Addenda to the original list in subsequent editions brought the account down to 1955.

The most recent compilation has been contributed by Gordon E. Dunn and Banner I. Miller of the National Hurricane Research Center at Miami, Florida, in their *Atlantic Hurricanes* (1960).[17] "Hurricanes Affecting the United States, by Sections" forms an appendix to this authoritative book. Attention is confined to those hurricanes which moved close to the shoreline of the present United States from Brownsville, Texas, to Eastport, Maine. The following are the number of pre-1815 storms listed: New England (8), Middle Atlantic States (4), South Atlantic States (21), Florida (13), Louisiana-Mississippi-Louisiana (17), and Texas (1). Dunn and Miller have eliminated a number of questionable storms from the Poey and Tannehill listings, and properly they cautiously include several with question marks, indicating that no reference to the historical source came to light in their researches.

THE TERM: HURRICANE

Perhaps at the outset it would be well to define our subject—hurricane. The following definitions appear in the *Glossary of Meteorology* published by the American Meteorological Society in 1959:

* **Hurricane**—(Many variant spellings.) A severe *tropical cyclone* in the North Atlantic Ocean, Caribbean Sea, Gulf of Mexico and in the Eastern North Pacific of the west coast of Mexico. For more complete discussion, see *tropical cyclone.*

Tropical Cyclone—The general term for a cyclone that originates over the tropical oceans. At maturity the tropical cyclone is one of the most intense and feared storms of the world; winds exceeding 175 knots (200 mph) have been measured, and its rains are torrential. . . .

By international agreement, tropical cyclones have been classified according to their intensity, as follows: (a) tropical depressions, with winds up to 34 knots; (b) tropical storm, with winds of 35 to 64 knots; and (c) *hurricane* or *typhoon*, with winds of 65 knots or higher.

Glossary of Meteorology. (Ralph E. Huschke, ed., 1959), 286, 593.

A judgment as to the intensity of an early American tropical storm, whether of full hurricane strength or not, must be mainly subjective as instruments to measure wind speed did not come into general use until the 1860's and 1870's. In general, wind force of sufficient strength to tear up trees by the roots, to cause major structural damage to buildings, or to dismast or put a ship on its beam-ends is considered sufficient evidence to meet the requirements of hurricane wind force. Whether the wind actually exceeded 65 knots (73 mph) limit must be speculative. A local distinction between storms, whether hurricanes or tropical storms, must rest mainly on the amount and type of damage caused. Admittedly, this method is subject to serious error since a particular locality might only be on the periphery of a strong hurricane and not experience the full wind fury. But the lack of a widespread observational system, communications, and daily weather maps precludes a more scientific approach.

A serious effort has been made to eliminate from the text and chronology all storms not of tropical origin since all previous lists contain one or more occurrences of land tornadoes, waterspouts, line squalls, and extra-tropical storms, largely a result of the prevailing loose usage of meteorological terminology over the years. Only within the last fifteen years has the American press been educated to the correct, restricted meaning of *hurricane*.

DATES

During our period of study two calendar systems were employed in the British colonies in North America. It is, therefore, necessary to correct dates from the Julian Calendar (Old Style) to the Gregorian Calendar (New Style) so that meteorological events will be attributed to their proper day and climatological season in accordance with our present method of reckoning. In the English-speaking world, the change took place after 2 September 1752 (Old Style). The next day was designated 14 September (New Style), eleven days being skipped to bring the position of the Sun and man's calendar into agreement. It is interesting to note that the Great Hurricane at Charleston occurred on 15 September 1752, the day following the change. No doubt some local commentator attributed the catastrophe to meddling with the calendar!

For the period prior to 3 September 1752, all dates in the text have been given in a double form, such as 12 (21) October 1492, in order to eliminate doubt as to which system was employed. This may appear a bit awkward to the general reader, but it is deemed essential for both the historian and the meteorologist to make the record correct according to our present climatological calendar. In the footnotes only, references to newspaper datelines prior to 1752 appear as they were actually printed in old style.

For the years from 1400 through 29 February 1500, nine days have been added to the Julian dates—from 1 March 1500 to 18 February 1700, ten days have been added—and from 19 February 1700 to 3 September 1752, eleven days. Concisely, in the XV Century, nine days, in the XVI and XVII, 10 days, and for the first half of the XVIII, eleven days have been added.

GUIDES TO THE SOURCES

A brief mention is in order as to the basic reference works employed. Probably the main reason why no historical volume has appeared previously about the early American weather scene from the national perspective lies in the enormity of the task of searching out the meteorological facts that are scattered throughout such a diversity of printed works and manuscript sources. To do a complete job of research in this field one should literally survey every document extant on American history. But in recent years the historian's task has been greatly eased by compilations of bibliographies of two types of sources vital to our particular subject of the weather. Without these, the time consumed in the search would have been multiplied many times, and the end result would have certainly been less complete and satisfactory.

The contemporary newspapers of our period offer, by far, the best source. Some of these contain actual meteorological observations, and most of those along the seaboard were edited by weatherwise writers who usually presented the essential facts about a storm. The first American periodical to appear on a regular basis was issued at Boston in 1704. Since then the present location of surviving files of newspapers for our first period will be found in the magnificent compilation of Charles S. Brigham: *History and Bibliography of American Newspapers, 1690–1820* (2 vols., Worcester, Mass., 1947). For our second period Winifred Gregory's *American Newspapers, 1821–1936* (New York, 1937), though less detailed, is extremely helpful.

The second major source for weather material lies in personal diaries. Here there are two worthwhile guides. William Mathews: *American Diaries: An Annotated Bibliography of American Diaries prior to . . . 1861* (Berkeley, 1945) mentions whether the diaries contain references to weather events. H. M. Forbes: *New England Diaries, 1602–1800* (Topsfield, Mass., 1923) is useful for a restricted area and period.

The admirable *Harvard Guide to American History* (Cambridge, Mass., 1954) comprises another recent compendium of references to our historical past that speeds one's way to pertinent sources and secondary works. To keep dates and events in their proper order and place, the *Encyclopedia of American History,* Richard B. Morris, ed. (New York, 1953) proves a trusted companion.

Finally, the advances in library techniques through microfilming and photostating of distant manuscripts and newspaper files permit one now to study a wealth of meteorological data never before assembled in one place. A great mass of pre-1870 weather records on micro-film is located at the National Weather Records Center at Asheville, North Carolina, with duplicate copies in the National Archives in Washington. There is an adequate guide to these: *List of Climatological Records in the National Archives,* Lewis J. Darter, Jr., ed. (Washington, 1942). Also helpful in the early period for locating manuscript weather records and journals is *An Annotated Bibliography of Meteorological Observations in the United States, 1731–1818.* James M. Havens, ed. (Washington, 1956).

None of the above, however, would have been of much avail without the close proximity of a first-class reference collection as afforded by the Firestone Library of Princeton University where the major part of this research was pursued.

[1] Sidney Perley. *Historic Storms of New England.* Salem, Mass., The Salem Press, 1891. 341 p.
Marjory Douglas. *Hurricane.* New York, Rhinehart, 1958. 393 p.

[2] W. Armstrong Price. *Hurricanes Affecting the Coast of Texas from Galveston to Rio Grande.* Technical Memorandum No. 78. Beach Erosion Board, Corps of Engineers. Washington, 1956. 52 p.
Analysis of Hurricane Problems in Coastal Sections of Florida. Survey Report. Jacksonville, Fla., U. S. Army Engineer District, 1961. Appendix A.
John C. Purvis. *Notes on Hurricanes in South Carolina.* Wash., Weather Bureau, 1956, 12 p.
C. B. Carney and A. V. Hardy. *North Carolina Hurricanes.* Wash., Weather Bureau, 1962, 26 p.

[3] Ivan Ray Tannehill. *Hurricanes.* Princeton, N. J., Princeton University Press, 1938. 304 p.
Gordon E. Dunn and Banner I. Miller. *Atlantic Hurricanes.* Baton Rouge, La., Louisiana State University Press, 1960. 326 p.

[4] William C. Redfield. Remarks on the prevailing storms of the Atlantic Coast, of the North American States. *American Journal of Science and the Arts* (New Haven). 20, July 1831, 17–51.

[5] William Reid. *An attempt to develop the Law of Storms by means of facts, arranged according to place and time; and hence to point out a cause for the variable winds, with a view to practical use in navigation.* London, John Weale, 1838. 431 p.

[6] M. Moreau de Jonnes. *Histoire physique des Antilles Francaises.* Paris, 1822. 1, 346.

[7] "Stormy Jack" Evans. A chronological list of hurricanes which have occurred in the West Indies since the year 1493; with interesting descriptions. *Nautical Magazine* (London). 1848. 397, 453, 524.

[8] Robert H. Schomburgk. *The History of Barbados.* London, 1848.

[9] Alexander Keith Johnston. *The Physical Atlas of Natural Phenomena.* Edinburgh, 1855.

[10] Andres Poey y Aguirre. A chronological table comprising 400 cyclonic hurricanes which have occurred in the West Indies and in the North Atlantic within 362 years, from 1493 to 1855, with a bibliographical list of 450 authors, books, etc. and periodicals where some interesting accounts may be found, especially on the West and East Indian hurricanes. *The Journal of the Royal Geographical Society* (London). 25, 1855, 291–328. Also translated into French: Paris, Paul Dupont, 1862, 49 p.
Andres Poey y Aguirre. Bibliographie cyclonique: catalogue comprenant 1,008 ouvrages, brochures et ecrits qui ont paru jusqu'a ce jour sur les ouragans et les tempetes cyclonique. *Annales Hydrographiques* (Paris). 1865. Published separately: Paris, Paul Dupont, 1866. 96 p.

[11] Lorin Blodget. *Climatology of the United States, and the temperate latitudes of the North American Continent.* Philadelphia, J. B. Lippincott & Co., 1857. 397–403.

[12] Increase A. Lapham. List of the great storms, hurricanes and tornadoes of the United States (1635–1870). *Journal of the Franklin Institute* (Philadelphia). 63, March 1872, 210–216.

[13] Elias Loomis. *Amer. Jour. Sci.* 3d ser., 12–67, July 1876, 15.

[14] Frank H. Bigelow. *Storms, Storm Tracks and Weather Forecasting.* U. S. Dept. of Agriculture, Weather Bureau, Bull. 10. Wash., 1897. 87 p.
Edward B. Garriott. *West Indian Hurricane.* Bulletin H. Washington, Weather Bureau, 1900.
William H. Alexander. *West Indian Hurricanes.* Bulletin 32. Washington, Weather Bureau, 1902. 79 p.
Oliver L. Fassig. *Hurricanes of the West Indies.* Bulletin X. Wash. Weather Bureau, 1913. 28 p.
Charles L. Mitchell. West Indian Hurricanes and other tropical cyclones of the North Atlantic Ocean. *Monthly Weather Review. Supplement No. 24.* Washington, Weather Bureau, 1924. 47 p.
Isaac M. Cline. *Tropical Cyclones.* New York, Macmillan & Co., 1926. 301 p.

[15] Alfred J. Henry. The frequency of tropical cyclones (West Indian Hurricanes) that closely approach or enter continental United States. *Monthly Weather Review.* 57, Aug 1929, 328–332.

[16] Ivan Ray Tannehill. *Hurricanes.* 9th ed. Princeton, N. J., Princeton Univ. Press, 1956. 308 p.

[17] Dunn and Miller. *Atlantic Hurricanes.* 313–318.

COLUMBUS

FIRST VOYAGE 1492–93

There is no better starting place for a discussion of early American hurricanes than in the New World experiences of Christopher Columbus, for it was on his Second Voyage in September 1494 that a European first encountered and described a true West Indian storm. Since then much has gone into print concerning the experiences of the Genoese sailor with hurricanes, though most of this writing shows little discernment as to what constitutes a hurricane, certainly not in accord with present-day usage of the term. Also a great deal of confusion and contradiction as to the dating of these meteorological events have appeared, even in our best authorities.

In recent years several writers have treated the subject—Tannehill (1938), Brooks (1940), Morison (1942), and Douglas (1958).[1] Ivan Tannehill of the U. S. Weather Bureau largely followed Andres Poey's compilation of 1855, with considerable updating by Tingley, in his listing of Columbian storms, though some important corrections appeared in later editions. The late Dr. Charles F. Brooks of Blue Hill Observatory, an experienced meteorologist, worked in collaboration with Prof. Samuel E. Morison, the eminent historian and skilled mariner, and the researches of these Harvard colleagues does much to set the record straight. Miss Douglas, writing in a more popular vein, followed Morison faithfully without adding anything new in this regard.

The outstanding meteorological fact of the First Voyage is simply that no hurricanes or severe storms were encountered in the West Indies despite the fact that the fleet of three small vessels traversed an area of tropical storm activity at the season of their most frequent occurrence.

Columbus saw the last of the Canary Islands on 9 (18) September 1492. Our best authority for the First Voyage is Bishop Las Casas of Hispaniola who wrote his *Historia de las Indias* about 1550–63 and appears to have studied a copy or abstract of the original, now long-lost, personal log of the Admiral.[2]

The only semblance of a storm on the outward leg occurred on 23 September (2 Oct.). Las Casas has Columbus remark that "the sea made up considerably, and without wind, which astonished them." Very probably this was the swell of a disturbance several hundred miles distant, but no mention of the direction from which the swell came was indicated.[3]

The highest winds on the outward voyage occurred on the day preceding the landfall when northeast trades mounted in the evening to near-gale force and drove the ships swiftly along at a nine-knot clip to their historic landing on Watling Island, or San Salvador as Columbus called it, in the eastern Bahama group. Morison tells us that the sighting of land at 0200 on the 12th (21st) took place under moonlight with planets and constellations visible.

One must conjecture at this time what the course of history in the West Indies might have been if, in the autumn of 1492, a full-blown tropical storm had dashed the frail craft of the Admiral's fleet to the bottom of the sea or flung them shipwreck on some tiny cay. The Old World might have long remained in doubt as to the fate of these sailors and further expeditions for a time might have been discouraged, thus continuing the ignorance of the New World for many more years. But, as we shall see, Columbus was favored by good weather during most of the days of the First Voyage to the "Indies," and in his few encounters with foul weather God intervened to preserve him and his ships. In addition to this dual blessing, he was a skilled mariner, an accomplished navigator for his period, and he learned the weather ways of the new waters in due time.

The subsequent early winter cruise of the *Nina, Pinta,* and *Santa Maria* through the southeastern Bahamas and along the northern coasts of eastern Cuba and Hispaniola were without major weather incident also, an abnormal situation which led Columbus into an early misconception as to the meteorological character of the Antilles. "In all the Indies, I have always found weather like May," he wrote to friends in Spain.[4] And Las Casas further comments: ". . . in the Indies he navigated all that winter without dropping anchor, and had always good weather, and not for a single hour did he find the sea such as he would have been unable to navigate."[5] It should be noted here that the wreck of the *Santa Maria* on a reef early Christmas morning of 1492 was not caused by unfavorable weather conditions.

The return voyage to Spain in the two remaining ships commenced on 16 (25) January 1493 from the northern coast of Hispaniola. By steering a northeasterly course for 18 days, Columbus reached the 35°N latitude of Cape Hatteras and Gibraltar where the

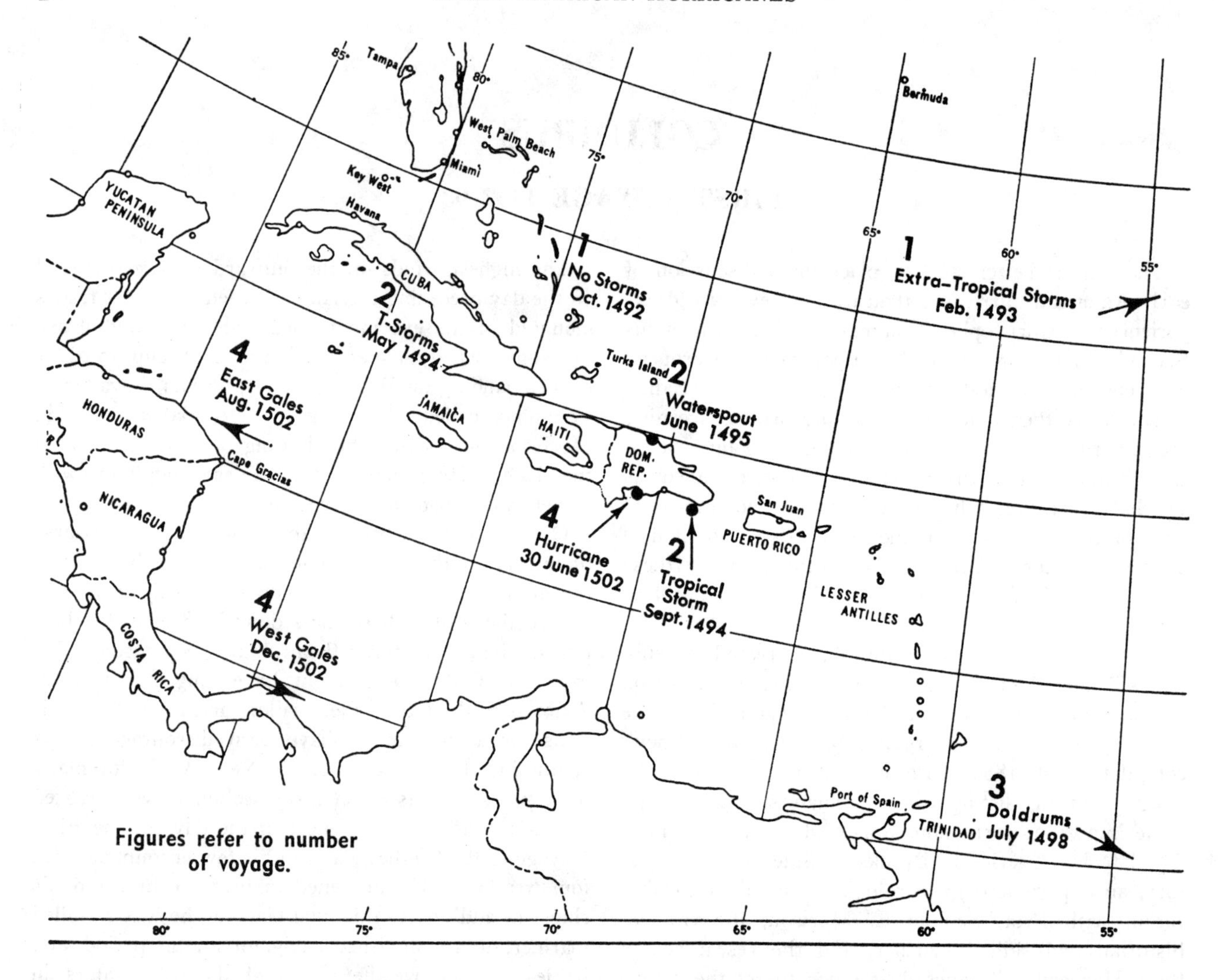

westerlies carried the ships homeward at an average of nearly 100 miles per day. The Admiral's weather luck, however, deserted him when approaching the usually smooth and navigable waters surrounding the Azores. In this part of the Atlantic, once or twice in a normal winter season, the atmospheric circulation pattern takes a radical change as a polar trough of low pressure extends southward, accompanied by clashing air masses, to replace the serene atmosphere of clear skies and light winds usually associated with the semi-permanent Azores high pressure system. On these occasions violent cyclonic storms may generate at low latitudes near the Azores, develop into gigantic vortices, and then whirl eastward to batter the Iberian and North African coastlines.[6]

Dr. Charles F. Brooks has translated the wind observations from the First Voyage log into a modern weather analysis and constructed simplified weather charts for the period. Dr. Brooks concluded that the two storms which Columbus encountered on 12–15 (21–24 February) and on 27 February–3 March (8–12 March) 1493 were not actually hurricanes, as previous authors had so often assumed. Rather these represented well-developed examples of extra-tropical cyclonic storms found in the Azores when the above described trough conditions exist.[7] Nevertheless, the legend that Columbus encountered a unique mid-winter hurricane on his First Voyage dies a hard death. Recently, when a mid-winter storm developed in the Leeward Islands, the press quoted misinformed meteorologists to the effect that such a phenomenon had not occurred since the time of Columbus.

Dr. Brooks has shown how the last storm overtook the *Nina* and held her in its grasp for nearly six days during which three distinct sections of the storm passed over the ship. Mounting southeast to south gales with heavy seas continued for three days as the frontal system of the storm approached. On the night of 1–2 (10–11) March the warm front overtook the ship and the wind went into the southwest in the warm air. This proved only a prelude to the passage of the cold front the next night which brought northwest squalls of whole gale strength. March 3d, the last day before sighting land, was the worst day of the entire voyage as the small ships, now long at sea and in need of complete overhauls, struggled with the northwest gale and the cross swell. But land was sighted that night and a safe landing made on 4 (13) March 1493.[8]

Certainly, Columbus should have considered himself lucky during his first trip under two meteorological scores: (1) He did not meet a tropical storm when in American waters, (2) In his one tough weather experience, he found himself in the southerly sector of the storms which generated north of the Azores. This is usually the less tempestuous part of such a storm. The world might never have known of his discoveries and honored the name of Columbus if his navigation had taken him a couple of hundred miles farther north where he would have encountered the easterly and northeasterly gales found in the eastern sector. These might have driven him much closer to the center of this exceptionally well-developed disturbance where the shifting gales would have been of even greater force.

[1] Ivan Ray Tannehill. *Hurricanes Their Nature and History*. Princeton, 1938. (9th ed., 1956.)
Charles F. Brooks. Two Winter Storms Encountered by Columbus in 1493 near the Azores. *Bulletin Amer. Meteor. Soc.* 22, 1941, 303–309.
Samuel E. Morison. *Admiral of the Ocean Sea A Life of Christopher Columbus.* Boston, 1942.
Marjory S. Douglas. *Hurricane.* New York, 1958.

[2] Morison, I, 67–72, discusses the sources for the First Voyage.

[3] Cecil Jane. *The Voyages of Christopher Columbus.* London, 1930. 142.

[4] Morison, I, 297–298.

[5] Jane, 252.

[6] Hans U. Groening. Azorentiefs . . . (The Azores lows). *Meteorologische Abhandlungen.* 5–1, 1957.

[7] Brooks, 306–307.

[8] Ferdinand Columbus. *The Life of the Admiral Columbus.* (1571). Benjamin Keen, trans. & ed. New Brunswick, 1959. 97–98.

SECOND VOYAGE 1493–96

After only a six-months stay at home Columbus set sail again for the New World from Cadiz on 25 September (4 October) 1493 with a grand fleet of 17 ships and some 1200 men. After a stop in the Canary Islands early in October, a landfall was made at Dominica in the Leeward Islands on 3 (12) November. The late-season arrival in the West Indies proved fortunate for it was close to the tailend of the normal hurricane period. The only meteorological experience of note on the outward voyage centered around a thundersquall of considerable violence which struck during the night on 26 October (4 November). The four-hour blow split sails and broke a few spars. During the electrical disturbance St. Elmo's fire, always fascinating to mariners prone to superstition, played around the masts, but by sunrise the wind had moderated and the sea appeared "smooth as polished marble." [1] This relatively brief storm certainly could not be classed as a hurricane. Such squally areas are usually associated with easterly wave situations which regularly move westward in the general east-to-west circulation at this time of year. Sometimes easterly waves can trigger a hurricane development when circumstances are favorable, but more often result in merely thundersqualls or rain showers which are of much less consequence to mariners.

The principal exploration cruise of the Second Voyage carried Columbus to Jamaica and along almost the entire southern coast of Cuba in the late spring and early summer of 1494. There were several weather incidents on the cruise which Poey, following others, included in his list of hurricanes, namely on 19–21 (28–30) May and 16 (25) June (July?). Bishop Las Casas, however, calls these thunderstorms, and Morison from personal experience in sailing the area points out the frequency of violent squalls and thunderstorms along the Cuban coast at this season.[2]

The first squall encounter occurred in mid-May (close to the 19th (28th) of May when the moon was full), and the second on the 16th (25th) of July. Either Poey or his printer erred in making this read June. The second squall almost ended the expedition, as Las Casas relates: "among many things that he suffered was a thundersquall so sudden, horrible and perilous that it threw the flagship on her beam-ends." Such a sudden blow shows that there were a variety of weather menaces to safe navigation in the realm of the hurricane.[3]

On the return to Hispaniola the fleet encountered what Morison has described in translation as a "September gale." The Admiral had been forewarned of the imminent storm by the appearance of a sea monster on the surface of the ocean, a sure sign of foul weather to come according to ancient mariners. In view of this, and probably backed up by a glance at an increasingly menacing cloud cover, Columbus took heed and anchored his fleet behind Saona Island, the bit of land just off the southeastern tip of Hispaniola. There his ships, protected from all quarters but the east, rode out the blow in safety.[4]

John Boyd Thatcher, a voluminous student of Columbus' ways, states that the fleet arrived at Saona on the day preceding an eclipse of the moon which took place on 14 (23) September. Hence the gale occurred on the 15th (24th) or later. Morison declares that Columbus observed the eclipse while waiting for the weather to clear up. It is impossible to give the "September gale" a definite date other than to state that it occurred in mid-September, old style.[5]

We have, unfortunately, only scanty details about the nature of the storm. There is no wind data or damage account which would justify calling this a hurricane with a definite circular wind pattern, nor do we know the relative wind speed. It would seem logical to classify this as a tropical storm of less stature than a full hurricane, though there is always the possibility that Columbus was on the fringe of a distant hurricane and did not sustain its full force. But without positive evidence, this will have to go into my books as a tropical storm, but an important tropical storm from which Columbus gained valuable experience. His knowledge of the early portents and the behavior of the full blow were to be put to good advantage on the Fourth Voyage.

The following year brought the most interesting of Columbian weather experiences in the New World, one which certainly caused the Admiral to modify his earlier opinion that "all the weather was like May" in the Indies. The following passage appears in *The Decades of the New World or West India,* written by historian Peter Martyr and published in 1511:

> This same year in the month of June, they say there rose such a boisterous tempest of wind from the SE, as hath not lately been heard of. The violence hereof was such that it plucked up by the roots whatsoever great trees were within the force thereof. When this whirlwind came to the haven of the city, it beat down to the bottom of the sea three ships which lay at anchor, and broke the cables in sunder: and that (which is the greater marvel) without any storm or roughness of the sea, only turning them three or four times about. The inhabitants also affirm that the same year the sea extended itself further in to the land, and rose higher than ever it did before in the memory of man, by the space of a cubit.
>
> The people, therefore, muttered among themselves that our nation had troubled the elements, and caused such portentous signs. These tempests of the air (which the Grecians call Tiphones, that is, whirl winds), they call Furacanes: which they say, do often times chance in this Island: But that neither they nor their great grandfathers ever saw such violent and furious Furacanes, that plucked up great trees by the rootes: Neither yet such surges and vehement motions of the sea, that so wasted the land. As indeed it may appear, for as much as, where soever the sea banks are near to any plain there are in manner everywhere, florishing meadows reaching even unto the shore.

Admiral Morison errs, in my opinion, in calling this disturbance a hurricane.[7] My reading of the Peter Martyr account, and this is the only contemporary reference to the event available, leads me to believe that a tornado, spawned somewhere in the back country of the island, moved in a northwesterly direction from the interior. When it struck the waters of the harbor of Isabella, the whirl became a waterspout and struck a devastating blow at the four ships which happened to lie directly in its path on that June day in 1495. My reasons follow:

(1) The only direction mentioned is that the disturbance moved from the southeast, and the wording "when this whirlwind came to the haven of the city" would indicate that it had a definite forward motion—a distinct characteristic of the tornado and waterspout.

(2) Circular motion, so evident in the tornado or whirlwind, is present here from the statement that the three doomed ships were whirled about three or four times before going to the bottom. This is quite unlike a sinking resulting from the rectilinear force of a hurricane wind.

(3) That the whirl had definite limits is in accord with the relatively small areal extent of a tornado. Martyr's account affirms that there was no general disturbance on the surface of the sea, such as a general hurricane would raise. And further, on land there was a definite limit in the destruction of trees as the statement that only those trees "which were within the force thereof" were destroyed.

(4) Such a violent whirl of this nature had not occurred before, according to the testimony of the natives' "grandfathers." This leads me to believe that this storm was unique in their experience and unlike the many normal hurricanes which had previously struck the area. Inland tornadoes and waterspouts along the shores of Hispaniola are not unprecedented, but are relatively rare phenomena.

In addition to my argument over the exact nature of the disturbance (whether a combined tornado-waterspout or a true hurricane), our authorities show disagreement, confusion, and contradiction as to the dating of the event.

Thatcher follows Las Casas in relating that the four supply ships sent out from Spain, which arrived at Isabella in October 1495, were destroyed by a hurricane "that fall." Morison believes that the ships in question arrived even later, toward the end of 1495, casting reasonable doubt as to the occurrence of a second storm and to the destruction of these same four ships. Later evidence which he brings forth indicates that some of these four vessels showed up in Spain two years later.

To complicate further matters, Morison, in a later passage when discussing a hurricane on the Fourth Voyage in 1502, states that Columbus "had witnessed a second when ashore in October 1495 on Hispaniola," without having made any previous mention of such in his regular discussion of October 1495 events. There is no positive evidence that a storm occurred in October 1495 as so many compilations claim.

We have no alternative in this case but to return to Peter Martyr's basic statement that the phenomenon was a "whirlwind" and that it occurred in "June 1495." Thus, we must erase one more "hurricane" from the traditional listings of Poey and Tannehill.

Peter Martyr's reference to the native word "Furacanes" to describe a tropical storm provides the basic clue for etymologists in tracing the origin of the word hurricane. Its entrance into the English language appears to have come through Eden's 1555 translation of Martyr's account. It was about 1600 that various forms of "hurricane" came into restricted use by Englishmen, and by 1650 it was in general use and the form became stabilized before the end of the century.

Of the four ships in the harbor at the time of the storm, only "stout little *Nina*" weathered the blow, according to Morison. The other three were salvaged from the harbor, broken up, and their parts reworked. Fortunately shipwrights were at hand among the colonists to construct a new ship, officially called *Santa Cruz* but nicknamed *India* by the sailors. She was completed, and ready to sail in March 1496. Onto these two small ships 255 disillusioned persons crowded for the voyage to Spain. A southerly route was taken, leaving Guadeloupe on 20 (29) April and arriving at Cadiz on 11 (20) June. This was quite a slow and tedious voyage. The only report about the weather was the lack of it—the winds being very light. The main scientific interest centered around the navigation whereby Columbus did some remarkable dead reckoning to the amazement of his professional pilots who were seemingly lost at sea.

[1] John B. Thatcher. *Christopher Columbus His Life His Works His Remains.* New York, 1903. II, 247.

[2] Morison, II, 129.

[3] Las Casas in Morison, II, 153; Thatcher, II, 335.

[4] Morison, II, 158–159.

[5] Thatcher, II, 338.

[6] Peter Martyr. *The Decades of the New World or West India.* Eden translation. (1555) in *The First Three English Books on America.* (E. Arber, ed.). Birmingham, 1885. 81.

[7] Morison, II, 172–173; Thatcher, II, 356n.

THIRD VOYAGE 1498–1500

The Third Voyage produced little of meteorological interest aside from the Admiral's first encounter with the doldrums. The small fleet of caravels left Spain on 30 May (8 June) 1498, stopping at the Canaries and the Cape Verde Islands on the way. Soon after leaving the latter, at latitude 8°30′N the four ships were becalmed on Friday the 13th (22) July, and for eight days drifted west-southwestward in the equatorial current. "The wind stopped so suddenly and unexpectedly and the supervening heat was so excessive and immoderate that there was no one who dared go below to look after the casks of wine and water, which burst, snapping the hoops of the pipes; the wheat burned like fire; the bacon and salted meat roasted and putrefied." [1]

The distaste of Europeans for tropical and equatorial heat was well expressed by Ferdinand Columbus: "He held on a southwest course until he found himself five degrees from the Equator, where, after they had sailed continually in the aforementioned haze, the wind fell; this calm lasted eight days, with a heat so excessive that it scorched the ships. None could endure staying below deck, and but for an occasional rain that obscured the sun I believe they would have been burned alive together with their ships. The first day of that calm being clear, the heat was so great they must have perished if God had not miraculously relieved them with the aforementioned rain and mist. So having turned somewhat northward and gotten seven degrees above the Equator, he decided not to sail any further south but instead proceed due west, at least until he could see how the weather was shaping up. For on account of the heat he had lost many casks of wine and water, which burst, snapping their hoops, and all their wheat and other provisions were scorched." [2]

On the 22d (31st) "at the lengthe an Eastsoutheaste wynde arose, and gave a prosperous blaste to his sayles," in the words of an 1555 English translator of Peter Martyr.[3] With this fortunate intrusion of the southeast trades north of their normal position, Columbus was able to steer westward to a landfall at Trinidad on the 31st.

The continent of South America was soon discovered and the coast of Venezuela explored during the following two weeks. This latitude is well below the usual area of hurricane formation and movement. But during the next two weeks Columbus crossed the open Caribbean to Hispaniola and made his way along the coast to Santo Domingo, arriving on 31 August (9 Sept.), and again he was favored with the absence of stormy weather while in hurricane territory.

For the next two years Columbus was beset with the problems of trying to organize and govern the "Terrestrial Inferno," to use Morison's apt phrase, made up mainly of frustrated treasure-seekers and ambitious political intriguers. We know little of meteorological events on the island during these two years as the scanty writings about this period are almost completely filled with self-justifying pronunciamentos. Poey quotes Moreau de Jonnes as citing a hurricane occurrence at Santo Domingo in August 1500, but a search has revealed no confirmation of this.[4]

Unable to put down the rebellious subjects, Columbus found himself replaced by an envoy of Ferdinand and Isabella in August 1500. To impress the local populace, the Admiral was placed in chains and sent home to Spain in October to plead his case with his sovereigns. Fortunately for the troubled passenger, the ship met "fair weather and favoring breezes" to make a quick voyage and reach home port within a month.

[1] *Raccolta di Documenti et Studi pubblicati dalla R. Commissione Columbiana* (Rome 1892–94) quoted by Morison, II, 241–242.
[2] Ferdinand Columbus, 179.
[3] Morison, II, 243.
[4] Alexandre Moreau de Jonnes. *Historie Physique des Antilles.* Paris, 1822. 386.

FOURTH VOYAGE 1502–04

The final voyage of Columbus to the New World commenced on 11 (21) May 1502. Though now divested of his original sweeping authority in the Indies, the now fifty-year-old Admiral wished to explore far southward and westward from his old haunts, sure that he could find a passage westward to the Cathay of Marco Polo. After a quick and uneventful outward trip, his fleet of four vessels reached Martinique on 15 (25) June, having been on the open seas only 21 days from the Canaries. This placed him well ahead of the date for the onset of the first tropical storm in a normal season.

An unauthorized stop was made at the harbor of Santo Domingo in an attempt to replace by trade a vessel which the Admiral deemed unsuitable for the exploring purposes in mind. Mindful of his past treatment and wary for his personal safety, Columbus sent one of his captains ashore on the mission and at the same time instructed him to convey a weather warning. The weatherwise Genoese observed the harbor packed with a large fleet of some thirty sail, many of them laden with treasure and slaves, ready to put out on the homeward voyage to Spain. From the hazy appearance of the atmosphere, the direction of high-flying cirrus, and the presence of an ominous southeasterly swell, Columbus sensed an imminent storm. He warned the unfriendly governor of Hispaniola that the large fleet should remain in the safekeeping of the harbor a few days longer. But scorning the Admiral's trading proposal and mocking his weather fears, the governor refused to deal with Columbus and even denied him shelter in the harbor. Forthwith the order was given for the Spain-bound flotilla to set sail although the seas outside were already making up.

In less than two days sail from Santo Domingo, just as the laboring fleet rounded the eastern tip of Hispaniola into Mona Passage, an increasing northeast wind caught the ships. Soon they were laboring heavily in a whole gale. So severe did the blow become, probably reaching full hurricane force, that about 20 vessels went to the bottom with all hands and six others were lost but had some survivors. Three or four managed to struggle back to the point of departure. Ironically, only one, that carrying gold belonging to Columbus which his factor had dispatched home, survived the storm's buffeting to continue on to Spain. Over 500 men lost their lives needlessly in this failure to heed the reading of the weather signs, and practically all the treasure which had been exacted from the natives with such toil and cruelty went to the bottom.[1]

Thanks to some weatherwise maneuvering, the Admiral's small fleet escaped major damage, though not without experiencing some dire moments. After being denied the protection of the best harbor on the coast, Columbus sailed westward along familiar shores and took up a station where he knew he would have shelter from north and west winds. The storm struck from the northeast on 30 June (10 July), increasing throughout the day till by night it was raging at full hurricane strength, if we may judge from the damage soon inflicted. The anchor chains held only on the Admiral's ship. The others were propelled seaward by wind and waves where they fought for their lives.

The Admiral described the scene: "The storm was terrible and on that night the ships were parted from me. Each one of them was reduced to an extremity, expecting nothing save death; each one of them was certain the others were lost." [2] With some family pride, Ferdinand Columbus noted that the two ships skippered by his family fared by far the best. Bartholomew commented: "experienced seaman that he was, had weathered that great storm by going out to sea, while the Admiral had saved his ship by lying close to the shore, like a sage astrologer who foresaw whence the danger must come." [3]

It would require a northerly wind component to drive the ships away from the shore; their position must have been in the western quadrant of the counterclockwise spinning hurricane, usually the most favorable area for a ship to weather such a blow. The center of the storm must have been close to Mona Passage with the trajectory of the eye over extreme western Puerto Rico. One can speculate as to what might have been the fate of the Admiral's fleet if the storm had passed just to the south, putting the nearby land to his lee.

By prearrangement, after the gales had subsided, the scattered fleet made for a rendezvous at a small land-locked bay farther along the coast to the west. They were assisted to this goal by a southeasterly wind which followed the passage of the hurricane. After a few days for repairs and recuperation, the fleet was seaworthy and again continued westward.

It would be a year and a half before Columbus would return to Santo Domingo and learn that the entire settlement of thatched-roofed buildings had been laid low by the hurricane about which he had given timely warning.

Columbus reached the mainland of Central America at Cape Honduras soon after August 1st (11th). Then commenced the most difficult experience, weatherwise, of the entire voyage. From the Cape the coast of Honduras trends east and then east-southeastward. Winds at this season are almost invariably from an easterly quarter, and these bring continual showers and squalls as the area lies close to the atmospheric equator where a low pressure trough features convergence and convection, a dual process resulting in continual overturning of the atmosphere. Beating to windward in such weather, it took 28 days to gain 170 miles easting. "During all this time, I did not put into harbor, nor could I, nor did the storm from heaven cease; there was rain and thunder and lightning continuously, so that it seemed as if it were the end of the world." [4] Not until 14 (24) September did the fleet round Cape Gracias a Dios and find a southward-trending coast along which they could sail steadily on one tack.

October and November spent along the shores of Panama seem to have been more pleasant months. But early in December the Admiral, wearied of continual headwinds along the eastward-trending coast of Panama, turned about to the westward to investigate a rumor of gold. The very next day the wind whipped around to the opposite quarter when the westerlies dipped down from temperate latitudes, as they often do at this season, to displace the tropical easterly flow which predominates in the warm season. With the current running with the wind also, the caravels were batted about, unable to make any westward progress, tacking back and forth in the area close to the present entrance of the Panama Canal. Columbus writes of this dreadful month briefly but eloquently:

> The tempest arose and wearied me so that I knew not where to turn; my old wounds opened up, and for nine days I was as lost without hope of life; eyes never beheld the seas so high, angry and covered by foam. The wind not only prevented our progress, but offered no opportunity to run behind any headland for shelter; hence we were forced to keep out in this bloody ocean, seething like a pot on a hot fire. Never did the sky look more terrible; for one whole day and night it blazed like a furnace, and the lightning broke forth with such violence that each time I wondered if it had carried off my spars and sails; the flashes came with such fury and frightfulness that we all thought the ships would be blasted. All this time the water never ceased to fall from the sky; I don't say it rained, because it was like another deluge. The people were so worn out that they longed for death to end their dreadful suffering." [5]

"Besides these different terrors," wrote Ferdinand, "there befel one no less dangerous and wonderful, a waterspout that on Tuesday, December 13th (23d), passed between two ships. Had the sailors not dissolved it by reciting the Gospel according to St. John, it would surely have swamped anything it struck; for it raises the water up to the clouds in a column thicker than a water butt, twisting it about like a whirlwind." [6]

These storms, of course, were not hurricanes, nor even tropical storms in the accepted sense. They represented one of the many varieties of atmospheric misbehavior that the tropics can muster on occasion along the zone of intertropical convergence. Surely, these experiences were a far cry from the Admiral's original concept of the climate of the Antilles.

The return from the mainland toward Hispaniola occupied the spring months of 1503. Morison has called this "at sea in a sieve" as the worm-eaten boats were in such wretched condition that only 24-hour pumping kept them afloat. Fortunately, the only serious weather encounter came off the south coast of Cuba about May 14th (24th) when a heavy squall struck and knocked the leaking ships about most of the night. "But it pleased God to deliver us there, as already He had from many other dangers."

The remaining period of the Fourth Voyage holds little of weather interest. For 369 days the band of about 100 mariners were marooned on the northern coast of Jamaica to await a rescue which came on 29 June (9 July) 1504.

The Admiral embarked at Santo Domingo for home on 12 (22) September at the peak of the hurricane season. The final return trip was long and troublesome. One storm on 19 (29) October broke the mainmast in four pieces, but a jury was rigged; and this was broken in a second wind encounter.[7] But on 7 (17) November 1504 the Admiral of the Ocean Sea, after a passage of 56 days, stepped ashore for the last time to end his epic voyages.

[1] Morison, II, 324–326, 330; Thatcher, II, 579.
[2] Jane, 293.
[3] Ferdinand Columbus, 229.
[4] Jane, 294.
[5] Raccolta, I, ii, 186 in Morison, II, 358; Jane, 297.
[6] Ferdinand Columbus, 246.
[7] Morison, II, 410.

THE FIRST TWO CENTURIES: 1501–1700

SIXTEENTH CENTURY

We must now take leave of Columbus and the sometimes stormy isles of the West Indies. Our attention henceforth, of necessity, will be focused on the shores of the Gulf of Mexico and the North Atlantic Ocean, from the Rio Grande in Texas around to Florida and northeastward to Maine. We shall concentrate our studies, not on the origin and early tracks of hurricanes, but on their impact on the shoreline of the present United States and their meteorological character at that time.

It is well to keep in mind that 115 years were to pass after Columbus made his landfall on Watling Island (San Salvador) until the English made a permanent settlement in what is now the United States. The Sixteenth Century was the great age of discovery and exploration when many European nations competed on the high seas to control the new lands to which Columbus had found the key. In 1513 the Spaniards, sailing forth from Hispaniola, first visited Florida, and another expedition in the same year discovered the Pacific Ocean. The Narvaez Expedition in 1528 visited the Tampa and Tallahassee areas; soon after their departure from Apalachee Bay in west Florida on 22 September (2 October), they were shipwrecked, perhaps by a tropical storm, and only ten of more than 400 survived.[1]

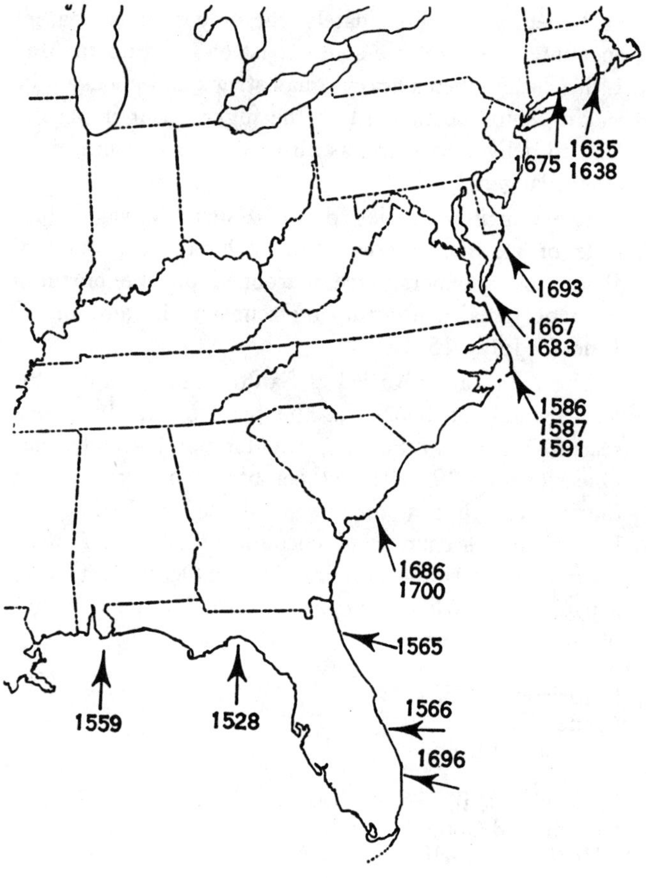

The Spaniards with their small ships appear to have been especially unfortunate in their voyages around the bays and shoals of the Florida peninsula. Woodbury Lowery, historian of the period, tells of wrecked Spanish fleets in 1545, 1551, 1553, 1554, and 1559.[2] The *Luna Papers* reveal that the latter storm caught a Spanish fleet of 5 vessels in the Bay of Santa Maria Filipina of Arellano (present Pensacola?) on 19 (29) August 1559. A northerly gale set in first, and then "the wind came from all directions." The blow continued for 24 hours. All ships were driven ashore. This is the first account at hand of a long line of tropical storms which have ravaged the Pensacola area.[3]

Spanish activity shifted to the East Coast of Florida in the 1560's. In 1565 we hear a familiar testimony from the oldest inhabitant: "The day he departed, which was the 10th (20th) of September, there arose so great a tempest, accompanied with such storms, that, the Indians themselves assured me, it was the worst weather that was ever seen on the coast." [4]

The founding of St. Augustine and its survival as a Spanish settlement has a hurricane-connected angle. In 1565 both the French at Ft. Caroline and the Spanish at St. Augustine (both close to the mouth of the St. Johns River) were engaged in a struggle to see who would massacre the other. Under Jean Ribaut the French sailed on 10 (20) September to attack the Spanish fortifications, but a storm, when three days aboard ship, caught the fleet off St. Augustine Bay and sent the ships to the bottom along with most of the attackers. The Spanish then immediately retaliated by attacking Ft. Caroline with its weakened garrison, and soon dispatched the defenders to another world. Thus, a tropical storm, possibly a hurricane, determined the immediate political fate of East Florida.[5]

Very intense hurricanes which cross Cuba often ravage the coasts of Florida and the Gulf States—those of 1846 and 1915 serve as examples of the most destructive type. Desiderio Herrera in his *Memoire sur les Oragans des Cuba* (1847) describes memorable hurricanes of this type which ravaged the island in 1527, 1557, and 1558, the latter being "more destructive than 1557." Herrera also mentions a storm on the East Coast of Florida on 16 (26) September 1566, but gives no details—this might possibly be confused with the storm of the previous year which wrecked Ribaut's fleet.[6]

Farther north hurricanes, no doubt, smashed across the Outer Banks of North Carolina with the same regularity as they have in subsequent centuries, though in the Sixteenth Century there were no permanent settlers to keep tabs on them. We hear of a Spanish brigantine lost in a storm near Cape Fear in 1526 while enroute to the temporary Spanish Colony of Chicora in the Cape Fear country.[7]

The ill-fated English attempts to found a colony in the Virginia country owed much discouragement to storm activity and the lack of sheltered harbors. Sir Humphrey Gilbert's first expedition in 1583 ran into a storm in September on the homeward voyage during which the leader's ship was lost. Gilbert's half-brother, Sir Walter Raleigh, accompanied an expedition the following year and entered Albemarle Sound on the North Carolina coast, naming the country Virginia in honor of his Queen. Next year, 1585, a small colony was established on Roanoke Island, but was abandoned the following summer after a three-day June storm. Sir Francis Drake had arrived offshore in early June, but "there arose a great storm (which they said was extraordinary and very strange) and lasted three days together, and put our fleet in great danger." [8] Ralph Lane, the leader of the colony, related in his history of the expedition that the storm occurred on 13–16 (23–26) June 1586.[9] "The 13th of June our ships were forced to put to sea. The weather was so sore and the storme so great that out anchors would not hold, and no ship of them all but either broke or lost their anchors. And our ship the *Primrose* broke an anchor of 250 lbs. weight. All the time we were in this country, we had thunder and rain with hailstones as big as hen's eggs. There were great spouts at the seas as though heaven and earth would have met." [10]

The results of the storm, together with mounting Indian hostility, caused Drake to evacuate the settlers and take them back to England, departing the Carolina shores on the 19th (29th) of June. Sir Richard Grenville arrived about two weeks later with supplies and new settlers. Though he was unable to find any trace of the former colony, he left 15 men with provisions for two years to maintain English sovereignty over the land.

When a fourth expedition under John White arrived in 1587 no trace could be found of the Englishmen who had planned to winter at Roanoke. The colony was again reestablished by White in July 1587 and again a great storm came. "August 21 (31)—there arose such a tempest at northeast, that our Admiral (Drake), then riding out of the harbor, was forced to cut his cables, and put to sea, where he lay beating off and on six days before he could come to us again. . . ." [11] White left a small group of colonists to winter on Roanoke Island and returned to England with the intention of bringing supplies and more settlers to the New World.

The constant threat of a Spanish invasion and the actual battle with the Armada occupied most of Britain's attention in 1588. An abortive attempt to send two small vessels was made in that year, but it was not until 1590 that an adequate expedition could be mounted to succor the infant Roanoke Colony. Two storm periods, on 1–9 (10–19) August and again on 17–18 (27–28) August 1590 impeded the landings. "We had very foul weather, with much rain, thundering and great spouts, which fell around us, high unto our ships," noted White.[12] No trace of the 1587 colonists was found. Their fate has always remained an intriguing mystery in American annals.

The season of 1591 may serve as an indication that hurricanes were as frequent in the Sixteenth Century as they are today. A final expedition visited the Carolina shore in August 1591 and once again had to contend with a strong northeast gale on the 16th (26th). "For at this time the wind blew at Northeast and direct into the harbor so great a gale, that the sea broke extremely on the bar, and the tide went very forcibly at the entrance." [13] Out on the broad Atlantic no less than four destructive storms occurred in the four weeks from 10 (20) August to 6 (16) September 1591, with the destruction of at least 27 ships reported. The Grand Fleet making its annual treasure run from Havana to Old Spain was caught in the 10 August gale and over 500 sailors went to the bottom with most of their loot.[14]

This brief and fragmentary survey of storm activity in the Sixteenth Century indicates that the hurricane posed a major problem to those bent on exploration, treasure-seeking, and settlement in that exciting period. No doubt, an intensive search of Spanish archives and a survey of British Admiralty records would reveal a list of many more hurricanes which churned the seas in those days. But in the absence of any permanent English settlement in America in this century, our meteorological data for the shoreline must always remain sketchy and unsatisfactory.

[1] Woodbury Lowery. *The Spanish Settlements in the United States, 1513–1561,* I, 198.

[2] *Ibid.,* 352–353, 360.

[3] Herbert Priestley. *The Luna Papers,* I, xxxvi, and II, 245.

[4] *Historical Collections of Louisiana and Florida.* n.s., 336; Lowery, II, 167–168.

[5] Lowery, II, 190.
[6] Desiderio Herrera. *Memoire sur les Ouragans des Cuba*, 46.
[7] David Stick. *Graveyard of the Atlantic*, 5.
[8] *The Roanoke Voyages 1584–1590.* (David Beers, ed.). For the Hakluyt Society: Ser. 2, vol. 104, I, 302.
[9] *Ibid.*, 291–292.
[10] *Ibid.*, 307–308.
[11] *Ibid.*, II, 532.
[12] *Ibid.*, 608.
[13] *Ibid.*, 611.
[14] Thomas Southey. *Chronological History of the West Indies*. London, 1827. I, 212; Tannehill, 242.

THE GREAT COLONIAL HURRICANE

Rev. Increase Mather in his *Remarkable Providences* (1684), a review of early natural and supernatural events in New England, thought he had heard "of no storm more dismal than the great hurricane which was in August 1635." [1]

This was the great meteorological event of the colonial period in New England, coming only 15 years after the settlement of Plymouth Plantation and in the 5th year of the Massachusetts Bay Colony. It would be 180 years before another great hurricane of similar importance would strike the area with equal force—The Great September Gale of 1815.

Perhaps the historical stature of the 1635 event owes its prominence to the unusual severity of the storm itself, since the accounts of whole forests being leveled would indicate that it was a hurricane of exceedingly great force. But to perpetuate its reputation for future generations, the Great Colonial Hurricane had two able eyewitnesses whose contemporary writings form the solid foundation upon which most of the history of early New England has been based: John Winthrop of Massachusetts Bay Colony and William Bradford of Plymouth Plantation.

The occurrence of the Great Colonial Hurricane seems to have been unknown to the historians of hurricanes of the past generations. Poey did not list this storm in his compilation of some 400 hurricanes occurring prior to 1855, and he studied William Redfield's manuscript notes very carefully.[2] Tannehill writing in 1938 mentioned a storm in August in the Windward Islands as "a violent gale between St. Kitts and Martinique," but did not indicate any further history of this or any other disturbance in New England in 1635. In his second edition (1942) Tannehill, in a separate list of severe tropical storms which have affected New England, correctly identified the Great Colonial Hurricane as occurring on August 15th, though he did not mention that this was Old Style and that ten days should be added for the present seasonal equivalent of August 25th.[3] Sidney Perley, however, commenced his *Historic Storms of New England* (1891) with a non-meteorological account of the hurricane of 1635 which was drawn mainly from the standard sources of New England's history which we shall employ also.[4]

John Winthrop, governor of the Massachusetts Bay Colony, kept a running journal of the principal events in the Boston area. Although his manuscript was available to colonial writers, it was not published in full until the early Nineteenth Century as *The History of New England from 1630 to 1649.* In all of his writings Winthrop showed himself a very weather-conscious individual, noting the course of the winds each day on the voyage across the Atlantic, chronicling the variety of the seasons he experienced during his two decades in New England and comparing them with those in the old country, and describing in detail all the unusual weather and storms on land and sea which afflicted the early settlers.

Winthrop's account of the Great Colonial Hurricane remains our best source and reveals some essential details as to the meteorological character of this celebrated storm:

> Aug. 16 The wind having blown hard at S. and S.W. a week before, about midnight it came up at N.E. and blew with such violence, with abundance of rain, that it blew down many hundreds of trees, near the towns, overthrew some houses, and drove the ships from their anchors. The Great Hope, of Ipswich, being about four hundred tons, was driven aground at Mr. Hoffe's Point, and brought back again presently by a N.W. wind, and ran on shore at Charlestown. About eight of the clock the wind came about to N.W. very strong, and, it being then about high water, by nine the tide had fallen three feet. Then it began to flow again about one hour, and rose about two or three feet, which was conceived to be, that the sea was grown so high abroad with a N.E. wind, that, meeting with the ebb, it forced it back again.
>
> This tempest was not so far as Cape Sable, but to the south more violent, and made a double tide all that coast . . . The tide rose at Naragansett fourteen feet higher than ordinary, and drowned eight Indians flying from their wigwams.[5]

Two of Winthrop's statements are most important in trying to recreate the meteorological conditions attending the hurricane. The wind in the Boston area and along the North Shore of Massachusetts Bay commenced at northeast, but backed to northwest—a sure indication that the center passed to the south and east of Boston. This is confirmed by Winthrop's statement that the storm was more severe to the south (in the direction of Cape Cod and Narragansett Bay in Rhode Island). In that direction, fortunately, we can turn to

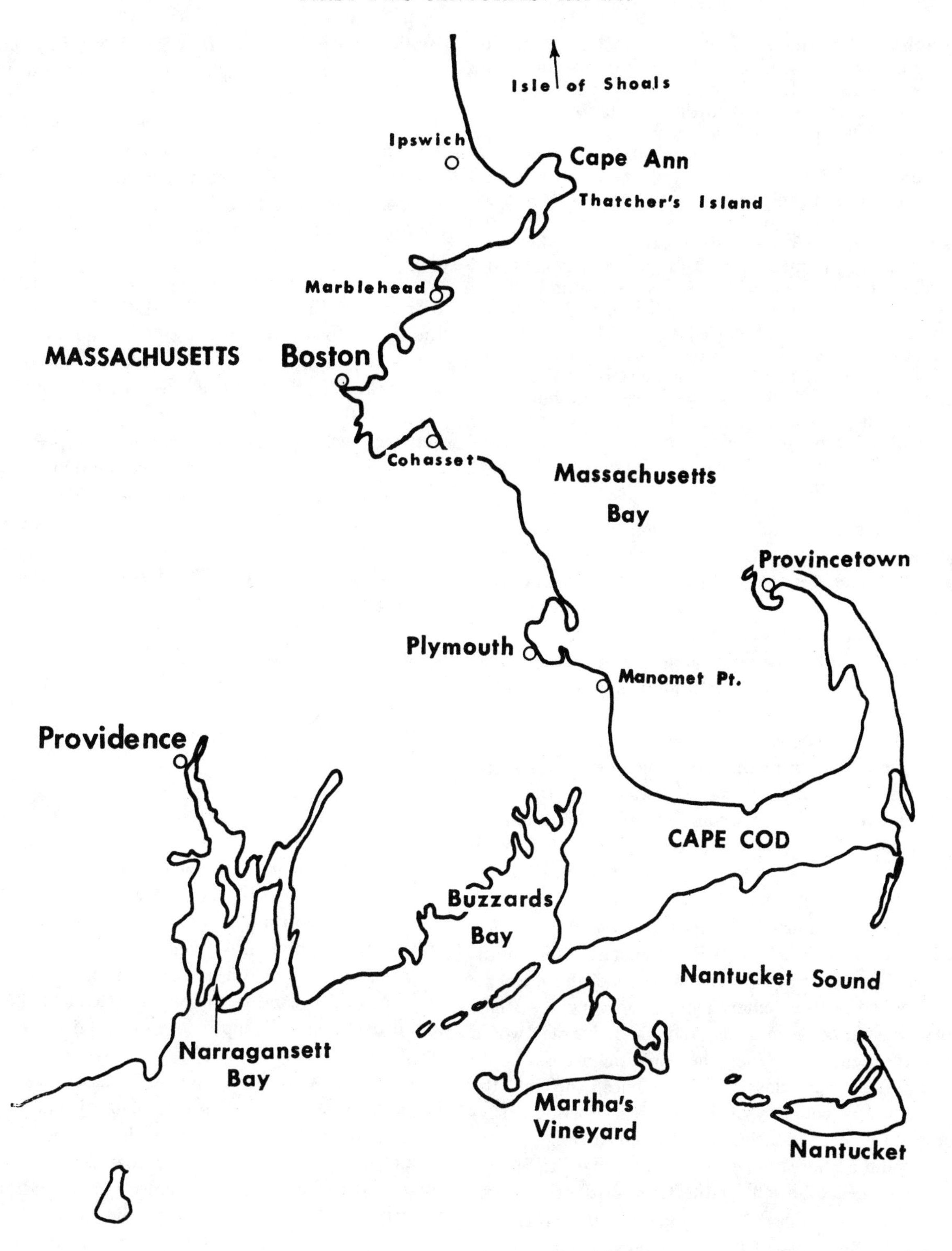

the Plymouth area, 45 miles southeast of Boston and to William Bradford's *Of Plymouth Plantation, 1620–47*. Though a more formal history than Winthrop's personal day-to-day journal, Bradford contributes several salient points about this justly famous storm:

This year, the 14th or 15th of August (being Saturday) was such a mighty storm of wind and rain as none living in these parts, either English or Indians, ever saw. Being like, for the time it continued, to those hurricanes and typhoons that writers make mention of in the Indies. It began in the morning a little before day, and grew not by degrees but came with violence in the beginning, to the great amazement of many. It blew down sundry houses and uncovered others. Divers vessels were lost at sea and many more in extreme danger. It caused the sea to swell to the

> south wind of this place above 20 foot right up and down, and made many of the Indians to climb into trees for their safety. It took off the boarded roof of a house which belonged to this Plantation at Manomet, and floated it to another place, the posts still standing in the ground. And if it had continued long without the shifting of the wind, it is like it would have drowned some part of the country. It blew down many hundred thousands of trees, turning up the stronger by the roots and breaking the higher pine trees off in the middle. And the tall young oaks and walnut trees of good bigness were wound like a withe, very strange and fearful to behold. It began in the southeast and parted toward the south and east, and veered sundry ways, but the greatest force of it here was from the former quarters. It continued not (in the extremity) above five or six hours but the violence began to abate. The signs and marks of it will remain this hundred years in these parts where it was sorest. The moon suffered a great eclipse the second night after it.[6]

Bradford's statement that "it began in the southeast and parted toward the south and east" requires some study. It would seem to me that he meant that the wind in the first phase of the storm blew from the southeast, and in the concluding phase there were northwest winds carrying the storm seaward off to the southeast. This interpretation would place Plymouth and southeastern Massachusetts in the dangerous eastern quadrant of the storm at its commencement, have the center pass just to the west, and bring in strong northwest following winds as the storm area moved away from the coast to the northeast. The mention of unusual tidal activity in Narragansett Bay would also indicate such a track as it requires a southerly component (southeast to south) to raise a great tide in that body of water.

If there were northeast winds at Boston preceding the arrival of the center, as Winthrop states, and southeast winds at Plymouth at the same time, there can be little doubt that the center, moving on a northeasterly course, passed between the two points. This track would bring the center in from the Atlantic across upper Narragansett Bay close to Providence, through the Massachusetts counties of Bristol and northern Plymouth, to enter Massachusetts Bay in Norfolk County on the South Shore somewhere near Cohasset. Such a track has been followed by the Great Atlantic Hurricane of 14 September 1944, and recently Hurricane Edna on 11 September 1955 moved on a parallel course, but about 25 miles to the east. The other *great* hurricanes of New England (23 September 1815 and 21 September 1938) followed paths far to the west of the above.

The meteorological conditions preceding the hurricane are supplied by Rev. Richard Mather when north of Cape Ann enroute from England to Boston. On both the 13th and 14th his ship experienced strong southsouthwest winds, an air flow which would indicate the presence of a well-established trough of low pressure over western New England or New York State. Such a trough is a prime prerequisite for a New England hurricane since it provides a congenial atmosphere for a tropical storm to speed northward and to maintain, or even increase, its intensity. Mather further states that on the night of the 14–15th the wind was "somewhat milder" so that his ship lay quietly until about break of day when commenced "a most terrible storm of rain and easterly wind."[7] Mather's comment as to the sudden onset of the easterly gale agrees with Bradford's observation on this point. This condition would be repeated in the Great September Gale of 1815, and suggests the presence of some type of front existing between the cool northeasterly stream prevailing before and the advancing tropical circulation.

The onslaught of the hurricane came at a season when a number of ships were enroute to Massachusetts Bay carrying new settlers and much-needed goods to the infant communities. The Great Hurricane caused many shipwrecks and several near-disasters. The *Great Hope* out of Ipswich, England, was driven aground by the northeast wind on Charleston Point and then blown off again when the gale shifted to northwest.[8] A similar occurrence near the Isle of Shoals endangered the lives of one hundred passengers, including the Rev. Richard Mather, when the *James* of Bristol drove close to the rocks near Piscataqua on the heels of a northeast gale. Then a seemingly miraculous shift of wind to northwest saved the vessels and passengers.[9]

The hurricane of 1635 gave rise to one cherished New England legend. A small coastal shallop with 21 persons aboard, enroute from Ipswich to Marblehead, was frustrated by southwesterly winds on 14 August in attempting to round Cape Ann. About midnight the sudden shift of wind to the northeast caught the vessels and drove the small ship before the wind until it lodged on a rocky ledge about 100 yards off a small island. One mighty wave lifted ship and passengers onto the ledge smashing ribs and staves of the boat.

First to reach the safety of solid land was Anthony Thatcher who had boarded the boat with his wife and four children for what was usually a short and uneventful trip. After some time ashore he discerned the remains of the shallop washing ashore with a body entangled in the wreckage. To his relief and delight, it turned out to be his wife who succeeded in freeing herself and struggled to the beach alive. Anxiously, they kept a vigil on the shore for sight of their children and other shipmates, but none ever made it alive. After a sorrowful 36 hours on the island, a passing boat spotted them and proceeded to the rescue. The place has ever since been known as Thatcher's Island, in memory of that fateful night during the Great Colonial

Hurricane. Thatcher later put his experiences into a manuscript which was republished several times and became a favorite New England legend.[10]

[1] Increase Mather. *Remarkable Providences* (1684), (1856 ed.), 220–221.
[2] Poey, 2.
[3] Tannehill. (2nd ed.), 222.
[4] Sidney Perley. *Historic Storms of New England* (1891), 3–10.
[5] John Winthrop. *The History of New England from 1630–1649.* (James Savage, ed. 1853), I, 195–198.
[6] William Bradford. *Of Plymouth Plantation 1620–1647.* (S. E. Morison, ed., 1952), 279–280.
[7] Richard Mather's Journal. *Chronicles of the First Planters of the Colony of Massachusetts Bay.* (Alexander Young, ed., 1846), 472–475.
[8] Perley, 5.
[9] Winthrop, 196.
[10] Anthony Thatcher's Narrative of His Shipwreck. Young, 485–495.

THE TRIPLE STORMS OF 1638

Winthrop described two noteworthy eastern New England storms in the year 1638. On 3 August (13) "in the night was a very great tempest, or hiracano at S.W. which drave a ship on the ground at Charlestown, and brake down the windmill there, and did much other harm. It flowed twice in six hours, and about Naragansett, it raised the tide fourteen to fifteen foot above the ordinary spring tides, upright." Without the reference to the extraordinary rise in tide in Rhode Island, one might be inclined to classify the blow as the cold front passage of an extra-tropical storm, in view of the mention of only southwest winds at Boston. Other hurricanes, however, in that locality have been accompanied by very damaging southwest winds following the passage of the center. Furthermore, the season of the year would be favorable for a hurricane occurrence and quite unfavorable for the development of a southwester strong enough to do structural damage.[1]

Eight weeks later another storm of probable tropical origin brushed the New England coast, to be followed by still a third much farther offshore. Winthrop relates: "Sept 25 (Oct. 5)—Being the third day of the week and two days before the change [equinox?], the wind having blown at N.E. all day, and rainy in the night, was a mighty tempest, and withal the highest tide, which has been since our coming into this country; but through the good providence of God, it did little harm. About fourteen days after, (Oct. 19) the wind having been at N.W. and then calm here, came in the greatest eastern sea, which had been in our time. Mr. Peirce (who came in a week after) had at that time a very great tempest three days at N.E." [2]

Confirmation of the first of these late season hurricanes will be found in the writings of John Jocelyn. In his *Two Voyages to New England* he made the following note in 1638 while at Scarborough, Maine, a coastal town just south of Portland: "Sept 24—Monday about 4 o'clock in the afternoon, a fearful storm of wind began to rage, called a hurricane. It is an impetuous wind that goes commonly about the compass in the space of 24 hours. It began from the N.N.W. and continued till the next morning—the greatest mischief it did us, was the wracking of our shallops, and the blowing down of many trees, in some places a mile together."

That the season of 1638 was prolific in the production of hurricanes would follow from the above, and is further bore out by another reported occurrence in the West Indies. St. Christopher Island experienced a severe tropical storm on 5 (15) August 1638. In what is a good candidate for the first imprint on an American weather event, John Taylor told the story in his *Newes and strange Newes from St. Christophers of a tempestuous Spirit which is called by the Indians Hurry-Cano or whirlewind.*[4] If our dating of the other hurricane occurrences above are corrected, it would appear that the St. Christopher storm was not connected with those which struck at the New England shore.

[1] Winthrop, I, 320.
[2] *Ibid.*, 330.
[3] *Mass. Hist. Soc. Coll.*, III–3, 1833, 227; also *Coll. Maine Hist. Soc.*, III, 1853, 88.
[4] Reprinted in *Two Tracts.* Oxford, 1946.

THE DECADE OF THE 1640'S

Winthrop's history continued through 1649, the year of his death. Although there are accounts of other meteorological phenomena, the only major storms which might possibly have been of tropical origin were three which occurred at the tail end of the normal hurricane season in modern November. Both Poey and Tannehill lack material for this decade as only three storms, all occurring in the single year 1642 and all confined to the West Indies, are mentioned.

On 12 (22) November 1641 Winthrop tells of "a great tempest of wind and rain from the S.E., as fierce as a hurricane. It continued very violent at N.W. all the day after."

On 29 October 1645 (8 November) a storm moved

across the New England area with evidence that the center might have passed right over Boston. "The wind E.N.E. with rain, so great a tempest that it drave three ships upon the shore . . . and the night after for the space of two hours the tempest arose again at S. with more wind and rain than before."

Just a year later another severe November storm did much damage in Massachusetts Bay on 4 November (14). Winthrop relates that the northeast wind, among other things, lifted the roof of Lady Moody's house in Salem and deposited the debris on the ground.

THE YEARS 1651-1666

Following the death of Winthrop, our sole meteorological reporter is John Hull of Boston who maintained a diary of local events covering 1634 to 1681. Although he details the characteristics of each winter, Hull makes mention of very few storms occurring at other seasons. Our conclusion must be that no noteworthy hurricanes, certainly none of great character such as 1635 and 1638, struck the area. This opinion is reinforced by discussions of the major hurricane of 1675 when people always harked back to the 1635 occurrence for a blow of comparative severity. Tannehill lists nine West Indian hurricanes in the decade 1651–60 as does Poey. In the decade of the 1660's Tannehill has nine again while Poey numbers seven. None of these is mentioned as crossing or approaching the American coastline. No doubt these years experienced the usual number of tropical storms and hurricanes, but their historical trail cannot be followed.

STRANGE NEWS FROM VIRGINIA: THE DREADFUL HURRY CANE OF 1667

The London Company established what was destined to be the first permanent English settlement in America at Jamestown in 1607. But for the next sixty years we have little historical information about storms or possible hurricane activity in the Virginia colony. Numerous accounts of shipwrecks appear in the annals, but no essential meteorological details are ever given. On land no writer of the stature of a Winthrop or a Bradford took up his pen to leave us a record of the vagaries of the Southern weather such as we have for New England.

A hurricane in the Bahamas in 1609, however, played an important role in guiding the immediate destinies of the young colony, then beset with internal political dissension, economic uncertainty, hunger, and disease. Sir Thomas Gates, destined to become governor of Virginia, met with a severe hurricane between Cuba and the Bahamas while enroute from England to his new post, and a near-disaster occurred. The story was told at the time by William Strachey in *A true reportory of the wracke, and redemption of Sir Thomas Gates, Knight; upon, and from the Islands of the Bermudas* (1612): "Upon Saint James day [25 July–4 Aug.] being about one hundred and fifty leagues distant from the West Indies, in crossing the Gulf of Bahama, there happened a most terrible and vehement storm, which was a tail of the West Indian hurricane." [1]

The 44-hour gale scattered the fleet of small vessels and sent one to the bottom immediately. Four of the remaining vessels collected again after the cessation of the blow and reached Virginia shortly thereafter, to be followed in a few days by three others. The remaining ship *Sea Adventure,* with the highest in command aboard, disappeared and was presumed to have been lost. By happy chance the ship made a landfall on Bermuda; though the vessel was wrecked in the landing through the coral reefs, all the passengers and crew struggled ashore. They remained there until a small boat could be constructed which took them to Jamestown the following May after a ten-month stay. It is said that this incident, as reported by Strachey, provided William Shakespeare with the background material for *The Tempest.*

A tremendous hurricane late in the summer of 1667 provided another major event in Virginia's early history. Our main source of information comes from a contemporary pamphlet which rates as the first imprint on an American mainland meteorological subject. *Strange News from Virginia, being a true relation of the great tempest in Virginia* was published in London sometime before the end of the year 1667.

The first 47 lines of *Strange News* concern the Virginia hurricane, while the addenda of 33 lines calls attention to various natural calamities which had afflicted diverse parts of the world. The Virginia account follows:

> The Copy of a Letter from Virginia, Containing the Relation of a Violent Hurricane, which happened the 27th of August and continued (without intermission) twelve dayes together.

Sir,

Having this opportunity, I cannot but acquaint you with the Relation of a very strange Tempest which hath been in these parts (with us called a Hurricane) which began Aug. 27. and continued with such Violence, that it overturned many Houses, burying in the Ruines much Goods and many people, beating to the ground such as were any wayes employed in the Fields, blowing many Cattle that were near the Sea or Rivers, into them, whereby unknown numbers have perished, to the great affliction of all people, few having escaped who have not suffered in their persons or Estates, much Corn was blown away, and great quantities of Tobacco have been lost, to the great damage of many, and utter undoing of others. Neither did it end here, but the Trees were torn up by the roots, and in many places whole Woods blown down, so that they cannot go from Plantation to Plantation. The Sea (by the violence of the winds) swelled twelve Foot above its usual height, drowning the whole Country before it, with many of the Inhabitants, their Cattle and Goods, the rest being forced to save themselves in the Mountains nearest adjoyning, where they were forced to remain many dayes together in great want, till the Violence of the Tempest was over, which while it continued, was accompanied with a very violent rain that continued twelve dayes and nights together without ceasing, with that fury, that none were able to stir from their shelters, though almost famished for want of Provisions. The ships that were in the Rivers have sustained great damage, but we hope there is none of them lost. This Tempest, for the time, was so furious, that it hath made a general Desolation, overturning many Plantations, so that there was nothing that could stand its fury. We are now with all the industry imaginable, repairing our shattered houses, and gathering together what the Tempest hath left us. Although it was not alike Violent in all places, yet there is scarse any place in the whole Country where there is not left sufficient marks of its ruines. By the ships you will hear a particular of all our losses.

Such Hurricanes on the Land are seldome heard of, but Hurricanes upon the Sea are common in those parts, which are many times very prejudicial and dangerous to the ships Trading there. It was by a Hurricane that excellent Commander Lord Willoughby perished, with divers others in his Company: By these kind of Tempests the King of Spain hath lost at several times near 1000 sail of ships.[2]

More information on the meteorological characteristics of the storm is given in a letter from Secretary Thomas Ludwell to Lord Berkeley on the subject of the "Dreadful Hurry Cane" of 27 August (6 September) 1667:

This poore Country . . . is now reduced to a very miserable condition by a continual course of misfortune. In April . . . we had a most prodigious storm of hail, many of them as big as turkey eggs, which destroyed most of our young mast and cattle. On the fifth of June following came the Dutch upon us, and did so much mischief that we shall never recover our reputations. . . . There were not gone till it fell to raining and continued for 40 days together, which spoiled much of what the hail had left of our English grain. But on the 27th of August followed the most dreadful Harry Cane that ever the colony groaned under. It lasted 24 hours, began at North East and went around northerly till it came to west and soe till it came to South East where it ceased. It was accompanied with a most violent raine, but no thunder. The night of it was the most dismal time I ever knew or heard of, for the wind and rain raised so confused a noise, mixed with the continual cracks of falling houses. . . . The waves were impetuously beaten against the shores and by that violence forced and as it were crowded into all creeks, rivers and bays to that prodigious height that it hazarded the drowning many people who lived not in sight of the rivers, yet were then forced to climb to the top of their houses to keep themselves above water. The waves carried all the foundations of the fort at Point Comfort into the river and most of our timber which was very chargably brought thither to perfect it. Had it been finished an a garrison in it, they had been stormed by such an enemy as no power but Gods can restrain. . . . Had the lightning accompanied it we could have believed nothing else from such a confusion but that all the elements were at strife, which of them should do most towards the reduction of the creation into a Second Chaos. It was wonderful to consider the contrary effects of the storm, for it blew some ships from their anchors and carried them safe over shelves of sand where a wherry could difficultly pass, and yet knocked out the bottom of a ship . . . in eight foot more water than she drew. But then morning came and the sun risen it would have comforted us after such a night, had it not lighted to us the ruins of our plantations, of which I think not one escaped. The nearest computation is at least 10,000 houses blown down, all the Indian grain laid flat on the ground, all the Tobacco in the fields torn to pieces and most of that which was in the houses perished with them. The fences about the corn fields were either blown down or beaten to the ground by trees which fell upon them & before the owners could repair them the hogs & cattle got in and in most places devoured much of what the storm had left.[3]

The above behavior of the wind would place the center as passing to the east and north of Jamestown, probably on a path that carried inland over northern Virginia. We have no accounts of the storm from New England, but a reference is made in Stokes' *Iconography of Manhattan Island* to a severe storm at Manhattan on 29 August (8 September).[4] No doubt this referred to the Virginia storm and provides further evidence for placing the track inland over the present Middle Atlantic States.

To the South the hurricane also struck the Outer Banks of North Carolina where a settlement on present Colington Island near Kitty Hawk had just been established. The Colleton papers mention "a great storme or reather haricane" on 27 August (6 September) which destroyed the corn and tobacco crops, blew down the eighty-foot-long hog house on Powells Point, and severely damaged other buildings.[5]

In the West Indies, Barbadoes and Nevis were devastated by a hurricane on 19 (29) August, possibly the same which struck the Carolina-Virginia coast a week later. This storm also had the distinction of being one of the first hurricanes to be predicted by a European. Captain Langford, an experienced skipper of the Royal Navy, had already passed through and described five storms in the previous decade. Now with this knowledge, he forewarned the British fleet in the vicinity of Nevis of the impending storm and assisted them in maneuvering to escape the full fury of the blow.[6]

News of the disaster in Virginia soon spread to the other settlements. Samuel Mavericke in Boston wrote on 16 (26) October to the Secretary of State: ". . . and in Virginia on the 23rd of August (sic.) there was such a dreadful haracana as blew up all the rootes that was on the ground, overturned many houses and an abundance of trees, and drove up some vessels of burthen above high water mark many foote. . . ." His failure to mention any effects of the storm in the Boston area is significant.[7]

The only further meteorological information available states that a twelve-day rainy period followed to cause high floods throughout Virginia. Poey lists a second hurricane in the West Indies at this time, on 1 (10) September at St. Kitts, "the most violent ever known," and this again may have moved close enough to the American mainland to produce further tropical downpours along the tidelands of Virginia.

[1] C. W. Sams. *The Conquest of Virginia, The Second Attempt.* 637–638.
[2] *Strange News from Virginia,* 3–5.
[3] Sec'y Thomas Ludwell to Lord Berkeley. P.R.O. C01–21 quoted in T. J. Wertenbaker. *Virginia under the Stuarts 1607–1688,* 131–132.
[4] I. N. P. Stokes. *Iconography of Manhattan Island,* 584.
[5] David Stick. *The Outer Banks of North Carolina,* 266.
[6] Capt. Langford. Observations of his own experiences upon hurricanes and their prognosticks. *Phil. Trans.* (London), 1698, xx, 407.
[7] Samuel Mavericke, Boston, Oct. 16, 1667 to Sec'y of State. *Documents Relative to the Colonial History of the State of New York,* III, 161.

THE NEW ENGLAND HURRICANE OF 1675

Just forty years after the Great Hurricane of 1635, a storm with many similar characteristics swept the New England coastal regions. There are reports from Connecticut, Rhode Island, and Massachusetts.

At New London a "dreadful storm of wind and rain at East" occurred in August according to the journal of Simon Bradstreet which chronicled the outstanding events in Connecticut and New England.[1] At nearby Stonington many ships were wrecked and "much loss of hay and corn. Multitudes of trees blown down."[2] Peter Easton in Rhode Island, after recalling the hurricane of 1635, related that "much the like storm blew down our windmill and did much harm the 28 August (7 September) 1675."[3]

The occurrence of the 1675 storm was noted by John Hull of Boston who wrote in his diary under 29 August: "a very violent storm, that exceedingly blew down the Indian corn and the fruit of trees; did much spoil on the warves, and among the ships and vessels in Boston, to value supposed a thousand pounds."[4]

Poey lists only one hurricane in the West Indies in 1675, at Barbados on 31 August (10 September). If his dating is correct, there could be no connection between the New England and the Barbados storms.

The evidence of trees blown down along the Connecticut shore and Hull's description of damage at Boston would place the 1675 storm high on the list of New England's most destructive hurricanes, just a notch below the great storms of 1635, 1815, 1938, and 1944.

[1] Simon Bradstreet. Bradstreet's Journal 1644–1683. *N.E. Hist & Gen. Reg.,* 9, 1855, 47.
[2] *The Diary of Thomas Minor, 1653–1684,* 131.
[3] Peter Easton's Notes. *R.I. Hist. Soc. Coll.,* 11, 1918, 79.
[4] John Hull. Public Journal. *Trans. & Coll. Amer. Antiq. Soc.,* 3, 1857, 241.

HURRICANE AND FLOOD OF 1683 IN CONNECTICUT

Increase Mather in his *Remarkable Providences* (1684) stated that a hurricane took place in Virginia at the same time that a vast flood inundated the Connecticut Valley. And Thomas Minor's diary related that at Stonington on the Connecticut shore of Long Island Sound there was a "great storme that blasted all the trees, Wednesday the 15th (25th) August."[1] The date is apparently in error by two days as are many of Minor's listings. Confirmation of the storm's occurrence is found in the journal of John Pike of New Hampshire who recorded succinctly on 13 (23) August: "exceeding high tide and stormy weather."[2] Also, at

Weatherfield, Connecticut, a local historian wrote that on 13 (23) August the Connecticut River rose 26 feet above its usual level as a result of torrential rains.[3] Mather's account probably formed the basis of the above and follows:

Some remarkable land floods have likewise happened in New England. Nor is that which came to pass this present year to be here wholy passed over in silence. In the spring time, the great river Connecticut used to overflow, but this year it did so after midsummer, and that twice; for, July 20 (30), 1683, a considerable flood unexpectedly arose, which proved detrimental to many in that colony. But on August 13, a second and more dreadful flood came: the water were then observed to rise twenty-six foot above the usual boundaries: the grass in the meadows, also the English grain, was carried away before it; the Indian corn, by the long continuance of the waters, is spoiled, so that the four river towns, viz. Windsor, Hartford, Weathersfield, Middle-Town, are extreme sufferers. They write from thence, that some who had hundreds of bushels of corn in the morning, at night had not one peck left for their families to live on.

There is an awful intimation of Divine displeasure remarkable in this matter, inasmuch as August 8, a day of public humilitation, with fasting and prayer, was attended in that colony, partly on the account of Gods hand against them in the former flood, the next week after which the hand of God was stretched out over them again in the same way, after a more terrible manner than the first. It is also remarkable that so many places should suffer by inundations as this year it hath been; for at the very time then the flood happened at Connecticut, there was a hurricane in Virginia, attended with a great exundation of the rivers there, so as that their tobacco and their Indian corn is very much damnified.

. . . There are those who think that the last comet, and those more rare conjunctions of the superior planets happening this year, have had a natural influence into the mentioned inundations. Concerning the flood at Connecticut, as for the more immediate natural cause, some impute it to the great rain which preceded; others didst imagine that some more than usual cataracts did fall amongst the mountains, there having been more rain than what now fell sometimes when so such flood followed. It is not impossible but that the wind might be a secondary cause of this calamity. . . . With us in Boston it was then at first an Euroclydon (northeaster); but in the afternoon the wind became southerly, when it blew with the greatest fierceness. If it were so at Connecticut, it seems very probable that the fury of the wind gave a check to the free passage of the river, which caused the sudden overflowing of the waters.

That season at sea William Dampier, the sailor and buccaneer, captured a vivid account of a hurricane which has become a classic in marine literature. His storm occurred on 26–31 August (5–10) September somewhere in the North Atlantic three days after leaving Virginia. The account forms an interesting section of his chapter: *Discourse of the Trade-Winds, Breezes, Storms, Seasons of the Year, Tides and Currents of the Torrid Zone throughout the World* published in 1703–05.[5] Poey lists no hurricane occurrences in 1683.

[1] *The Dairy of Thomas Minor, 1653–1684,* 179.
[2] Journal of John Pike. *New Hamp. Hist. Coll.,* III, 62.
[3] H. R. Stiles. *History of Wethersfield,* I, 72.
[4] Increase Mather. *Remarkable Providences.* (George Offor., ed., 1856), 232–234.
[5] William Dampier. *A new voyage round the world.* Reprinted in *Dampier's Voyages.* (John Masefield, ed., 1906), I, 98–99.

THE SPANISH REPULSE HURRICANE OF 1686

(See account under Hatteras South, p. 41.)

THE GREAT STORM OF 1693 AT ACOMACK

A great storm which changed the course of many rivers and modified the beachline from Virginia to Long Island occurred on 19 (29) October 1693. A letter appeared in the *Transactions of the Royal Society* of London in 1697 "giving a relation of the effects of a violent storm at Acomack in America, 19th October 1693, on the rivers of that country." *Ye Kingdome of Acomack* was part of the Virginia settlements on what is now known as the Delmarva peninsula between Chesapeake Bay and the Atlantic Ocean. The letter was contributed by a Mr. Scarburgh: "Oct. 19, 1693. There happened a most violent storme in Virginia, which stopped the Course of the ancient Chanels, and made some where there never were any: So that betwixt the Bounds of Virginia and Newcastle in Pennsylvania, on the Sea-board side, are many navigable Rivers for Sloops and small Vessels." [1]

Another possible reference to this storm lies in a statement of M. B. Flint in his *Early Long Island.* He relates that tradition has it that the Fire Island Cut, east of New York City, was broken through in 1691.[2] Tradition might be wrong by two years!

[1] *Philosophical Transactions* (London), 19, August 1697, 659.
[2] M. B. Flint. *Early Long Island,* 25.

THE STORMY AUTUMN OF 1698

This was a stormy autumn along the New England shore. We are indebted to John Pike of Dover, New Hampshire, who maintained a diary of events from 1678 to 1709; for many of these years he had an addenda "Observable Seasons" which gave a brief summary of the outstanding weather happenings. The successive storms chronicled for 1698 came within the normal hurricane season and might have been of tropical origin:

Sept. 30 (Oct. 10)—Was a violent south-east storm that blew down many fences & Shattered tops of some houses and barns.

Oct. 13 (23)—A violent north-east storm produced ye like effects, nearer ye sea it fell rain, higher up in the country snow.

Oct. 19 (29)—A violent north-east storm—melted snow—caused freshets higher than ever known.[1]

The same sequence of events would occur again in October 1783.

[1] Journal of John Pike. *New Hamp. Hist. Coll.*, III, 64–65.

HATTERAS NORTH: 1701–1814

THE YEAR OF THE BIG WINDS—1703

The early years of the Eighteenth Century brought memorable storms to both sides of the Atlantic Ocean. It was on 26–27 November (7–8 December) 1703 that the *Great Storm* raged across the British Isles creating what was probably the most powerful wind force ever experienced in the modern annals of southern England.[1] The storm's fame has been recorded and immortalized by its historian, Daniel Defoe, whose *The Storm: or, A Collection of the most remarkable Casualties and Disasters which happened in the late Dreadful Tempest, both by Sea and Land,* published in 1704, is a meteorological classic.[2]

A month and a week earlier another severe storm had struck the Middle Atlantic coast of America; some writers have associated these two as parts of the same storm system, but the time interval separating them is much too long, from a meteorological point of view, to associate them together as emanations of the same disturbance. Nevertheless, the two trans-Atlantic occurrences indicated that the late fall of 1703 was a storm-breeder of the first order.

Our main account of the American hurricane of 7 (18) October 1703 is taken from printed excerpts of the manuscript diary of Rev. Sandel, a Swedish clergyman then living along the shores of the lower Delaware River: "On the 18th of the same in the evening, a hurricane arose which caused great damage. In Maryland and Virginia, many vessels were cast away, several driven to sea, and no more heard of. Ten tobacco houses belonging to one man were overturned. In Philadelphia, the roof of a house was torn off. A great number of large trees blown down." [3]

Rev. Sandel also mentioned in a previous sentence that an unusual early season snowstorm had taken place just eight days before the hurricane, on Michaelmas Day (29 September or 10 October, new style). This fact greatly assists now in identifying the year and the date of the storm's occurrence, as the *Journal of Rev. John Pike* of Dover, New Hampshire, also mentioned the early snow of 1703 and further remarked: "Oct 6, 7, and 8 (17, 18, and 19) very cold storm of rain." [4] Though there is no mention of wind damage in southeastern New Hampshire, the track of the storm probably moved close to the New England shore and induced a cold northeasterly air flow over New England. No other material on this storm has come to light since the first regular newspaper in America did not commence until the following year, 1704. Perhaps additional in-

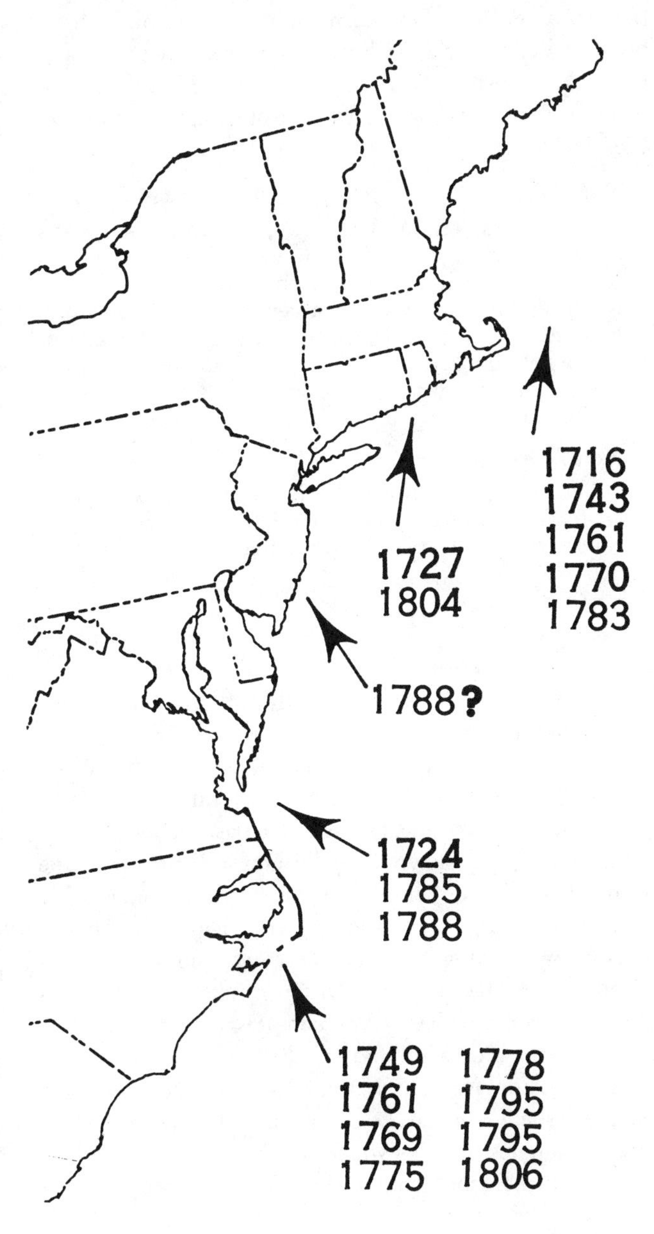

formation lies buried in the manuscript of some other unpublished diarist.

[1] Carr Laughton and V. Heddon. *Great Storms.* London, 1927.

[2] Daniel Defoe. *The Storm: or, A Collection of the*

most remarkable Casualties and Disasters which happened in the late Dreadful Tempest, both by Sea and Land. London, 1704.

[3] Nicholas Collin. Observations made at an Early Period, on the Climate of the Country about the River Delaware. Collected from the records of the Swedish Colony. *Trans. Amer. Phil. Soc.,* 1, n.s., 1818, 350–351.

[4] Journal of Rev. John Pike. *Collections of the New Hampshire Historical Society,* 3, 1832, 65.

THE STORMY SEASON OF 1706

The Boston *News Letter* carried several accounts of storms in the late season of 1706. Our meager information on a mid-October storm hints that it might have been similar to the 1938 hurricane in respect to causing floods in eastern New York and western New England. Another blow two weeks later would appear to have been well offshore:

New Haven, 16 Oct.

On the 3d, 4th, and 5th (14, 15, and 16th) days of this instant, there fell an excessive rain, which caused an unusual flood in the Connecticut River, which was accompanied by a very hard Gale of Wind between the East and South, to the great damage of the people on that river, it covered all the meadows, and carried away near 1000 loads of hay, besides the loss of the second crop of greass, and other losses." [1]

New York, 14 Oct.

We have great rains here and mighty Floods; They write from Albany, that the late great Rains has caused the greatest Flood there that ever was known; they say that Renslaers Island was 6-foot under water, and that it drowned their low lands.[2]

In the following spring the Boston *News Letter* carried an account of a severe storm which had raged offshore from Virginia to New England on or about 26 October (6 November) 1706. An England-bound fleet sailing from New England was scattered by a great wind when two days at sea. A Virginia fleet, also sailing about the same time, met a storm soon after its departure. Fourteen of the ships were known to have foundered, and others were given up for lost. Many returned to Virginia ports to refit after extensive damage to masts and sails. The first news of these twin disasters was apparently carried back from England by a ship arriving at Boston in March 1707.[3]

[1] Boston *News Letter,* 28 Oct 1706.
[2] *Idem.*
[3] *Ibid.,* 24 March 1707.

THE 1716 HURRICANE

A violent storm of wind and rain visited eastern Massachusetts on 13–14 (24–25) October 1716, apparently the western part of a severe gale which swept the shipping channels offshore with devastating effect. Samuel Sewall, the eminent Boston diarist, noted "many trees, fences, etc. blown down." [1] Out at Truro on Cape Cod, Moses Paine, witnessing the storm, marked in his diary: "Oct. 14, 1716—being Lords Day, and an excessive wind so great that there was no meeting in Eastham." [2] The hurricane was also noted on Martha's Vineyard where Rev. William Homes called it a "violent storm of wind and rain." [3]

The Boston *News Letter* did not mention the storm in the issue following its occurrence, but later carried several reports of ships which had met its virulence offshore. One Europe-bound ship out of Boston on the 11th encountered high winds about 30 leagues at sea.[4] Another enroute from Jamaica to New York met the gale just before putting into New York harbor.[5] From the New Hampshire port of Piscataqua came the report: "since the last great storm we have 6 or 7 vessels come in here, and every one has lost their masts." [6] Neither Poey nor Tannehill list any hurricane occurrences in 1716.

[1] Samuel Sewall. *Diary.* 3, 107.
[2] Diary of Moses Paine. *N.E. Hist. & Gen. Reg.* 54, 1900. 88.
[3] Diary of Rev. William Homes of Chilmark. *Ibid.,* 49, 1895. 414.
[4] Boston *News Letter,* 29 Oct 1716.
[5] *Ibid.,* 5 Nov 1716.
[6] *Ibid.,* 29 Oct 1716.

THE GREAT GUST OF 1724

Though the severe tropical storm which struck the Chesapeake Bay area of Virginia and Maryland in August 1724 produced a long-lasting impression, very few meteorological details have been preserved. The *Virginia Gazette* twenty years later in describing a December storm of 1744 remarked: "The like has not been known in the Memory of Man, not even in the great Gust in the year 1724." [1]

The occurrence of the Great Gust in Virginia has been definitely dated through a letter of Lt. Governor Drysdale to the Council of Trade and Plantations in London. Drysdale comments: ". . . had it not that violent storm which happened the 12th (23rd) of August almost wholly destroyed all the tobacco on the ground . . . nor has the storm affected only one crop of tobacco, but the country suffers very much for want of corn. . . ."[2] In a second letter Drysdale pointed out the economic stress in the community as a result of a poor crop in 1723 and the disruption of prices resulting from the damaged crop in 1724. A temporary prohibition in export of Indian corn was put into effect following the damaging hurricane.[3]

A contemporary witness, John Custis, who lived on the James River, wrote with a date line 12 (23) August 1724: "We have had such a violent flood of rain and prodigious gust of wind that the like I do not believe never happened since the universal deluge." He added that most of the tobacco in the region had been destroyed, some homes wrecked, and several vessels driven ashore.[4]

If our dating of the occurrence of the hurricane in Virginia is correct, there must have been a second one which followed closely on the heels of the Great Gust. There are two bits of contemporary evidence for this surmise: (1) Governor Nicholson of South Carolina wrote to the Duke of Newcastle on 25 August (5 Sept.) 1724): "We had on ye 17th instant a sort of hurricane which I thank God did no damage to the shipping here, but ye violence of the rain and wind I hear hath damnified some of the Indian corn and rice and ye flood hath ruined some of the bridges."[5] (2) A dispatch in the Boston press from Rhode Island described the arrival of a ship which had lost both masts in a great storm offshore on the 19th (30th).[6] Another press report from Philadelphia mentioned that a sloop had been driven ashore recently at Lewis Town on Delaware Bay in a storm (no date given).[7]

The continued rains in Virginia would give credence to the belief that at least two tropical storms were involved in the floods of 1724. Unfortunately, neither Tannehill or Poey make any mention of even a single hurricane as occurring in 1724.

[1] *Va. Gaz.*, 12 Dec 1744, in *Penna. Gaz.*, 29 Jan 1745.
[2] *Calendar of State Papers, America and West Indies, 1724–25*, 29, 318.
[3] *Ibid.*, 379.
[4] John Custis to (William Byrd II?), 12 Aug 1724. Ms. Custis Letterbook, 1717–41, LC, 18, quoted in Arthur R. Middleton, *Tobacco Coast,* 48.
[5] *Calendar State Papers,* 29, 214.
[6] Boston *News Letter,* 23 Sept 1724.
[7] *Idem.*

"A GREAT RAIN AND HORRIBLE WIND"—THE HURRICANE OF 1727

The hurricane of 1727 lived long in the memory of residents of eastern Massachusetts. Some octogenarians about Boston as late as 1804 recalled the similarities between the storm of 16 (27) September 1727 and the small destructive hurricane of 9 October 1804. It was recalled then that the Rev. Samuel Phillips had preached a sermon in December 1727 about the calamities which beset New England that year: "Then the Lord sent a great rain and horrible wind; whereby much hurt was done, both on the water and on the land."[1]

The *Weekly News Letter* of Boston in September 1727 carried the following account:

> On Saturday last the 16th Instant, we had here a violent storm of wind and rain, which lasted till about midnight, whereby great damage was done to the wharfs and shipping in the harbor, as also to the fruits of the earth, and to many buildings: Many trees blown up by the roots, and chimnies blown down: a kitchen chimney of Mr. Jacob Sheaf's at the bottom of the Common, blew down and beat on the roof, which killed a child of about 7 years of age, wounded two others, and broke the bone of Mrs. Sheaf's leg. But it is pleasant to behold very early in the next morning, which lasted a considerable length of time, a very fair rain-bow, a token of the covenant between G O D and the earth that the waters shall no more become a flood to destroy all flesh. Gen.9.11–17."[2]

From Marblehead came a more enlightening meteorological report: "The last Saturday we had a most terrible storm of wind and rain, such as has not been known among us, which began about noon and continued to about twelve at night. The wind was at N.E. and N.E. by N. which brought in a very high tide. . . . It blew up many trees by the roots."[3] Great damage occurred among the ships and vessels in the ports northeast of Boston. The main fleet at Marblehead was fortunately at sea; otherwise a major disaster might have occurred.

An idea of the ferocity of the storm on land appears in the diary of Stephen Jacques, as quoted by Newbury's historian: "In the month of September on Saturday in the afternoon the wind began to be strong and increased more in the night. It blew down and brake six trees in my old orchard and trees all over the woods. There never was the like known. It twisted young walnut trees in the midst. It raised a great tide, which swept away near two hundred loads of hay that was in swath."[4]

William Douglass, a Bristol physician who had settled

in Boston with the expressed intention of studying meterological conditions there, made a survey of the storm's activity and damage in much the same manner as William C. Redfield would do a century later. He interviewed captains of ships which arrived at Boston during the days and weeks following the blow. By November he was able to send the details to his friend and fellow physician, Cadwallader Colden in New York. The storm extended from the latitude of Bermuda northward to 43°N., but was not felt farther than that, and extended out to sea about 200 leagues (600 miles). The wind at Boston came out of the east-northeast with it main violence occurring toward evening. "The strength of its central current passed through the county of Essex" (northeast of Boston). He thought it the most violent wind in years: "it drove all of our fishing shallops at Marblehead in Essex ashore, brought down chimnies, overset barns, rooted up a vast number of trees to the ruin of some orchards. . . ."[5]

The hurricane also struck coastal Connecticut and Rhode Island a severe blow. At New London, according to the farmer-diarist Joshua Hempstead, the northeast wind was accompanied by rain, and many trees were uprooted.[6] At Swanzey on Narragansett Bay, perhaps closer to the center, "it blew up trees by the roots in abundance; blew down several chimneys, and blew off the roof of a house, and blew sundry vessels on shore."[7]

The center of the storm must have passed over some land area of southeastern Massachusetts. The east-northeast wind at Boston would indicate a close proximity as would the severe destruction noted in Essex County. We have no reports from Cape Cod at this time which would give a clue to the exact track. The evidence of trees torn up by the roots from Connecticut to Cape Ann, northeast of Boston, would indicate that the storm packed a full hurricane punch even on its western side. Neither Poey or Tannehill record any hurricane in the tropics in 1727.

[1] Rev. William Bentley. *Diary,* III, 117.
[2] *The Weekly News-Letter* (Boston), 21 Sept 1727.
[3] *New England Weekly Journal* (Boston), 27 Sept 1727.
[4] Joshua Coffin. *History of Newbury,* 197.
[5] Douglass to Colden, 20 Nov 1727. *Mass. Hist. Soc. Coll.,* IV, 2, 172–173.
[6] Joshua Hempstead. Diary. *Coll. New London Hist. Soc.,* 1901, I–2, 188.
[7] Diary of John Comer. *Coll. R. I. Hist. Soc.,* VIII, 45.

BENJAMIN FRANKLIN'S ECLIPSE HURRICANE

The storm which raced northward along the Atlantic Coast on 22 October (2 November) 1743 deserves a unique place in the annals of American meteorology. Not only was this the first tropical storm in America to be measured accurately by scientific instruments, but it also provided Benjamin Franklin with a key to unlock for the first time the secret of a storm's forward movement.[1]

John Winthrop, professor of Natural Philosophy at Harvard College, had commenced a meteorological register at Cambridge in 1742. His notation for 22 October (2 November) 1743 follows: "NE by N. worst in years—great damage on land as well as at sea. Barometer 29.35″. Tide within 4″ of 20 years ago. Storm abated about 7 P.M Barometer lowest at 2 P.M."[2]

The Boston press gave the storm full coverage. The most informative account appeared in the Boston *Evening Post:*

> Last Friday night, soon after a total and visible eclipse of the Moon (which began about nine and ended past one o'clock) came on a storm of wind and rain, which continued all the following day with great violence, and the wind being at N.E, the tide was raised as high within a few inches, as that remarkable one about 20 years ago: And as Dr. Ames had given no hint in his almanack of these events, (For which Omissions let him answer) which might have put the people on guard, the greatest damage by far has been done here, that was ever known to be done by a storm in the memory of man.[3]

The Boston *Post Boy* described the damage about the harbor:

> The wind being excessive high vast damage was done to the wharves and shipping, some vessels that got loose were drove ashore higher up than was ever known before, and several small vessels were cast upon the wharves and boats floated into the streets. . . . Tis impossible to enumerate all the particulars the terrible effects of this storm or estimate the damage sustained by it.[4]

The editor of the Boston *News-Letter* added:

> At noon the wind seemed to blow in prodigious gusts, and with the greatest fierceness and brought in an exceeding high tide, which overflowed most of our wharves, and came up into several streets higher than has been known for these twenty years past.[5]

To the northward the storm also was violent. At Piscataqua, New Hampshire, the tide rose very high and overflowed the wharves. At Newbury much damage was done, as usual, to the fields and salt marshes.[6]

Southward along the coast the disturbance appeared earlier. A Philadelphia paper mentioned "a violent Gust of Wind and Rain attended with Thunder and Lightning" on Friday evening.[7] This stormy weather greatly dismayed Benjamin Franklin in the Quaker City since he had wished to observe the scheduled eclipse of the moon that evening, but was prevented from doing so by the cloud cover attending the northeast storm. Later he wrote his brother at Boston about his disappointment, assuming that he, too, had been prevented from seeing the spectacle since the cloud-bearing wind current seemed to be moving from the northeast. Much to his surprise, Franklin later learned that the eclipse had been seen at Boston, but that clouds arrived and a northeast wind set in soon afterwards, to be followed by a violent storm.

Franklin noted the difference in time between the onset of the storm at Philadelphia and Boston and reasoned that it must have been the same disturbance, traveling from southwest to northeast against the current of the surface wind. He embodied this thought later in a letter to Jared Eliot in 1750, and his correct induction is generally regarded as the first tangible progress in trying to understand what the next century would refer to as the "law of storms." [8]

In later years there was much speculation as to just what day and to what storm Franklin had made reference since his letter was indefinite on this point. In 1833 his great-grandson Alexander Bache traced by astronomical calculations the occurrence of the eclipse and correctly dated the storm.[9] And as late as 1906 some comments on the same subject appeared in the publications of the American Philosophical Society to establish definitely that Franklin was the originator of this basic principle of storm movement.[10] The above controversies could probably have been avoided if a scientifically-minded historian had previously noted the significance of the first sentence of the Boston *Evening Post's* write-up of the 22 October 1743 storm as given above.

[1] Tannehill states erroneously that this storm occurred in September 1743.

[2] John Winthrop. Ms. Meteorologic Observations at Cambridge in New England. Houghton Library, Harvard.

[3] Boston *Evening Post*, 24 Oct 1743.

[4] Boston *Post Boy*, 24 Oct 1743.

[5] Boston *News Letter*, 27 Oct 1743.

[6] Boston *Post Boy*, 31 Oct 1743.

[7] *Penna. Gaz.*, 27 Oct 1743.

[8] Franklin to Jared Eliot, 13 Feb 1750. *The Papers of Benjamin Franklin.* (Leonard W. Labaree, ed.). New Haven, 1961. 3, 463.

[9] A. D. Bache. Attempt to fix the date of Dr. Franklin's observation, in relation to the North-east Storms of the Atlantic States. *Journal of the Franklin Institute,* Nov 1833, ns., 12, 300–303.

[10] William M. Davis. Was Lewis Evans or Benjamin Franklin the first to recognize that our Northeast storms come from the Southwest? *Proc. Amer. Phil. Soc.*, 45, 1906, 129–130.

THE OCTOBER HURRICANE OF 1749

The Middle Atlantic Coast from North Carolina to New Jersey experienced a severe storm on 7–8 (18–19) October 1749. The center may have moved inside Cape Hatteras, but appears to have remained a short distance offshore from Virginia northward to New Jersey and then passed close to the islands of southeastern Massachusetts and Cape Cod. At Ocracoke on the Outer Banks of North Carolina, of ten vessels awaiting for a fair wind to put to sea, all were lost save one. Two ships were driven over the bar and sunk five miles to the northward when the tide rose ten feet higher than ever witnessed before, according to a report by Captain Rivers.[1]

Our most informative account is contained in a letter written at Annapolis, Maryland, on the 14th (25th) describing the storm's destruction at Norfolk. On Saturday night, the 7th (18th), the wind began to blow hard and by 0100 was "very violent at NE with rain." The hardest portion of the storm occurred from 1000 to 1400 on the 8th (19th). The bay waters rose 15 feet perpendicular, according to the witness: "The tide kept continually fluxing and run at the rate of five knots an hour, overflowing all their streets and has carried some small craft near a mile from common high water and left some in cornfields." Damage at Norfolk was estimated at £30,000.[2]

The storm struck hard at the entire lower Chesapeake Bay area "with a great gust of wind and rain" on Sunday. Near Williamsburg some houses were carried away by the flooding waters and one entire family drowned. At Hampton the water rose four feet deep in the streets; trees were torn up by the roots, others snapped off in the middle. "The like storm has not been known here in the memory of the oldest men," concluded the account.[3] In the upper country of Virginia and Maryland torrents of tropical rain caused great freshes on all the streams and rivers.

At Annapolis the tide ran high, but not so great as one which occurred during the previous summer. No damage occurred in the immediate Annapolis area.[4] At Lewes on Delaware Bay the raging ocean cut a passage through the beach near Cape Henlopen into the Bay with a five-foot clearance so that small boats could sail through. The force of the wind there uprooted trees and drove many small craft ashore.[5]

At New York City the "violent gale" out of the east and northeast left many small craft ashore high and dry.[6] From Boston the *Evening Post* reported: "We had for the time it lasted as violent a gale of wind as has been known, which did considerable damage to shipping in this harbor." Seven vessels were ashore at Martha's Vineyard.[7] Edward Holyoke, president of Harvard, entered in his diary: "a violent storm at NE on the 8th." [8]

It is interesting to note that Benjamin Franklin at Philadelphia had this storm under surveillance. From the fact that it was first reported in North Carolina and Virginia on the 7th and in New England on the 8th, he drew confirmation for his hypothesis that coastal storms moved from the southwest and were preceded by northeastly winds.[9]

[1] *Penna. Jour.,* 23 Nov 1749.
[2] Extract of a letter from Annapolis, 14 Oct 1749. *Post Boy* (Boston), 6 Nov 1749.
[3] *Va. Gaz.,* 12 Oct 1749.
[4] *Post Boy,* 6 Nov 1749.
[5] Boston *News Letter,* 2 Nov 1749.
[6] *Post Boy,* 23 Oct 1749.
[7] Boston *Evening Post,* 16 Oct 1749.
[8] *Holyoke Diaries,* 10.
[9] Franklin to Alexander Small, 12 May 1760. *The Complete Works of Benjamin Franklin.* (John Bigelow, ed.). New York, 1904. 3, 260.

THE SOUTHEASTERN NEW ENGLAND HURRICANE OF 1761

A major hurricane hit southeastern New England on 23–24 October 1761. President Edward Holyoke of Harvard thought it "as great a storm as I have ever known," [1] and historian Jeremy Belknap, then a student at Cambridge, judged the blow the "most violent storm in 30 years," [2] probably refering back to the September 1727 hurricane. The intensity of the wind in the Boston area may be gathered from the testimony of John Whiting of Dedham who tells of the northeast gales "tearing up whole trees by the roots," exhibiting a force which would put the storm in the full hurricane class.[3]

The meteorological history of this storm is not very complete. There is no mention of the disturbance in the press prior to its striking the Rhode Island and Massachusetts coastline. For the Boston area John Winthrop's journal indicates that the severe period of the storm commenced at 2100 on the 23rd, continued to 0200 the following morning, a five-hour period, with remote lightning and thunder lasting for two more hours. Unfortunately, Winthrop failed to record his barometer reading at 1215 on the 23rd, though noting the wind northeast at only force 1. As was his custom, he did take a reading during the height of the storm at 0145 on the morning of the 24th: the pressure stood at 29.57" and the wind was at northeast force 4, his highest rating.[4] The Boston *News Letter* account agrees with Winthrop as to the timing of the storm: "Last Friday evening between 8 & 9 o'clock came on the severest NE storm of wind and rain that has been known here for 30 years past and continued till between 2 and 3 o'clock next morning." [5]

At Salem, Dr. Edward A. Holyoke noticed very moist air on both the 22nd and 23rd with southerly winds. Late on the 23rd the wind went to northeast with "a most violent storm for six hours" following.[6] Farther north reports received from ships coming down from the Bay of Fundy, Casco, and Piscataqua indicated that the gale reached those points, but that no material damage had been done.[7] The high winds were also felt inland as far west as Rutland in Worcester County, Massachusetts.[8]

The area of greatest material damage lay in Rhode Island and southeastern Massachusetts. Around Providence "on both roads east and west so far as we have heard, the roofs of houses, tops of barns, and fences have been blown down, and it is said thousands of trees have been torn up by the roots by the violence of the above storm." [9] Weybosset Bridge across the head of Narragansett Bay at Providence was wrecked by the wind and tide. At Newport the steeple of Trinity Church, a favorite target for hurricane winds, crashed to the ground during the blow.[10]

[1] *Holyoke Diaries,* 24.
[2] Jeremy Belknap. Ms. Diary 1761. Mass. Hist. Soc., Boston.
[3] Diary of John Whiting. *N.E. Gen. & Hist. Reg.,* 63, 192.
[4] John Winthrop. Met. Obs.
[5] Boston *News Letter,* 29 Oct 1761.
[6] Edward Augustus Holyoke. Ms. Meteorological Journal. Houghton Library, Harvard.
[7] Boston *News Letter,* 29 Oct 1761.
[8] *Journal of Seth Metcalf,* 18.
[9] Boston *News Letter,* 29 Oct 1761.
[10] S. G. Arnold. *Hist. of Rhode Island,* II, 232.

THE SEPTEMBER HURRICANE OF 1769—II

A traveler who viewed the damage of the September Hurricane of 1769 around New Bern and then traveled northward to Norfolk thought that the North Carolina towns had suffered more greviously than the Virginia communities through which he passed.[1] No doubt, the storm tide on the Carolina rivers far exceeded anything farther north. Nevertheless, this hurricane struck a mighty blow at the Middle Atlantic and New England coasts and must be placed among the more destructive storms of the century there.

Williamsburg, the colonial capital of Virginia close to Chesapeake Bay, lay near the track of the center. About 0100 on Friday morning the 8th, the local *Virginia Gazette* related that there "came on at northeast a most dreadful hurricane." The gale blew violently until between 1000 and 1100 and "then shifted northwest, when the storm increased, and continued without any abatement, until about dinner time." [2] The increase in wind force after the wind shift appears to have been a characteristic of this storm's western sector as the same condition was mentioned in a New Jersey account.[2]

Heavy damage resulted throughout the Chesapeake Bay area. Four ships in York River were driven ashore; another, newly arrived from England, had to cut both main and mizzen masts and, thus stripped, rode out the gale in the river. The wind rush carried away the top of a wharf at Yorktown, and a schooner ran its bowsprit into a nearby storehouse.[3] The damage appeared confined to the tidewater areas, not being felt inland in the Winston-Salem region of North Carolina nor in the Staunton sector of the Blue Ridge Mountains of western Virginia.[4]

Farther up the Bay at Annapolis the residents on Thursday night and all the next day experienced "the most violent storm of wind and rain from the northeast." In the lower part of the province over 100 tobacco houses were blown down and the crops of corn and tobacco leveled. Rain was reported to have come through walls which were 14 inches thick.[5] The intensity of the rainfall in this "gust" aroused the scientific interest of Lewis Nicola. Next year he fashioned a rain gage and commenced a record of Virginia precipitation which he communicated to the American Philosophical Society in Philadelphia.[6]

At Philadelphia a northeast storm with rain on the 7th continued through the next day, "the wind increasing till it became a mere hurricane." The strongest blasts came from the north so shipping in the river did not suffer the damage that a southeast gale would cause.[7] But on land it was a different story with many trees torn up and all rivers in flood. Jacob Hiltzheimer relates in his diary that on the 9th he "walked to Centre woods, where I found above 100 trees blown down." [8]

A shipwreck occurred at Barnegat Beach in New Jersey when a sloop was driven on land by an east-northeast gale. The crew made it ashore soon before 1800 when the "wind suddenly shifted to north-northwest and blew a more violent storm than before, with heavy rain and hail." [9] The storm hit the New Brunswick area in central Jersey as a severe northeast blow with much damage from wind and flood—trees down, bridges carried away, and roads "hurt." [10] New York City was swept by a "violent gale of wind at NE accompanied by a heavy rain which continued without intermission till past 2000." Trees were torn up in the city and corn crops greatly damaged.[11]

The storm then roared along the New England coast being felt as a strong northeast gale at New London, Newport, Cambridge, Boston, Dover, and Portland. At the last place weather-watching Rev. Thomas Smith described "a dreadful N.E. storm." [12]

John Winthrop's barometer on the Harvard campus dropped to a reading of 29.57″ at 2215 on the evening of the 8th during "a great storm of wind and rain. Lightning." It was his custom to write down an observation at the time of lowest barometer during a severe storm. If this holds true in the present case, the storm center would have traveled from a point directly east of Williamsburg at 1030 to a point east of Boston at 2215, a period of 12 hours, at a forward rate of about 40 mph. Winthrop's rain gage caught a total of 3.69″ during the storm.[13]

[1] *Va. Gaz.,* 14 Sept 1769.
[2] *Idem.*
[3] *Idem.*
[4] *Records of the Moravians in North Carolina,* I, 383; *Va. Gaz.,* 21 Sept 1769.
[5] *Md. Gaz.,* 14 Sept 1769.
[6] Ms. Communication to Amer. Phil. Soc., Natural History, I, 10A.
[7] *Penna. Gaz.,* 14 Sept 1769.
[8] *Diary of Jacob Hiltzheimer,* 18.
[9] *Penna. Gaz.,* 28 Sept 1769.
[10] *Journal of Elizabeth Drinker,* 24.
[11] *Penna Gaz.,* 14 Sept 1769.
[12] Ezra Stiles. Ms. Thermometrical Register. Yale Univ. Library; Jeremy Belknap. Diary; *Journal of Rev. Thomas Smith,* 22. Boston *Evening Post,* 11 Sept 1769.
[13] John Winthrop. Met. Obs.

THE LATE SEASON STORM OF OCTOBER 1770

A late season storm in October 1770 struck the New England Coast from eastern Connecticut to Maine and achieved lasting reknown by bringing in a tremendous tide, said to be the highest since the much publicized harbor flood of February 1723. All winds reported from land stations were from northeast to north; since we do not have any data from Cape Cod or the Islands, we do not know whether the center of the storm cut across any outer land areas there.

The *New London Gazette* reported a northeast storm

on Friday night the 19th which continued into Saturday and drove two vessels on shore.[1] Farther east at Newport, Ezra Stiles, an avid storm-watcher, described the blow on the 20th as "a violent hurricane. Wind N or NE. Rain violent—hail—vane of church steeple blown off—cleared up after 1500." Again this was the much-abused Trinity Church spire which had suffered in the 1761 hurricane. Stiles' thermometer dropped to 35.5° by 1900 and a stiff west wind was blowing.[2] The mention of hail (probably actually ice pellets or modern sleet) would indicate that a very cold air mass lay to the west of the storm track over western and northern New England, a situation that would certainly intensify the storm circulation.

The Boston area lay within the track of high winds. "Last Saturday morning," reported the *Massachusetts Gazette,* "a most violent storm came on the wind about NNE, attended with rain and hail. The tide rose about noon to such a height that it overflowed most of the wharves in this town . . . it is said that the tide rose higher than has ever been known, excepting once about 47 years ago, it rose a foot higher." [3] The tide at noon overflowed King Street as far as Admiral Vernon's Head Tavern and also into Dock Street and around the drawbridge.[4] Fort William in the harbor also received a severe beating. Capt. John Montresor, on an inspection trip when the storm hit, reported the sentry-boxes carried away, wheelbarrows, framework, and timber scattered, chimneys downed, and most of boats stove. Wind was NNE there.[5]

At Cambridge, John Winthrop noted an extremely low barometer of 28.96 inches at 1500 on the 20th. Wind was northeast force 3, one below his highest rating. Temperature 67.7°. "A great storm does a vast damage," he wrote. Rain and hail continued into the evening for a total precipitation figure of 2.48 inches.[6]

The storm area extended northward and northeastward. At Portsmouth several buildings were blown down along with many fences.[7] Here, too, a high tide did much damage to wharves and goods. Inland at Bedford near Manchester, Matthew Patton noted a very great storm of rain and hail from the northeast that "worked backward to the north." It rained 24 hours steadily there.[8] And at Portland Rev. Thomas Smith thought it "an exceeding great N.E. storm." [9]

[1] *New London Gaz.,* 26 Oct 1770.
[2] Ezra Stiles. Ms. Thermometrical Register. Yale Univ. Library.
[3] *Mass. Gaz.* & Boston *News Letter,* 25 Oct 1770.
[4] Boston paper in *N. H. Gaz. & Hist. Reg.* (Portsmouth), 2 Nov 1770.
[5] Journal of Capt. John Montresor. *Coll. N. Y. Hist. Soc.,* 1881, 402.
[6] John Winthrop. Met. Obs.
[7] *Mass. Gaz.,* 29 Oct 1770.
[8] *Diary of Matthew Patton,* 255.
[9] *Journals of Rev. Thomas Smith,* 277.

THE INDEPENDENCE HURRICANE OF 1775

A savage hurricane swept out of the tropics in September 1775 just as the opening maneuvers of the War of Independence were in progress. Smashing over settlements ashore and overwhelming ships at sea, the storm raged from North Carolina to Newfoundland, exacting a toll of human lives higher than any previous American mainland hurricane.

A storm occurred at Martinique on 25 August and at Santo Domingo two days later—possibly the same disturbance which brushed the coast of South Carolina, then raced across eastern North Carolina on 2 September.[1] The *North Carolina Gazette* at New Bern reported: "We had a violent hurricane the 2d instant which has done a vast deal of damage here, at the Bar, and at Matamisket, near 150 lives being lost at the Bar, and 13 in one neighborhood at Matamisket." [2] Since the gales on the 2nd had destroyed all the corn growing in the fields of Pasquotank County in the northeastern part of the colony, the Legislature voted to allow 40 shillings extra to each militiaman for purchase of food while enroute to assembly points.[3]

A Williamsburg, Virginia, paper noted: "the shocking accounts of damage done by the rains last week are numerous; most of the mill dams are broke, the corn laid almost level with the ground, and fodder destroyed; many ships and others drove ashore and damaged at Norfolk, Hampton, and York." [4] A second press gave more meteorological data: rain every day during the week—Saturday it never ceased pouring down and about noon the wind began to rise and blew furiously from the northeast until near 2200 of the 2nd.[5]

The path of the center of the storm apparently lay inshore, probably traversing Chesapeake Bay from south to north and then passing through eastern Pennsylvania in much the same fashion as Hazel would do in 1954.

The reliable meteorological register of Phineas Pemberton at Philadelphia showed a northeast wind on September 1st and 2nd, but at observation time of 0800 on the morning of the 3rd, the wind had veered into the southeast. His barometer then read 29.50 inches. Thereafter the glass rose, and the wind by 1500 was in the southwest. His side notes are pertinent:

> Sept 3—Stormy & showery. A violent gale from NE to SE the preceding night with heavy rain, lightning and thunder—a remarkably high tide in the Delaware

this morning. Flying clouds & wind with sunshine at times P.M.[6]

The local press called this the highest tide ever known.[7] At Philadelphia it usually takes a strong southeasterly gale which continues over several hours to raise a great tide.

New England lay well to the east of the central track of the hurricane. At Newport, Rhode Island, the wind also went around from northeast at 1000 on the 3rd to southeast by 1430 with "plentiful rain." But two hours later the sky was fair though the wind continued in southeast.[8] At Salem Dr. Holyoke noticed much rain and moist air.[9] Up in New Hampshire at Bedford "the wind high at about ENE and in the afternoon it rained considerably" according to Matthew Patten's notation for the 3rd.[10]

Shortly afterwards another storm was reported to have struck Newfoundland with unprecedented force. A traveler from Halifax related to the Boston press that the blow occurred on 9 September, a full six days after the above hurricane passed through New England. It is possible that the Newfoundland storm could have been the same, but it is also quite logical to believe that it was a second storm which swarmed northward in the congenial atmosphere left by the earlier disturbance. The Newfoundland blow caught many fishing boats on the Banks and drove them ashore before they could seek shelter in harbors. Four thousand sailors were reported to have drowned, most of them from Irish and English fishing ports. Total losses in ships, fish, oil and merchandise ran to £140,000. On land roofs were torn off, chimneys crumbled, and houses collapsed from the force of the wind.[11] Certainly, this was a major hurricane of the period under study.

[1] Poey, 13; also *Monthly Weather Review,* Feb 1906, 72; *S. C. & Amer. Gen. Gaz.,* 8 Sept 1775.
[2] New Bern, 9 Sept 1775, *Va. Gaz.,* 21 Oct 1775.
[3] *Colonial Records of N. C.,* 10, 574.
[4] Williamsburg, 9 Sept 1775, *Md. Gaz.,* 14 Nov 1775.
[5] Williamsburg, 8 Sept 1775, *Penna. Gaz.,* 20 Sept 1775.
[6] Phineas Pemberton. Ms. Meteorological Register (Amer. Phil. Soc.).
[7] *Penna. Gaz.,* 6 Sept 1775.
[8] Ezra Stiles. Thermometrical Register.
[9] E. A. Holyoke. Met. Journal.
[10] *Diary of Matthew Patten,* 347.
[11] Cambridge, 7 Dec 1775, *Penna. Magazine,* Dec 1775, 581; Boyle's Journal, *N.E. Hist. & Gen. Reg.,* 85, 27.

THE ORDERING OF PROVIDENCE: THE HURRICANE OF AUGUST 1778—II

A hurricane of moderate size moved along the familiar tropical storm track close to the Georgia and South Carolina coasts on the 10th and 11th of August 1778. After making a landfall in eastern North Carolina and doing extensive damage to crops, the storm center apparently moved northeastward out to sea as there are no further reports of activity to the north. The hurricane's future course might have been lost to history if two powerful fleets, each anxious to do battle, had not been maneuvering for wind advantage off the New England shore.

The military situation at Newport, Rhode Island, in early August 1778 bore many resemblances to that which would arise three years later in the Chesapeake Bay area. The Continental Army under General Sullivan was attempting to seal off the Narragansett Bay area and hem the British Army on the southern part of the island upon which the town of Newport stood. Admiral D'Estaing, newly arrived in America with 12 men-of-war and four frigates, sailed past the British batteries into the bay to assist in assaulting the enemy position. At this moment Lord Howe out of New York arrived off Point Judith with a fleet somewhat inferior in fire power to the French. Howe hovered off the harbor entrance enjoying the weather-guage with a seasonal southwest wind blowing. D'Estaing, eager to give battle on this occasion, sailed out of Newport harbor early on the morning of the 10th when an unexpected shift of wind to the north enabled him to maneuver past the British batteries and gain the wind advantage over Howe.[1]

The British retired southward, hoping for a change of wind which would return them the weather-gauge. The French were unable to come up to the skillful withdrawal of Howe on the 10th, but next day with the wind now from an easterly quarter were rapidly closing on the British in the late afternoon when the first puffs of a hurricane approaching from the southward began to kick up the seas. In view of the lateness of the hour and the increasing storm, the French bore away to the southward, and by sundown all were under close-reefed topsails.[2]

Admiral A. T. Mahan has described the following scene:

> The wind now increased to great violence, and a severe storm raged on the coast until the evening of the 13th, throwing the two fleets into confusion, scattering the ships, and causing numerous disasters. The *Apollo* lost her foremast, and sprung the mainmast, on the night of the 12th. The next day only two

British ships of the line and three smaller vessels were in sight of their Admiral. . . . Many injuries had been received by the various ships, but they were mostly of a minor character; and on the 22nd the fleet again put to sea in search of the enemy.

The French had suffered much more severely. The flagship *Languedoc,* 90, had carried away her bowsprit, all her lower masts followed it overboard, and her tiller also was broken, rendering the rudder unserviceable. The *Marseillais,* 74, lost her foremast and bowsprit. In the dispersal of the two fleets that followed the gale, each of these crippled vessels, on the evening of the 13th, encountered singly a British 50-gun ship; the *Languedoc* being attacked by the *Renown,* and the *Marseillais* by the *Preston.* The conditions in each instance were distinctly favorable to the smaller combatant; but both unfortunately withdrew at nightfall, making the mistake of postponing to tomorrow a chance which they had no certainty would exist after today. When morning dawned, other French ships appeared and the opportunity passed away.[3]

Ashore at Newport the hurricane gave the land troops a severe beating. Fortunately, for this and other events during the war, Lt. Frederick MacKenzie of the Royal Engineers was at hand to record the military happenings along with his usual perceptive meteorological comments:

13 August—The rain continued very heavy from 4 o'clock yesterday evening, all last night, and all this day; with a very strong gale of wind at N.E. The wind fell about 8 this evening.

Most of the tents are blown down and torn to pieces; particularly those constructed by the Seamen of the ship's sails. Much damage is done to the Indian corn, great part of which is laid quite flat.

The troops are at present in a most uncomfortable situation, few of the officers having a tent standing, and every thing belonging to the men being perfectly soaked with the rain.

We are under great apprehension for the safety of Lord Howe's fleet. . . .[4]

On the night of the 11th and throughout the 12th and 13th the hurricane was felt up and down the Middle Atlantic and New England shores. At New York City, Captain Montresor described "a mere tempest most all day at northeast and rain which abated toward night. Rather cold than cool this evening." [5] Ezra Stiles at New Haven, where high winds from the northeast swept the college town, related that "the storm has been violent all day." His temperatures on the 12th ran close to 70°—the wind at 1600 at N.E. 4, his highest rating. At 0600 of the 13th the wind still came out of the northeast and continued to blow hard the rest of the day.[6] Montresor at New York again experienced "a violent gale at N.N.E. with heavy rain, weather cold for the season and very uncomfortably so." [7]

The blow of the 11–12–13th reached northward along the coast as there were gale reports from Boston, Salem, and Dover in New Hampshire.[8] But it appears that the worst of the blow remained offshore since all wind reports are northeast. Fortunately or unfortunately, as the case may be, the greatest reported damage occurred on the two opposing fleets which must have been rather close to the center of this northeastward-moving hurricane.

[1] A. T. Mahan. *The Influence of Sea Power upon History 1660–1783,* 361.
[2] W. M. James. *The British Navy in Adversity,* 105–106.
[3] A. T. Mahan. *The Major Operations of the Navies in the War of American Independence,* 75–76.
[4] *Diary of Frederick Mackenzie,* II, 352.
[5] Journals of Capt. John Montresor. *Coll. N. Y. Hist. Soc.,* 1881, 509.
[6] Stiles. Thermo. Register.
[7] Montresor, 509.
[8] Edward Holyoke. Met. Journal; Jeremy Belknap. Diary; *Boston Gaz.,* 24 Aug 1778.

THE STORMY OCTOBER OF 1783

October 1783 proved a tempestuous month along the Atlantic seaboard from the Carolinas to New England. Two major storms, both conceivably of tropical origin, swept up the coast, bringing destructive gales, heavy precipitation (some of which fell as snow over the northern interior), and destructive floods when a third rainstorm caused a rapid snow melt and runoff. A writer in the Bennington *Vermont Gazette* complained of the early fall weather, saying such a long-continued cloudy spell had never been known before.[1] Perhaps persistent trough conditions over the East Coast caused the cloudy skies and set the scene for the tropical storms to race northward.

The first disturbance seems to have been an authentic hurricane. It struck Charleston a severe blow on the 7–8 October, though no previous storm in the West Indies at this time is mentioned by Poey or Tannehill. The center passed to the east of the South Carolina port as the wind there backed from northeast to north and to northwest, the shift to north taking place at 1300 on the 7th when the center was probably directly east of the town.[2]

Extensive damage reports came from the Wilmington and Cape Fear area of North Carolina and also from inland points as far west as Winston-Salem where the storm "during the night assumed the proportions of a hurricane, damaging buildings, fences, and buildings, and blowing down many trees in the woods." [3]

All wind reports are from an easterly quarter. It appears that the center, after passing close to Charleston, moved northward, probably across eastern North Carolina, and then brushed the coast northeastward near enough to the shoreline to cause whole gale winds on land.

At Richmond, Virginia, on the 8th, "a most violent gust of wind came from the northeast which continued without intermission for 24 hours." No serious material damage occurred there, but at Norfolk, probably close to the path of the center, the tide was reported in the press to have risen an incredible "twenty five feet," causing damage there and across the river at Portsmouth to the tune of £8,000–10,000.[5]

The whirl of winds raked the southern New Jersey and Philadelphia area with northeast gales on the same day. At Cape Henlopen, on the southern side of the entrance to Delaware Bay, nine square-rigged, ocean-going vessels were driven ashore by the raging northeast gales. Later when the wind went into the northwest two additional ships, sheltering behind the lee shore of Cape May, were also driven onshore by the shift of the gale. Two brigs came ashore on New Jersey's Long Island near Egg Harbor a bit farther north.[6]

Around New York City on the 9th there arose an "uncommon high tide attended with a hard gale at northwest," flooding cellars along the waterfront and overflowing wharves.[7] At New Haven the press reported: "a severe northeast storm; and the fullest tide that has been known these forty years [probably a reference back to Franklin's Storm of October 1743], which has done considerable damage along shore by carrying off fences, lumber, etc." [8] At Providence the wind came out of the east-northeast for several hours, a compass point which would indicate that the center was not too far away to the southeastward.[9] From Boston: "Last Thursday night (9th) we had the severest north east storm that has been felt for a number of years." One large ship sank at its dock, while two schooners shared a similar fate.[10] Rev. Manasseh Cutler, a weather observer for the American Academy of Arts and Sciences, recorded for the 9th: "a very severe storm of rain." [11]

A second storm, packing winds of less violence than those of the 7–9th but producing a greater precipitation output, moved up the coast on the weekend of the 18–19th. A good reservoir of cold air lay over Canada. This contributed to a considerable deepening of the storm center off the New England coast and brought a mid-October snowfall almost to the shores of Long Island Sound.

At New Brunswick in central New Jersey on Saturday evening, the 18th, a "violent storm of wind and rain" destroyed a large number of small boats by driving them ashore.[12] At New Haven: "Saturday evening last there came on a violent storm of wind and rain, which continued for 24 hours; but happily the tide not being very high, there was little damage done in our harbor. In Goshen [northwestern Connecticut] and several of the adjacent towns, there was a considerable fall of snow." [13] Bostonians judged the storm more severe than the hurricane of the previous week though the tide damage there was not great.[14] Manasseh Cutler at Ipswich reported "a very severe storm of rain attended with excessive winds from the northeast" and "towards the close of the storm on the 19th a considerable quantity of snow fell, but soon melted." [15]

Inland at Bedford, near Manchester in New Hampshire, Matthew Patten noted the events:

> 18th Was a great rain all day and the whole night following.
> 19th Rained all the fore part of the day. Hard in the afternoon. It snowed until some time in the night. It fell two inches deep or more. The wind at NE. All this storm was violent and tedious.
> 20th The rain made a good freshet.[16]

Back in the hills of Vermont the snowfall reached a depth of 6 to 8 inches on the height-of-land to provide a good runoff on a soil already saturated from a wet September and an early October hurricane. Another heavy rainstorm struck the Vermont area on Wednesday, the 22nd, as reported in the Windsor *Vermont Journal,* "which raised streams to such a degree as produced the greatest flood ever known since the settlement of this country." Bridges and mills were washed out, and travelling was reported to be extremely difficult.[17]

[1] *Vt. Gaz.* (Bennington), 28 Sept 1783; *Journal of Rev. Thomas Smith,* 282.
[2] Charleston, 11 Oct, *Penna. Jour.,* 25 Oct 1783.
[3] *Moravians in N. C.,* 4, 1843.
[4] Richmond, 18 Oct, *Penna. Gaz.,* 29 Oct 1783.
[5] *Idem.*
[6] *Penna. Jour.,* 15 Oct 1783.
[7] New York City, 13 Oct, *Penna. Gaz.,* 22 Oct 1783.
[8] New Haven, 16 Oct, *Penna. Gaz.,* 5 Nov 1783.
[9] Providence, 11 Oct, *Penna. Jour.,* 29 Oct 1783.
[10] Boston, 13 Oct, *Penna. Gaz.,* 29 Oct 1783.
[11] Manasseh Cutler. *Trans. Amer. Acad.,* I, 369.
[12] New Brunswick, N. J., 21 Oct, *Penna. Jour.,* 1 Nov 1783.
[13] New Haven, 22 Oct, *Penna. Gaz.,* 5 Nov 1783.
[14] Boston, 23 Oct, *idem.*
[15] Manasseh Cutler. *Trans. Amer. Acad.,* I, 369.
[16] *Diary of Matthew Patten,* 473.
[17] *Vt. Jour.* (Windsor), 30 Oct 1783.

THE EQUINOCTIAL STORM OF 23–24 SEPTEMBER 1785

The year 1785 proved a productive period for hurricanes with no less than six, and possibly eight, tropical storms listed by Tannehill. The only gale which affected the American coastline swept in over the Carolinas and Virginia soon after the equinox on 23–24 September. The center appears to have passed over Ocracoke Bar at the entrance to Pamlico Sound in North Carolina as press reports spoke of a calm existing there. A major breach was made in the sand dunes which permitted the flood tide to drive the waters of the sea two or three miles inland. A large number of cattle drowned, and people took to the trees to save themselves.[1]

From the lower Chesapeake Bay area which lay in the path of the storm, there are two eyewitness reports which are presented in full:

> Norfolk, 25 Sept.—A higher tide and severer storm were never known than happened at this place yesterday; the damages sustained thereby are immense—almost all the ships in the harbor were drove from their moorings, and many warehouses were entirely carried away—vast quantities of salt, sugar, corn, lumber and other merchandise were totally lost—the lower stores of many dwellings were filled with water.[2]

> Portsmouth, 29 Sept.—On Thursday, Friday, and Saturday we had the most tremendous gale of wind ever known in this country, from N.E. to N.W. The whole town was overflown, and numbers of vessels driven into the cornfields and woods; storehouses drifted from their foundations, and every kind of property floated with the tide. It is supposed the damage amounts to £30,000. Not less than 30 sail of vessels are on shore, but it is expected the greatest part will be got off.[3]

Farther north there were reports of storm damage from Baltimore, Philadelphia, New York, and Boston.[4] At New York City it was called "a heavy equinoctial storm." The Swedish ship *Alstromer* was driven on Governor's Island; a newly-built house in the Bowery was blown down by the force of the wind.[5] At Boston it was "a violent equinoctial storm" which began at the Hub City about 1500 of the 24th and continued until noon of the next day.[6]

[1] *S. C. Gaz.,* 1 Nov 1785.
[2] Norfolk, 25 Sept, *Va. Gaz.,* 1 Oct 1785.
[3] Portsmouth, 29 Sept, *N. J. Gaz.,* 1 Oct 1785.
[4] Noah Webster. *Diary* (Ford, ed.), I, 140.
[5] New York, 27 Sept, *N. J. Gaz.,* 17 Oct 1785.
[6] Jeremy Belknap. Ms. Diary.

GEORGE WASHINGTON'S HURRICANE, 23–24 JULY 1788

An early season hurricane of great destructive power caused many ship disasters as it swept the ocean area south and west of Bermuda in late July 1788. It then roared inland on a northwesterly course over tidewater North Carolina and Virginia to pass directly over the Mt. Vernon plantation to which General Washington had retired to enjoy the rural pleasures of farming and watching the weather.

It seems likely that this hurricane had its origin farther north than is customary for most tropical storms. At least, the first notice of the blow came from Bermuda and from ships in the near vicinity. The following dispatch would indicate a close approach to Bermuda:

> Bermuda, 26 July—Saturday last (19th), in the forenoon began a heavy gale of wind from NE which soon got around to ESE when it blew with great violence all the afternoon and until twelve at night, at which time it moderated. Much damage was done by unslating houses, tearing up trees, to the boats, and destroying vegetables and provisions; two houses were also thrown down.[1]

The center of the storm appears to have moved west-northwestward and then northwestward during the next three days at a slow pace, for the press was filled with accounts of wrecks at this time in the area southwest of Bermuda.[2] An unusually large number of ship losses followed in the vicinity of Ocracoke Inlet, just southwest of Cape Hatteras, where six vessels were total wrecks, eleven driven ashore, and two dismasted. When the gale was at its height on the night of the 23–24th, the wind at the bar varied between northeast and northwest with the hardest blow coming from the northwest after the wind shift. Another report had 22 out of 30 ships either dismasted or driven ashore. The shift of wind to a northwest hurricane caught many ships sheltering behind the bar and drove them high and dry on the landward side of the sand barrier. At the same time the continuance of the northwest gale forced the water out of the Sound to strand many vessels on the shoals there.[3] The wind pattern clearly indicates that the center passed to the east and north of the Cape Hatteras area.

The Lower Chesapeake Bay with Norfolk and the smaller ports of Portsmouth, Hampton, and Yorktown lay in the direct path of the advancing whirl. A letter from Norfolk, dated 25 July 1788, gives a rather full account of the meteorological events there:

> Wednesday last (23rd) the most violent storm ever known commenced at 1700 and continued for 9 hours

—wind at start from NE—at 0030 it suddenly shifted to S and blew a perfect hurricane—tearing up large trees by the roots, removing houses, throwing down chimneys, fences, etc., and laying the greatest part of the corn level. The tide was not so high as in 1785. Only two ships in Hampton Roads survived the gale.[4]

The quick shift of the wind from northeast to south would indicate the passage of the center of the disturbance, and the same maneuver was noticed by George Washington at Mount Vernon some 140 miles to the northwest of Norfolk. Washington's diary usually paid great heed to the current weather situations and the present storm received full attention:

> Thursday—July 24th. Thermometer at 70 in the morning, 71 at noon, and 74 at night. A very high No. Et. wind all night, which, this morning, being accompanied with rain, became a hurricane, driving the miniature ship *Federalist* from her moorings, and sinking her, blowing down some trees in the groves and about the houses, loosening the roots, and forcing many others to yield, and dismantling most, in a greater or lesser degree of their bows and doing other and great mischief to the grain, grass, etca. and not a little to my mill race. In a word it was violent and severe—more so than has happened for many years.
>
> About noon the wind suddenly shifted from No. Et. to So. Wt. and blew the remaining part of the day as violently from that quarter. The tide this time rose near or quite 4 feet higher than it was ever known to do, driving boats, etc. into fields where no tide had ever been heard of before, and must, it is apprehended have done infinite damage on their wharves at Alexandria, Norfolk, Baltimore, etc. At home all day.[5]

Colonel James Madison, father of the future president and also a keeper of a meteorological register, noted the storm at Montpellier, his home near Orange, Virginia. There were easterly winds on the 23rd and northwesterly on the 24th: "great winds and rain all night."[6]

The *Maryland Gazette* at Annapolis noted the greatest tide in memory with northeast winds which gradually veered to southeast, but no abrupt shift to southerly took place, to put the Maryland capital east of the track of the center.[7] At Baltimore a violent storm from the east-northeast raged for 12 hours. Original estimates of damage ran to £50,000, but this was scaled down considerably in a later survey.[8]

The Alexandria reports confirmed Washington's account of the sudden wind shift:

> Alexandria, 31 July. The wind was at ENE when the storm began, but changed suddenly to the southward brought in the highest tide that was ever known in this river.
>
> The damage in the country to the wheat, growing tobacco, Indian corn, &c. is beyond description; and many planters and farmers, who flattered themselves with much greater crops than have been known for many years past, had their hopes blasted by the violence of the storm.[9]

There are no further notices of the hurricane from any location north of Maryland and Virginia. It is assumed that the momentum of the whirl kept it moving northwestward into the Appalachian Mountains of present West Virginia, Maryland, and Pennsylvania.

[1] Bermuda, 26 July, *New Haven Gaz.*, 4 Sept 1788.
[2] Bermuda, 2 Aug, *American Mercury* (Hartford), 22 Sept 1788.
[3] *Penna. Jour.*, 20 Aug 1788.
[4] Norfolk, 30 July, *Independent Gaz.* (Phila.), 8 Aug 1788.
[5] *Diaries of George Washington*, III, 394.
[6] James Madison at his Plantation, Ms. Amer. Phil. Soc.
[7] *Md. Gaz.*, 31 July 1788.
[8] *Md. Jour. and Balto. Adv.*, 25 July 1788.
[9] Alexandria, 31 July, *Penna. Jour.*, 9 Aug 1788.

HURRICANE, TORNADO, OR SQUALL LINE? 19 AUGUST 1788

From the scientific point-of-view the most fascinating of all the severe storms studied in this period occurred in mid-August 1788, just a month after a major hurricane had swept inland over North Carolina and Virginia with devastating effect. The August storm of 1788 presents the historian with an intriguing problem in his search for firsthand material to delineate the storm's path and effects, and after this the meteorologist must grapple with these tantalizingly sparse morsels of data in attempting to analyze its physical characteristics. Of all the storms studied, past and future, in this work, the mid-August 1788 occurrence appears unique, without known precedent.

A tropical hurricane lashed Martinique in the Windward Islands on 14 August, the gale commencing in the morning at northeast and continuing until about 2300 from quarters between northeast and north-northwest. At that time there was a sudden shift to southwest with redoubled violence, a wind maneuver which will again appear in the later history of this tempest.[1] Puerto Rico and Haiti were visited by a severe storm two days later, on the 16th, when 50 vessels were driven ashore at the latter place.[2] It would appear that these two were the same storm, and as the following will show there was a subsequent history.

The next land report comes from Philadelphia where on the 18–19th a tremendous downpour of tropical proportions deluged the city as a preliminary to a severe wind storm which swept over New Jersey, New York, and New England during the daylight hours of the 19th.[3] If we consider noon of the 16th the time of passage of the center near Puerto Rico, and noon of the 19th its arrival time at New York City, the rate of movement would be slightly in excess of 20 mph, certainly a not unreasonable average forward speed when one considers that the center later appeared to be moving about 40 mph over New Jersey, across New York, and into New England.

Perhaps it would be more satisfying to both the historian and the meteorologist if we approach the account of this storm in a different manner from the rest by presenting all of the known sources of information in their original form as they appeared to readers in 1788:

NEW JERSEY

Morristown, Tuesday 19th: Rain again this day until 2, and just before it cleared away the wind changed from S.E. and N.W. and blew violently so that it threw down the corn and buckwheat and fruit considerably. The waters were raised above the banks and many dams and forges &c. damaged. *Diary of Joseph Lewis,* 47.

NEW YORK STATE

New York City, 20 August: Yesterday between eleven and twelve o'clock A.M. came on a severe gale, wind at S.E. For upwards of 23 minutes it blew with incredible fury; and had the wind not shifted to N.W. every vessel in the harbor must have been drove on shore. The tide rose to a very great height, and most of the cellars in Front and Water streets, and a great number in Queen were filled with water. The West side of the Battery, which had stood the force of the elements for a great number of years, is almost laid in ruins. The damage sustained will be very considerable. *Daily Adv.* (New York) 20 Aug 1788.

Poughkeepsie, 26 August: But on Tuesday morning it rained moderately with a light wind from the southeast, which about eleven o'clock shifted to the N.E. somewhat increased. At 12 got about to N still blowing harder, and kept shifting westerly, increasing to one o'clock, when for about an hour it seemed to be fixed at W. and blew with such violence, that the largest oaks in the woods could not withstand its fury —and at which time rained so violently, that the creeks in this neighborhood as well as those twenty miles distant, were raised to such a degree as to take away bridges, and others, the water ran over the top, and floated away the planks with which they were covered. Great numbers of fruit trees were torn up by the roots. Every field of corn wherever the storm had reached was level'd to the ground. . . . A great plenty of apples, pumpkins, & squashes have been driving down the river. *Country Journal* (Poughkeepsie), 26 Aug 1788.

Hudson, 26 August: The 18th and 19th instant, we experienced in this country an excessive flood of rain, it continued for nearly 48 hours, almost without cessation, and toward the close very fast; much damage has been done on the low lands, by the carrying off of hay, corn, &c. . . . We learn that it has actually carried off a number of mills between this and Albany. Its effects will be severely felt, both at the northward and westward. *Hudson Weekly Gazette,* 26 Aug 1788.

1788

Hillsdale, N. Y., 8/20: Yesterday a scene was exhibited, in our neighborhood, which considering its approach, progress, and consequences, has not been equaled within the memory of the oldest inhabitants. The day had passed, from morning till about 4 o'clock P.M., attended by showers, the air remarkably thick and sultry, the wind at southeast, when it chopped around, as it were in an instant, and blew a most violent hurricane from the northwest, attended by an amazing deluge of rain, it could hardly, however, with propriety be called rain, it rather wore the appearance of large rivers precipitated from huge mountains, and driven through the atmosphere by the irrestible force of the wind. The fall of these aerial cataracts almost instanteously swelled our small domestic streams into dangerous torrents. The little brook which runs almost within our piazza, and was scarsely noticed by geese and children, arose to a Jordan like importance. The sudden and surprising rise and roar of water, attended by the continual crush of the stupendous pines, which, as it were, hung noddling over our houses; combined with ideas accompanying extraordinary phenomena, afforded me the sublime, the most striking solemn, and awful scene I ever before beheld—it was however short, the consequences not so alarming as might have been expected. A neighbor of mine, I hear, is badly hurt, some oxen and sheep killed, fences, mill dams, potash works suffered very much, but I believe not so much here as those lying lower down and further from the place where the devastation first began. Indian corn and almost every specie of domestic vegetation is laid level with the ground. *Hudson Weekly Gazette,* 26 Aug 1788.

CONNECTICUT

New Haven, 21 August: Last Tuesday morning came on a violent gale of wind from South, which about one o'clock P.M. veered to S.S.W. and blew a perfect hurricane. Several vessels were driven ashore, and very material damage done to the long wharf. *New Haven Gazette,* 21 Aug 1788.

New London, Tuesday 19th—Lowry morning, fresh breeze; meridian: squally rain wind S. Thomas Allen's Marine List in *Middlesex Gazette* (New London), 25 Aug 1788.

Norfolk, 27 August: On the 19th about 2 o'clock came on the most terrible gale of wind known in these parts. It began to blow with greater violence than usual from the S.E. but at once shifted to the S.W. and blew for about 20 minutes with tremendous fury, continually increasing, which spread a solemn amazement universally among the inhabitants. *Conn. Courant* (Hartford), 15 Sept 1788.

Litchfield, 25 August: On Tuesday last, a violent tornado, attended by heavy rain, laid waste many valuable fields of corn, and proved equally destructive to an infinite number of fruit and other trees:—Buildings unroofed, chimnies blown down, and fences laid level:—Many cattle were killed and wounded; and incredible damage done for a considerable extent of miles—Mr. Simeon Baldwin, of New-Milford, in passing through Washington, had his horse killed under him by the falling of a tree, and himself dangerously wounded. The damage sustained by the hurricane, in different parts of the country, must be great. *Weekly Monitor* (Litchfield), 25 Aug 1788.

MASSACHUSETTS

Pittsfield, 21 August: Tuesday last we had the most violent gust of wind which has been known in this town since its first settlement. The fore part of the day had been interchangeably rain and sun shine, until about noon, when it seemed to commence a settled rain, accompanied with a brisk breeze from the southwest, which continued with several variations and incessant rain, until about four o'clock, when the wind suddenly shifted to north west, and almost instantly exhibited a scene truly awful and magestick, the tallest trees fell before it. . . . *Berkshire Chronicle* (Pittsfield), 21 Aug 1788.

Lee, 19 August—Morning wind gently from south with rain. High wind from one to four o'clock. We were suddenly alarmed with an uncommon noise in the west like the roaring of thunder: the clouds were all in a most frightful commotion, and the prospect was indeed so tremendous, that we could hardly flatter ourselves with anything better than that we should be involved in one general destruction—The wind then shifted about into the west, and in a few minutes we saw roofs torn from buildings,—the fences blown away—the best of our timber broken down and scattered with the wind—our corn levelled with the ground—and our other fruits destroyed. The hurricane lasted about a quarter of an hour; four cows and several swine were killed, but no persons as we have yet heard. We have been informed that the storm was no less violent at Lennox, than it was here. Extract of a letter from a gentleman at Lee, Berkshire Co., Mass., 19 Aug, in *New Haven Gazette,* 4 Sept 1788.

Northampton, 27 August—About two o'clock the wind rose from the south and continued with increasing force until about three, it then veered to southwest, and for about 20 minutes was exceedingly violent here—three barns and a number of hovels were blown down, several barns unroofed, many apple trees were demolished, and a greater part of the apples blown from the trees which remained; considerable damage was done to fences, stacks of grain, and the Indian corn; but from the kindness of Providence no lives were lost. (A child was reported killed at Hatfield, and a man at Conway). *Hampshire Gazette* (Northampton), 27 Aug 1788.

VERMONT

Dummerston (Brattleboro) n.d: The wind blew about 15 minutes southwardly, and then suddenly varied southwestwardly, seemingly with redoubled violence. Trees two feet in diameter were broken off only a few feet from the ground and many were violently removed many yards distance. . . . *Mass. Spy* (Worcester), 28 Aug 1788.

Windsor, 25 August: Tuesday last Mr. Pearly Robarts of Plainfield, N. H., was killed by fall of limb from tree. *Vermont Journal* (Windsor), 25 Aug 1788.

Bennington, 25 August: The northern post informs that a great deal of damage has been done in the different towns on his rout, to the corn &, many cattle have been killed by the falling of trees during the late heavy storms. *Vermont Gazette* (Bennington), 25 Aug 1788.

Williston (near) 1789: "Almost half of ye trees in ye woods blown down by ye violence of ye wind last year." Nathan Perkins, *Narrative of a Tour* (1789), 17.

NEW HAMPSHIRE

Exeter, 6 Sept: We hear from Sanbornton that on the 19th they had a violent gale of wind from SSE to WNW attended with heavy showers of rain, which at about half past one blew a hurricane. Several hundred acres of thick woodland in different parts of the town, were blown up and hurled to the ground; even the stoutest oaks and rock-maples were either blown off or hurled to the ground. Many barns were unroofed and hardly a shed escaped being blown to pieces: one in particular 80 feet long, was taken up and carried 150 rods. Many young and beautiful orchards were suddenly destroyed.—cornfields blown flat to the ground, and even pole fences did not escape its fury. Many cattle were killed in the woods, but we hear that no person was materially hurt.

But its fury did not stop here, in Meredith and Newhampton, adjoining towns, and as far to the northward as we can learn, its effects were equally dreadful. *Freeman's Oracle* (Exeter), 6 Sept 1788.

Hanover, n. d.: New dwelling house of Capt. McCluse blown down. Shed with 2 men and 14 horses blown down and all killed. People in woods crushed by trees. *Freeman's Oracle* (Exeter), 6 Sept 1788.

The question arises as to what was this: Hurricane? Tornado? Line Squall? First of all, we can get a fair estimate of its forward speed. The airline distance from New York City to Northampton, Mass., is about 130 miles and the center apparently took about three hours, if we can depend on the press figures, to make the journey at a rate of about 45 mph. This forward progression of the storm over New England from southwest to northeast and the fact that the severe part of the blow came from different directions at different places would rule out the possibility that this was merely a line squall. Further, the great amount of rainfall accompanying the western section of the track would lessen the utility of the line-squall hypothesis.

The forward rate of 45 mph would not be unusual for a well-developed tornado, but nowhere do we find the rotary wind movement so characteristic of the tornado. The wind at a single location was reported as coming from a single direction. One eyewitness of the storm, writing at a later date, described the wind as "not circling like a whirlwind but right onward, leaving the trees, large and small, prostrate in one direction." [4]

The consensus of the newspaper reports would also confirm the straightline nature of the wind and the forward progress of the meteor.

We then return to the original assumption that this was a true hurricane, directly related to the tempest which moved through the eastern West Indies on 14–16 August. It is probable that this tightly-knit storm of very small diameter made no landfall before the Delaware Bay area. No ship reports of damage at this time have been noted, and no mention appeared in the Norfolk or eastern Carolina papers. My opinion is that the storm traversed the length of New Jersey from south to north, passing very close but just to the west of New York City. Then it followed a slightly-curving course with a northeast bearing over the Hudson River highlands near Bear Mountain, into the central Berkshires where the present Massachusetts Turnpike crosses the ridges near Stockbridge and Lee, downslope across interior Hampshire County, just cutting southeastern Vermont, and then over south-central New Hampshire on a northeasterly track which took it near Lake Winnepesauke, from whence it passed beyond the area of settlement.

The wind action is certainly like no other hurricane which has ever affected this area. The extremely short period of severe wind force, estimated from 15 to 25 minutes at several points, would indicate a very small center, and also a rather fast forward movement. Destruction in eastern New York and extreme western Massachusetts came on the west or northwest wind, while at New York City, New Haven, and generally in the Connecticut Valley it came with a southwest wind. It is probable that the area of high winds did not exceed 100 miles in breadth, but in about a fifty-mile-wide path the speeds must have been well in excess of 75 mph to cause such destruction, especially in the forests. One can contemplate the damage a repetition of such a power-laden storm would cause in this now heavily-populated strip from New Jersey to interior Maine.

[1] *New Haven Gaz.*, 18 Sept 1788.
[2] *Conn. Courant* (Hartford), 22 Sept 1788; Luis A. Salivia. *Historia de los temporales de Puerto Rico*. San Juan, 1950. 106; *Monthly Weather Review*. 34–2, Feb 1906. 70–72.
[3] *Penna. Jour.*, 27 Aug 1788.
[4] Hiram Barrus. Town of Goshen. *Hampshire County, Mass.*, 112.

THE TWIN NORTH CAROLINA HURRICANES OF 1795

Two hurricanes accompanied by extremely heavy rainfall struck the Carolinas and Virginia early in August 1795 creating a serious flood situation throughout the Piedmont. Rivers crested at the highest level in years: on some the greatest since the Regulator Flood of 1771, on others the greatest since the Great Gust of 1724. Thomas Jefferson at Monticello complained of the unusual wetness of the season: ". . . during the summer months of this year there were probably twice as many wet days as in common years, for nothing like it has ever been seen within the memory of man. . . ." [1]

On 2 August six vessels were driven ashore at Ocracoke Bar by an easterly gale, and the tidewater ports of New Bern and Washington in eastern North Carolina experienced damaging high tides.[2] The same gale caught a Spanish fleet of 18 vessels bound from Havana to Old Spain while in the latitude of Charleston and drove them on the Hatteras shoals.[3]

The blow reached inland where the *North Carolina Journal* at Halifax, then the state capital, reported: "Yesterday week we had one of the most violent storms of wind and rain remembered for many years past. The wind rose about two o'clock in the morning, and continued with unabated fury for about 12 hours, accompanied with vast torrents of rain." Corn was leveled to the ground and the incessant downpours put the Roanoke River into its third fresh of the year. In Granville County, farther west, the violence of the wind tore down many trees and blew fruit to the ground.[4]

The storm of the 2nd also struck the Norfolk area, but the wind coming from the southeast was not as dangerous in the harbor area as a northeast blow. A ship sank off Cape Charles and others went ashore on the Carolina Outer Banks.[5]

In northwestern Virginia the heavy rains attending the hurricane caused floods at Winchester and at Martinsburg in the upper Potomac watershed. Much corn and hay were ruined in the vicinity.[6] The only notice of the storm from a point farther north came from New York City where Henry Laight, destined to be Manhattan's longtime weatherman, mentioned "a very heavy gale of wind and a great deal of rain from the East" on Monday, August 3rd.[7]

The exact track of this hurricane remains in doubt due to the lack of wind direction data. It certainly passed south and east of New York City, and there are no reports of storm damage from New England. The center probably orbited in over eastern North Carolina like a comet and then headed east-northeastward out to sea again after dousing the entire tidewater and Piedmont in a tropical deluge.

Ten days later North Carolina again experienced a major hurricane along with additional torrential downpours. The high winds reached farther inland than previously as the Winston-Salem area in the west-central part of the State had a major blow. At Salem the Moravians were attending a religious service on 13 August when the storm struck:

> We were not a little disturbed by the terrible storm which lasted from three o'clock in the afternoon to near midnight, and resembled a West Indian hurricane. The storm came in great gusts from the east, and one could hear the roaring from a distance; some of the fruit trees were broken off at the roots; and the heavy rain was driven against the houses with such force that it penetrated the walls, even when built of stone. The downpour caused our creek to rise so that some of the planks were torn from the bridge, though it stands very high. The flood was twenty-one and a half inches higher than the one of last month, and that was higher than anything seen since Salem was founded. Our grist mill again suffered no small injury. During the following days we heard from various places that the flood swept away most of the mills in this section, and some dwellings, and that many head of cattle were drowned.[8]

Thomas Jefferson inland at Monticello near Charlottesville in Virginia felt the double blow of these storms. "I imagine that we never lost more soil than this summer," he complained in his *Garden Book,* "it is moderately estimated at a year's rent." [9]

At Petersburg, Virginia, "a most powerful torrent of rain" fell on the 13th, and the creeks were reported higher than at any time for the past 70 years, or since the memorable Great Gust of 1724.[10] In Granville County, North Carolina, a planter wrote that the inundation had put all his low ground under water, swept away 300 panels of fence, drowned 15,000 corn hills, and destroyed 2000 pounds of tobacco—a staggering economic loss! [11]

The second hurricane may have been related to the one which Poey mentioned at Jamaica on 10 August. A steady forward course might have brought it ashore in the Cape Fear area on the 13th. The damage in the vicinity of New Bern, where 100 dwellings were overthrown and a three-masted ship grounded, suggested an easterly wind, and at Bethabara near Winston-Salem the wind direction was reported at southeast.[12] These facts would suggest that the storm moved on a northwesterly course close to the North Carolina-South Carolina border and lost its tropical characteristics in the central Appalachians. We have no reports of its presence northward of southern Virginia.

[1] Edwin M. Betts. *Thomas Jefferson's Farm Book,* 47.
[2] *North Carolina Journal* (Halifax), 17 Aug 1795.
[3] *Norfolk Herald,* 12 Aug 1795.
[4] *N. C. Jour.,* 10 & 24 Aug 1795.
[5] *Norfolk Herald,* 12 Aug 1795.
[6] *Md. Gaz.,* 20 Aug 1795.
[7] Henry Laight. Ms. Met. Register. National Archives.
[8] *The Records of the Moravians in N. C.,* 6, 2534.
[9] Edwin M. Betts. *Thomas Jefferson's Garden Book,* 238.
[10] *N. C. Jour.,* 24 Aug 1795.
[11] *Ibid.*
[12] *Moravians,* 6, 2547.

NEW ENGLAND'S SNOW HURRICANE OF 1804

A tropical storm of small size, but packing full hurricane force, swept the Middle Atlantic States and New England on Tuesday, 9 October 1804. Its path ran very close to New York City and then across southern New England, with the center probably passing directly over Boston and Salem. Though this storm did great damage and was considered at the time to be the severest blow on record, its memory has all but been submerged in the flow of notoriety given the much larger hurricane a decade later, the Great September Gale of 1815. Nevertheless, in its limited area, the Snow Hurricane of 1804 probably carried as strong a local punch as its big cousin, and from the meteorological standpoint offers many interesting and unusual facets for study.

Tannehill and Poey make no mention of a hurricane in the West Indies at this time, nor do Southern papers carry any notices that might be tied in with the storm under consideration. Meteorological data from Norfolk northward indicates that there was a deep trough present along the Atlantic seaboard. It is possible, of course, that the storm could have formed in the southern Appalachians and sped northward in the congenial atmosphere of the trough, deepening all the time, until in the vicinity of New York City it became a violent circulation since it is only over southern New England that we find damage on a hurricane scale. Nevertheless, it did exhibit some of the characteristics of a tropical storm circulation and occurred well within the hurricane season.

The first evidence of storm activity came from the Virginia Capes area. The Norfolk meteorological observations published weekly in the *Public Ledger* merely indicate that a trough passed through on the morning of the 9th with a shift of wind from SW force 3 to WNW force 6 by 1400.[1] A ship just off Cape Henry reported a "dreadful squall" from the northwest struck at 1100.[2]

In the upper Chesapeake Bay the violence of the following gale on the 9th prevented the mail boat from crossing at Havre de Grace, Maryland, since the wind "was incessantly varying from west to northwest, and for its duration exceeded anything we have witnessed at this season of the year." [3] Baltimore: "For 2 days a very smart gale from W & N—set against tide, left many ships dry at wharves." [4]

At Philadelphia a violent squall about 0800 on the 9th struck and submerged a newly arrived ship. Farther up the river the violence of the wind on the same morning upset the ferry at Trenton.[5] To the east on the Jersey shore we have an important report of a ship being driven on Absecon Beach (Atlantic City) by a *southeast* wind.[6]

The next check-point is New York City where William M'Intosh operated the City Observatory for the Board of Health. Fortunately his records were published weekly in the *New York Post.* At sunrise on the 9th rain was falling from a cloudy sky with the wind in the *southeast.* His barometer had dropped 0.44″ overnight. By 1400 the wind shifted to NNW and was high —weather cloudy—the barometer off 0.18″ more to an uncorrected level of 28.87″. We do not have information as to the exact hourly character of the wind shift at New York, but the temperature remained high at 55°, off only three degrees from the morning reading. His barometer remained at this low point until the sunset observation, at least; his thermometer, however, had then slumped to 42° and a hard rain from the north was sweeping the city. The rain gage on the 9th showed a total of 2.27″.[7]

At nearby Belleville, New Jersey, Gerard Rutgers had rain and northeast winds all day the 9th, but it was not until dark that mention was made of "a violent gale of wind from the NE. It continued blowing a great part of the night." [8] In New York harbor on the 10th there were no arrivals of ships since a 24-hour gale from the north had been blowing down the Bay.

Farther up the Hudson Valley the storm lashed severely at both Poughkeepsie and Hudson. At the first place a northeast wind was mentioned in the press which "intermitted for an hour, when it renewed its violence, and continued until morning." [9] At Hudson the main force of the wind was from the north and blew "with a violence not experienced in this city since its settlement." [10]

More satisfactory meteorological details are available at New Haven where both President Day of Yale College and Thomas Beers, a bookseller at College and Chapel Sts., kept meteorological registers. The latter's comments are more complete and are given in full:

October 9—
0600—Hard rain in early part, noon heavy black clouds wild and dark, very heavy rain most of forenoon with considerable heavy distant thunder—wind highest at SE till about 1030, round to SW, W and NW and blew very hard with heavy rain—slacked toward noon
1300—appears to be clearing off
1800—wild heavy black clouds driving rain, clouds fly quick
2100—a high gale of wind this evening and for part of night [11]

Professor Day's observations on the Yale campus also showed a southeast wind at 0600, but by 1300 it had shifted to northwest. By sunset, with the gale now out of the north, his thermometer read 38°, a drop of 17 degrees since morning. He reported "frequent lightning and heavy thunder. Wind very violent through the night." Rainfall measurement for the period totaled 3.66 inches.[12]

At New London: "On Tuesday morning last, a violent gale commenced here, wind at S.E.—at noon it became suddenly calm, but in the afternoon it recommenced with greater force, wind NE and blew heavy until 10 P.M." [13] The wind at Providence varied at points between northeast and southeast during the day, and "during the course of the night, this town experienced the heaviest gale within the recollection of any of its inhabitants," though the direction was not mentioned.[14]

From Boston we have a very significant report. The wind began rising from the south-southeast on Tuesday morning, later shifting to east and increasing in power until about 1500. Then the gale abated for a few minutes, veering to northeast, and blew "with a violence and fury unprecedented in the annals of this town." The steeple of the North Church fell, and the tower roof of the Stone Church sailed 200 feet through the air.[15]

Farther northward along the coast the storm hit Salem a severe blow and glanced at Portsmouth, New Hampshire, where "in the town no damage was done, except blowing down some trees and fences—nor was any material injury to vessels in the harbor." It appears that the New Hampshire port area comprised about the northern fringe of the high winds.[16]

Back at Salem, some 15 miles northeast of Boston, it was a different story. Here we have Edward Holyoke's full meteorological observations, good coverage in the press, and a most complete diary account. On Tuesday morning about 0900 the wind shifted into northeast with rain and thunder following the whole day. Holyoke's barometer dropped to 29.29″ at sunset and the thermometer was down 8 degrees, at 44°, from morning. After sundown the wind rose violently and continued through the night. Mr. Atwater Phippen reported that 4 inches of rain fell during the day, and 3 inches the following night—a total of 7 inches, "a greater quantity than has ever been known in the same space of time."

At Salem also dwelt Rev. William Bentley who maintained a daily chronicle of massive size concerning home and national events. He was also an experienced weatherman, keeping a meteorological register of local events and also operating a clipping service for weather items from all over the country. In his four-volume diary covering the years 1785–1817, he gave an unusually full and interesting treatment of the 1804 storm:

This morning the wind was in the South and the weather uncertain. About 7 it shut down and it began to rain at S.E. and soon the wind rose and the wind changed to N.E. Its violence increased till sundown and continued all night. The barn belonging to Perkins on the Neck, was blown down and one horse killed. Beckets barn down, all vessels drove from their anchors. Chimnies were blown down, roofs and windows injured and trees destroyed in great number. The fences suffered so much that in the eastern part of the Town which I visited it was easy to pass over any lot in that part of the Town. The damage is so equally divided that few have special cause to complain. It was the heaviest blow ever known in Salem and it will be remembered as the Violent Storm of 9 Oct. 1804. We had thunder and lightning all day. We lost the Railing from the top of the house in which I live. It was totally destroyed.

We are every moment receiving accounts of the injuries done by the Storm. The Vessels in Cape Ann and Marblehead that were at anchor are ashore. The damage done in Boston is great. The celebrated Steeple of the North Church is blown down. Mr. Atwater Phippen who for many years has noticed the fall of rain, distinguished the rain of yesterday as the greatest he ever knew, four inches fell in the day and three inches in the night.

Continued account of the Storm. From the Coast, accounts general only as yet. Roads everywhere much obstructed by the fall of trees, etc. Revere's Buildings over his furnace destroyed. Not of great value. Covering of Chapel Church tower blown down. Mr. Eaton at Boston, new brick walls tumbled upon his old house from which he had just time to escape. The woman who lived with him killed, servants wounded. The spire of Charleston steeple bent down. The top of Beverly steeple blown off. The dome of the Tabernacle in this town uncapped, and shattered the Lantern. A vessel from Cape Ann harbour, belonging to Kennebunk, lost her anchor and split her sails and drove up over our Bar into the Cove within the Beacon upon Ram Horn rock. This is the only Vessel ashore on our coast not in the Harbour. The Boston account is an almost total destruction of all small boats at the wharves. The damage to Houses, buildings, trees, fences, etc. is incalculable, but such losses not heavy to individuals, but a distressing loss to the public.

This day I rode through south fields and Marblehead farms to Nahant. Every where trees are blown down and barns unroofed and the road in several places would have been impassable had it not been cleared. Even at Nahant Great head, Wood lost part of the roof of his new Barn erected this year. The reports are endless, but we cannot distinguish truth from falsehood at any distance from home at present. But the reports shew the state of the public mind. The quantity of seaweed driven up is beyond any former example. I had a good opportunity of examining a rich variety on every part of Nahant. The most common there in deep water is the Kelp, the seagrass and the wrack as they are called. The Dulce Conpici, etc. were in less abundance. It would not have been imagined that the beaches over which we passed had ever been used for pleasure had they been seen only after the late storm.

I cannot refuse to adopt the belief that the late storm was the most severe ever felt in this part of America. All the accounts which I have seen represent nothing like it. In Boston, the old people are said to represent that a storm like it happened 16 September 1727. As yet I have found no tradition of such a storm among our old people or upon record or any report of its consequences. I suspect as our winters have less horrour we partake more of a southern climate from the great quantity of heat and consequently have more stormy weather of this kind and therefore may expect more of it in future years. I can find no history of wharves, ships, trees, houses, fences, out houses which lead to suspect great calamities from high winds. From Cape Ann we learn that many of their boats were lost entirely and some greatly injuried by the storm. But we have hopes from the news from Plymouth and Portland, that the storm was much more limited than we have expected from its great severity here and near Boston.[17]

Tracing the exact path of the storm presents the usual difficulties of the period with its lack of hourly observations and detailed wind data. It appears that the center passed between Philadelphia and Atlantic City (Absecon Beach), very close to New York City, south of Poughkeepsie, certainly north of New Haven and New London, and probably right over Boston. Such a parabolic curve over the Middle Atlantic States and then eastward across southern New England is not unknown in hurricane annals. It would seem that the small-diameter whirl on the morning of the 9th entered a swiftly-moving trough, oriented from northeast to southwest along the coastline of the Middle Atlantic States. As the northern part of the trough sped eastward more rapidly than the southern, the hurricane track curved northeast and then east-northeast over southern New England as the strength of a strong westerly flow behind the trough dominated. Certainly a great mass of unseasonably cold air was poised over northern New York and northern New England that morning as evidenced by the commencement of heavy snowfall in those areas about dawn.

Our data showed that only the northwesterly and northerly winds following the passage of the center were of whole gale and hurricane force. No doubt the cold air mass was responsible for the steepening of the pressure gradient and the intensification of the whole system. An interesting phase of the storm structure lies in the comments from Poughkeepsie, New Haven, and Boston to the effect that the wind abated for a while or "intermitted" before the heavy blow from the northerly quarter commenced. The trough of low pressure must have maintained a sort of corridor of light winds to the right of the path of the storm's center now in the process of transforming into an extra-tropical disturbance. The "eye" perhaps lost its circular shape becoming an elongated oval as the storm center progressed to higher latitudes and met the resistance of the high pressure over Canada.

This storm exhibited an unusually wide precipitation pattern. Timothy Dwight, the travel-minded natural scientist who later became president of Yale, found himself at Bemis, near Rochester, New York, on the day of the New England hurricane. There a storm from the northeast raged accompanied by heavy rain and "a considerable flight of snow, which, however, dissolved as it fell." The storm continued at Bemis from 2100 on the 8th until 1300 the next day—indicative of a trough passage and the usual following snow squalls.[18]

The mention of snowfall at low levels in western New York associated with this storm pattern is most significant since at the same time the hill country of eastern New York and most of New England was experiencing the greatest early season snowstorm known at that time. From the hills of southern Connecticut northward into Canada snow fell throughout October 9th and well into the 10th. Noah Webster relates that the hills of Woodbridge, Connecticut, just north of the Yale Bowl and near where the Wilbur Cross Parkway tunnels through West Rock, were white with snow.[19] At Litchfield three inches were reported, and at the higher elevation of Goshen the depth amounted to 12 inches.[20]

In western Massachusetts the press reported amounts up to 24 and 30 inches in the Berkshires, and northward in Vermont the Windsor *Vermont Journal* estimated depths of three to four feet along the height of land of the Green Mountains where temperatures were a bit lower.[21] In the Catskill Mountains west of the Hudson River in New York State accounts told of amounts up to 18 inches.[22]

A rather complete account of the snow situation in the Connecticut Valley was given by the editor of the Walpole, New Hampshire, *Farmers Cabinet*. Writing on Saturday the 13th, he recounted the commencement of the snowstorm on Tuesday in the middle of the forenoon when a temperate rain changed to snow, accompanied by thunder and a high wind. This continued with some intermission until Wednesday morning. He estimated that the bottom of the Connecticut Valley received a mean depth of 15–18 inches, most of which dissolved upon falling, but still measured 4–5 inches at the conclusion of the storm and remained on the ground for 30 hours before completely melting. On the nearby highlands, back from the river, the snow was still to be seen four days later where it had originally drifted over the fences during the height of the blow. Since leaves were still on the trees, many branches and tree trunks were shattered by the unusual weight of the loose, clinging snow. Great damage to orchards and sugar groves through the hill country of New England resulted.[23]

From northern Vermont a reliable report concerning the Lunenburg area in the upper Connecticut Valley came from Dr. H. A. Cutting when later recalling famous storms in his "Natural History of Essex County":

> October 9, 1804 brought with its dawn a great snow storm. The weather had been cloudy and extremely cold for the season for a number of days; and on the morn of this day it commenced snowing, and continued almost without intermission until full 20 inches of snow had fallen.[24]

Many New England town histories, especially those of Vermont and New Hampshire hill towns, recount this unusual, early October snowstorm. In many cases, they state that the snow cover remained until next spring, but a study of temperature conditions for the remainder of the month casts doubt as to the accuracy of this legend. In fact, only in 1843 does an October snow in low-level towns of northern New England appear to have remained throughout an entire winter. The Snow Hurricane of 1804 deservedly lived long in the memory of New Englanders until a somewhat similar hurricane situation brought an even earlier general snow on 3–4 October 1841.

1 *Public Ledger* (Norfolk), 18 Oct 1804.
2 *New York Post,* 22 Oct 1804.
3 *Ibid.,* 13 Oct 1804.
4 *Amer. Adv.* (Phila.), 13 Oct 1804.
5 *U. S. Gaz.* (Phila.), 11 Oct 1804.
6 *True American* (Phila.) in *New York Post,* 17 Oct 1804.
7 *New York Post,* 15 Oct 1804.
8 *Gerard Rutgers.* Ms. Diary. Rutgers University.
9 Poughkeepsie, 16 Oct., *New York Post,* 20 Oct 1804.
10 Hudson, 16 Oct., *Md. Gaz.,* 25 Oct 1804.
11 Thomas Beers. Ms. Thermometrical Register. Yale Univ. Library.
12 Jeremiah Day. Ms. Meteorological Register. Yale.
13 *Conn. Gaz.* (New London), 17 Oct 1804.
14 Providence, 11 Oct., *New York Post,* 15 Oct 1804.
15 *Boston Gaz.,* 11 Oct 1804.
16 *Portsmouth Oracle,* 13 Oct 1804.
17 William Bentley, *Diary,* III, 116–117; E. A. Holyoke. Met. Journal; *Salem Gaz.,* 11 Oct 1804.
18 Timothy Dwight. *Travels,* IV, 94.
19 *Notes on the Life of Noah Webster* (E.E.F. Ford, comp; E.E.F. Skeel, ed.). New York, 1912. 1, 563.
20 Jeremiah Day, Met. Reg.
21 *Vermont Journal* (Windsor), 16 Oct 1804.
22 *Amer. Adv.* (Phila.), 26 Oct 1804.
23 *Farmers Cabinet* (Walpole, N. H.), 13 Oct 1804.
24 H. A. Cutting. Natural History of Essex County. *The Vermont Historical Gazetteer* (A. M. Hemenway, ed.). 1868. I, 1056.

THE GREAT COASTAL HURRICANE OF 1806—II

A hurricane of great size and destructive power raged along the Atlantic Coast on 21–24 August 1806. Except for a short transit over eastern North Carolina it remained offshore during a rather leisurely journey northeastward, with its gales restricted to the bays, sounds, and islands which fringe the coastline. Noah Webster tells us that it was not felt 100 miles inland, but along the coastal shipping lanes it exacted an enormous toll of lives and vessels.[1]

Again, as is so often the case in American mainland hurricanes of the early period, neither Tannehill nor Poey make mention of a tropical disturbance which could be connected with the one under study. Our first report came from a ship at 26°30′N, 74°W (225 miles northeast of Nassau) which encountered a four-day storm on 19–22 August. This position would indicate that the track lay to the east and north of all the Bahama Islands.[2]

The first land report came from Charleston where the wind, which had been at easterly points for several days, freshened from the north-northeast on Wednesday the 20th, and early the next morning at 0200 increased to a "complete gale with rain." [3] The center appears to have passed inland close to Cape Fear as both Smithville and Wilmington experienced a sudden 180-degree wind shift at hurricane force. At the later place it was the "most violent and destructive storm of wind and rain ever known here." [4] See full account in Hatteras South section.

The eye appears to have crossed eastern North Carolina to emerge again close to, but east of the Norfolk area, as indicated in the following report:

> The appearance of the weather, had for some days past, indicated a storm, which came on Saturday last attended with heavy rain. The wind thro' the day was from South to East. In the night it got more to Northward, and from midnight to about three in the morning, the wind blew with great violence, from North to Northwest, after which it abated. It was fortunate that the violence of the gale was of such short duration, for had it continued it must have done great damage.[5]

Several new buildings and chimneys were blown down along with the usual toll of trees and fences. Damage to shipping proved minor with two vessels reported ashore. If our press figures can be relied on, the eye of the hurricane required 36 hours to move from a landfall at Cape Fear to the latitude of Norfolk, a distance of 200 miles; but once clear of the shoreline the forward movement greatly accelerated as evinced by the relatively short duration of the peak winds at Norfolk. The hurricane caught many ships of the British and French war fleets, then engaged in the far-flung conflicts of the Napoleonic Wars, in our coastal shipping lanes. The port of Norfolk was treated to the spectacle of witnessing ships of two warring nations put in for repairs and refitting.[6]

As the hurricane gained forward speed and roared up the coast, shipping suffered severely. The most tragic incident involved the coastal ship *Rose in Bloom* which upset on Sunday morning, the 23rd, in a stiff northeast gale off Barnegat Inlet, on the central Jersey coast, with the loss of 21 of the 49 persons aboard. This disaster, first reported in a New York paper, received wide publicity in the press of the nation.[7]

The New York City area received a sound lashing from the fringe of the storm, with a northeast gale prevailing from Saturday morning to noon of Sunday, when it shifted to north-northwest, but still blew a gale until dark. A heavy, incessant rain accompanied the wind until the shift occurred.[8] At Belleville, New Jersey, Gerard Rutgers noted "a violent gale of wind" on Sunday morning which blew down several peach trees. His thermometer read 70°, 72°, and 70° at his three daily observations.[9]

The gale next gave Cape Cod and the Islands a severe sideswipe. The marine reporter for the *Boston Gazette* noticed a hard rain and blow in the harbor and mentioned that the wind raged much higher out on Massachusetts Bay on Sunday.[10] From Barnstable on Cape Cod and Edgartown on Martha's Vineyard, much closer to the track of the ocean storm, the reports spoke of a tremendous deluge and winds severe enough to cause structural damage.

At Edgartown an observer related that Saturday morning dawned with the wind in the southwest and a light rain falling, with flashes of lightning but no thunder. The rain increased, falling in torrents most of the day, but abated toward evening. Early in the night the wind went into the east, then veered to northeast "and increased to one of the severest gales I have ever experienced." His rainfall report would seem incredible —he noticed a barrel filled to a depth of 30 inches. From this and the sheet of water covering the ground from the extended heavy downpour he estimated that 36 inches fell during the storm. There was great crop destruction on the island. Five coasters were driven ashore there.[11]

This report received confirmation from Barnstable County on Cape Cod across Nantucket Sound. From Brewster came accounts of great destruction to crops and to the valuable salt works from the combined onslaught of wind and rain. The observer wrote: "It is supposed there is 18 inches of water on the level" in speaking of the deluge. One certainly wishes that a reliable rain gage had been exposed at one of these locations since this storm appears to have been a rain-producer of unprecedented proportions. From Boston, probably well to the west of the heavy rain belt, we have a rain gage measurement to the effect that 0.40-inch fell in the hour from 1200 to 1300 on Sunday, a respectable figure, but certainly not a record for Boston.[13]

[1] *Notes on the Life of Noah Webster*. I, 566.
[2] *Palladium* (Boston), 9 Sept 1806.
[3] *American Register* (Phila.), I, 8.
[4] Wilmington 26 Aug, in *Conn. Courant* (Hartford), 17 Sept 1806.
[5] *Public Ledger* (Norfolk), 25 Aug 1806; *Palladium* (Boston), 5 Sept 1806.
[6] *Nat. Int.* (Wash.), 19 Sept 1806.
[7] *N. Y. Daily Adv.* in *U. S. Gaz.* (Phila.), 1, Sept 1806.
[8] New York 25 Aug, in *Public Ledger* (Norfolk), 1 Sept 1806.
[9] Gerald Rutgers. Ms. Diary.
[10] *Boston Gaz.*, 25 Aug 1806.
[11] *Ibid.*, 4 Sept 1806.
[12] *Ibid.*, 1 Sept 1806.
[13] *Medical and Agricultural Register* (Boston), 1806, 143.

HATTERAS SOUTH: 1686–1814

THE SPANISH REPULSE HURRICANE OF 1686

Hurricanes and Spaniards were the twin scourges of the South Carolina area during the colonial period. From the day of the first English settlement there in 1670, a weather-eye to the sea had to be maintained for self-preservation. Charleston was established in its present location at the junction of the Ashley and Cooper rivers in 1680. It so happened that only six years later the twin scourges struck at the same moment with momentous results for the infant colony.

John Bartram, while visiting Charleston toward the close of the first 100 years of settlement, took up the study of hurricanes. From eyewitnesses and tradition, he compiled an interesting and rather reliable account of Charleston's battle against the tropical storm monster. He noted: "about the year 1686 they had a grievous hurricane; and the Indians told the English that they knew one that raised the water over the tops of the trees where the town now stands."[1] Perhaps this would refer to a total inundation of the island prior to 1680, possibly in 1667 at the time of the Great Hurricane in Virginia. A total inundation has not occurred in the years of settlement by the whites.

The Spaniards, based at St. Augustine, though certainly well aware of the presence of the hurricane season, chose late summer of 1686 to mount an attack on the lower Carolina settlements. They landed near North Edisto Island and struck toward Stuart Town near Beaufort on 25 August (4 Sept.) 1686. But that evening the wind picked up to a gale, driving two of the Spanish galleys so high on land that they had to be abandoned and the attack called off.[2]

There is only one contemporary description of the event by an unidentified writer to the Lords Proprietors in London:

> As the night came exceeding black and menacing clouds began to show themselves and were the next morning (being the 26th) succeeded by a hurricane wonderfully horrid and destructive whereof your Lordships shall hereafter receive a more particular relation. . . .
>
> . . . The expedition of the several parties (in pursuit) was much impeded by frequent contrary and false intelligences together with continual wet and windy weather which day by day succeeded the hurricane, the violent extremity whereof continued not above 4 or 5 hours yet was attended with such dismal dreadful and fatal consequences that the hand of Almighty God seems to concur with the malice of our enemies to hasten our ruin and desolation. Your Lordships cannot imagine the distracting horror that these united evils plunged us into. All the ships and vessels in the road and harbors were driven up on the land and whether any of them can be fitted out again may be yet a question.
>
> The whole country seems to be one entire map of devastation. The greatest part of our houses are blown down and still lie in their ruin, many of us not having the least cottage to secure us from the rigor of the weather. The long incessant rains have destroyed almost all our goods which lie intombed in the ruins of our houses. Our corn is all beaten down and by means of continued wet weather lies rotting on the ground.
>
> Our fences are laid flat so that the little corn that escaped the storm is devoured or destroyed by our hogs and cattle. Abundance also of them were killed in the tempest by the falls of trees which in infinite number are blown down and lie in confused heaps

all over the country so that most of our cattle are in great danger of running wild, there being scarse any probability of finding them out or possibility of driving them home when they are found.

In some places for 3 or 4 miles together there is scarse one great tree standing. All paths being so impassable that there is no traveling on horseback and scarse any on foot, whereby all society and communication with our neighbors, one of the greatest comforts of our lives, is for many years rendered extraordinary difficult. With the falls of trees the foods of our hogs is likewise destroyed which will cause them all to run wild; or which is as bad, they will all be starved from these and the like calamities which now attend us. We have too great reason to fear the near approach of famine to complete all our miseries which we pray God in his mercy to direct from us.

And now having given your Lordships a brief account of the invasion of the land by the Spaniards, a cruel and inveterate enemy, and the inconceivable detriment we have sustained by the terrible hurricane, we crave leave to acquaint your Lordships with the sad consequences thereof both at present and what we may reasonably expect for the future.[3]

[1] John Bartram. Diary of a Journey through the Carolinas, Georgia, and Florida from July 1, 1765 to April 10, 1766. (Francis Harper, ed.) *Trans. Amer. Phil. Soc.*, ns., 33–1, 1942, 20.

[2] *Report of the Committee . . . St. Augustine Expedition of 1740.* (J. T. Lanning, ed.), 4.

[3] Paper to the Lords Proprietor, c. 1686, quoted by J. G. Dunlop, *South Carolina Hist. & Gen. Mag.*, 30–2, 1929, 83–84.

THE *RISING SUN* HURRICANE OF 1700

In the last year of the Seventeenth Century a hurricane struck a savage blow at the Carolina coast with tragic effect. It so happened that on 3 (14) September 1700 the large Scottish ship *Rising Sun* was standing off the bar at Charleston harbor, unable to cross to a safe refuge and awaiting the return of about 15 of the ship's company who had gone ashore in quest of fresh provisions. The *Rising Sun* along with the *Duke of Hamilton* and several smaller vessels had evacuated the remnants of the ill-fated Darien settlement of Scots in Panama, and were presumably enroute back to the home isle.[1]

The increasing hurricane winds and storm tide caught the *Rising Sun* while anchored off the bar and hurled her onto the sand beaches where the ship broke up. All those then aboard were reported lost along with Captain James Gibson. Accounts vary as to the number actually lost. The figure 97, which appeared in a contemporary document, is perhaps a close approximate.[2] Most of the bodies were recovered the next day along James Island and buried there. The lucky voyagers who had gone ashore assisted in the task. The other vessels within the harbor were also completely wrecked, but the passengers were all able to get ashore safely as the rising winds foretold disaster.

The date of the occurrence is fixed by the following letter as 3 (14) September 1700, though Ramsay quotes another source as 5 (16) September.[3] The communication to the Lords Admiralty, dated New York, 15 October 1700, by Lord Bellomont follows:

> Some Scotchmen are newly arrived hither from Carolina that belonged to the ship Rising Sun (the biggest ship they sent out for their Caledonia expedition) who tell me that on the third of last month a hurricane happened on that coast, as that ship lay at anchor, within less than three leagues of Charles Town in Carolina with another Scotch ship called the Duke of Hamilton and three or four others; that the ships were all shattered in pieces and all the people lost, and not a man saved. The Rising Sun had 112 men on board. The Scotchmen that are come hither say that 15 of 'em went on shore before the storm to buy fresh provisions at Charles Town by which means they were saved. Two other of their ships they suppose were lost in the Gulph of Florida in the same storm.[4]

There is no actual meteorological data available for this storm. John Bartram merely adds that the storm "was very severe, overthrowing many houses and overflowing the town." [5]

[1] Charles L. G. Anderson. *Old Panama and Castilla Del Oro*, 498.

[2] *Calendar of State Papers, American and West Indies, 1700.* (845, xxxi), 598.

[3] Ramsay. *History of South Carolina*, II, 176.

[4] *Calendar State Papers*, 598.

[5] Bartram, 20.

THE CAROLINA HURRICANE OF 1713

The third hurricane of note in Charleston's early history occurred on 5–6 (16–17) September 1713. Three contemporary witnesses confirm this date as did John Bartram's interview in 1766 with a survivor.[1] Mark Catesby, who was in the Carolinas on a natural history survey from 1722–26, mentioned the 1713 event as an example of the severe type of storms which occasionally afflicted the young settlements.[2] Since Catesby returned to England prior to the great hurricane of 1728, that of 1713 must have stood in the minds of the Carolinians as the outstanding storm precedent.

Catesby's book was published in England in 1731 with parallel English and French text, and was soon translated into German. His remarks, therefore, are of interest in showing the state of knowledge of hurricanes in the 1730's:

> Usually once in about seven years, these rains are attended with violent storms and inundations, which commonly happen about the time of the hurricanes that rage so fatally amongst the Sugar Islands, between the tropicks, and seem to be agitated by them or from some cause, but are much mitigated in their force by the time they reach Carolina; and tho' they affect all the coast of Florida, yet the further they proceed, so much the more they decrease in their fury, Virginia not having often much of it, and north of that still less. Tho' these hurricanes are seldom so violent as in the more southern parts, yet in September 1713, the winds raged so furiously that it drove the sea into Charles-town, damaging much of the fortifications whose resistance it is thought preserved the town. Some low situated houses not far from the sea were undermined and carried away with the inhabitants; ships were drove from their anchors far within land, particular a sloop in North Carolina was drove three miles over marshes into the woods.[3]

There is practically no significant data extant on the behavior of the wind except for the assertion of Thomas Lamboll that the storm raged more severely northward of Charleston and was not felt violently at Port Royal, 50 miles to the southwest.[4] Mark Catesby mentioned a sloop being driven three miles inland over marshes and into the woods in North Carolina (probably the Cape Fear country). From the above, it appears that the landfall must have been to the north of Charleston. The violence of the wind is attested by Lamboll's description of buildings blown down, and a declaration that an inundation from the sea drowned 70 persons.

The origin and track of this hurricane cannot be determined. Poey mentioned a storm at Guadeloupe and St. Thomas in this year. Catesby some years later saw a boat stuck in a tree 10 or 12 feet above the ground on Eleuthera Island in the Bahamas; he attributed this to the 1713 storm.

We present the accounts of the three eyewitnesses who set down their impressions shortly after the event.

THOMAS LAMBOLL

1713—On September 5th (16th) came on a great hurricane, which was attended by such an inundation from the sea, and to such an unknown height that a great many lives were lost; all the vessels in Charleston harbor except one were driven ashore. the new Lookout on Sullivan's Island, of wood, built eight square and eighty feet high, blown down; all the front wall and mud parapet before Charleston undermined and washed away, with the platform and guncarriages, and other desolations sustained as never before happened in this town: To the northward of Charleston the hurricane was more violent, but at Port Royal it was not much felt. Thomas Lamboll (died 1775), Ms. Ramsay, *History of South Carolina,* 1858 ed., II, 175–176n.

DR. FRANCIS LE JAU

It is miraculous how any of us came to escape from the great Hurricane we felt Sept 5th (16th) last past. It continued for 12 hours, and had the two rivers on both sides of Charles Town been joined for some time that place would now be destroyed. There have been 70 persons drowned in the Province and much damage to our fortifications, houses, barns and Plantations. God of his Goodness has preserved us. Le Jau to the Secretary, 22 Jan 1714, South Carolina Parish of St. James near Goose Creek, 22 Jan 1713/14. *The South Carolina Chronicle of Dr. Francis Le Jau* (F. J. Klingberg, ed.), 1956, 136.

INDEPENDENT CHURCH

Memorandum—There was a former register kept belonging to the meeting-house and congregation, which by misfortune of the great hurricane that happened the 5th and 6th of September, 1713, was lost; when the house where the late Mr. William Livingston, deceased, then lived, and in whose possession it was, at White Point in Charleston, in this province, was washed and carried away by the overflowing of the sea. Ms. found in a blank leaf of the church book of the Independent Church, Ramsay, *History of South Carolina,* 1858 ed., II, 175n.

[1] Bartram, 20.

[2] Mark Catesby. *Natural History of Carolina, Florida, and the Bahama Islands,* ii.

[3] *Idem.*

[4] Lamboll Ms. quoted by Ramsay, II, 175–176n.

THE CAROLINA HURRICANE OF 1728

The fourth major hurricane to occur in Charleston's first half-century arrived in mid-summer 1728. David Ramsay in his summary of Carolina hurricanes, published in 1809, stated that very few particulars had been preserved since "newspapers, which are now so common, had then no existence in Carolina." He dated the occurrence through a family tradition of the Laurens family as taking place on September 3rd (14th). Alexander Hewatt in his 1779 account of South Carolina history merely gives the times as "late August." But in 1766 John Bartram transcribed a note "made by an accurate observer of veracity" dating the storm on 2 (13) August 1728.

Present-day historians with all the aids of modern

research could easily turn to a photostat or microfilm file of the Boston *Weekly News Letter* and find that the issue of 24 October 1728 (Old Style) clearly states, in confirming Bartram's note, that the storm occurred "on August 2nd (13th) about 10 o'clock at night. . . ."

That the storm was well remembered by those who experienced its fury is attested by the information given to Vincent Pearse during the War of Jenkins' Ear in 1740. When planning an attack to take place in July, the Carolinians informed Pearse that the last hurricane had occurred on 2 (13) August and that its early-in-the-season occurrence was considered quite exceptional. No mention of this storm was made by Lining, Chalmers, Milligen, or other writers in the 1750's and 1760's who chronicled Charleston's meteorological history. It seems odd that remembrance of this earliest in relation to season hurricane should have faded from memory so that Ramsay, in a diligent study of the subject, was unable to date the occurrence correctly when writing in 1809. But then Bartram misplaced erroneously by two years the Great Hurricane of 1752, even though only 14 years had passed since its occurrence.

The account of the 1728 event by John Bartram's "observer of veracity" is the most satisfying from a meteorological point of view. It indicated that the center passed near, yet to the south and west, of Charleston:

> Friday, August the 2nd 1728. We had a very violent Huricane, the wind being N.N.E. and continuing from N.N.E. to E.N.E. for about six hours, from 9 of the clock in the morning to three in the afternoon: and as the tide began to ebb the wind shifted to the S.E. and E.S.E., still blowing violently until four of the clock on Saturday morning; at which time it began to thunder, attended with violent showers of rain, and then it broke up. But the wind continued at S.E., or thereabouts for several days after.[6]

The news of the disaster did not reach New York until 14 (25) October, two and a half months after the occurrence.[7] All Boston newspapers gave space to the item. The *Weekly News Letter* a week later devoted a column of nine inches of type to include a listing of the 23 ships damaged or lost, their home ports and masters, and the content and extent of damage to their cargoes. The accounts of Hewatt and Ramsay appear to have drawn their data from the same dispatch which must have circulated widely in the colonial and British press.

> Boston, October 24—Since our last, we have received a more particular account of the late storm or hurricane at South Carolina, which was August 2nd about 10 o'clock at night, and lasted the next day until noon; It hath done a vast deal of damage on the land to the houses, wharfs and bridges, besides the destroying abundance of corn, and other fruits of the earth: Several lives were lost, both of White men and Negroes. . . . The great damage done to the shipping in Charlestown Harbor in South Carolina are as follows, viz." [8]

Of the 23 ships damaged or lost, it was thought that only eight were total wrecks and that the others would get off. The two men-of-war, the *Fox* and the *Garland,* there for the protection of trade, were the only ships which rode out the storm. A total of 1531 barrels of rice were lost on ships and 500 barrels on land from warehouses. Hewatt adds the further information that the hurricane, though it leveled many thousands of trees in the maritime parts, was scarcely perceived a hundred miles inland from the shore.[9]

The only additional information about the extent of the storm appeared in a follow-up note in the Boston press to the effect that a ship was lost with all its crew only six miles from Tokelcock Bar (possibly Ocracoke was meant here).[10] This interpretation would place the influence of the storm as extending well north along the coast to the Cape Hatteras area.

[1] Ramsay, II, 176n.
[2] Alexander Hewatt. *An Historical Account of the Rise and Progress of the Colonies of South Carolina and Georgia,* (1779), I, 316.
[3] Bartram, 20.
[4] See also: *New England Weekly Journal,* 21 Oct 1728; *Boston Gazette,* 21 Oct 1728.
[5] *Report of Committee to Investigate the St. Augustine Expedition of 1740* (J. T. Lanning, ed.), 82.
[6] Bartram, 20.
[7] *New York Gazette,* 14 Oct 1728.
[8] *Weekly News Letter* (Boston), 24 Oct 1728.
[9] Hewatt, I, 316.
[10] *Weekly News Letter,* 24 Oct 1728.

THE GREAT HURRICANE OF 1752

THE SOUTH–CAROLINA GAZETTE

September 19th, 1752. (Numb. 953.)

Charles-Town, Sept. 19.

The most violent and terrible HURRICANE that ever was felt in this province, happened on Friday the 15th instant in the morning; and has reduced this Town to a very melancholy situation: As the public doubtless will expect a particular account from the press, so we have endeavored to obtain the best information we possibly could of this deplorable calamity; for it is impossible, as yet, to tell all the damage and devastation we have sustained from the violence of the wind and waves.

On the 14th in the evening, it began to blow very

hard, the wind being at N.E. and the sky looked wild and threatening: It continued blowing from the same point, with little variation, 'till about 4 o'clock in the morning of the 15th, at which time it became more violent, and rained, increasing fast 'till about 9, when the flood came in like a bore, filling the harbor in a few minutes: Before 11 o'clock, all the vessels in the harbor were on shore, except the Hornet man of war, which rode it out by cutting away her main-mast; all the wharves and bridges were ruined, and every house, store, &c. upon them beaten down, and carried away (with all the goods, &c. therein), as were also many houses in the town; and abundance of roofs, chimneys, &c. almost all the tiled or slated houses, were uncovered; and great quantities of merchandize, &c. in the stores on Bay-street damaged, by their doors being burst open: The town was likewise overflowed, the tide or sea having rose upwards of Ten feet above the high-water mark at spring-tides, and nothing was now to be seen but ruins of houses, canows, wrecks of pettiauguas and boats, masts, yards, incredible quantities of all sorts of timber, barrels, staves, shingles, household and other goods, floating and driving, with great violence, thro' the streets, and round about the town. The inhabitants, finding themselves in the midst of a tempestuous sea, the wind still continuing, the tide (according to its common course) being expected to flow 'till after one o'clock, and many of the people already being up to their necks in water in their houses; began now to think of nothing but certain death: But, (Here we must record as signal an instance of the immediate interposition of Divine Providence, as ever appeared.) they were soon delivered from their apprehensions; for, about 10 minutes after 11 o'clock, the wind veered to the E.S.E., S., and S.W. very quick, and then (tho' it continued its violence, and the sea beat and dashed every where with amazing impetuosity) the waters fell about 5 feet in the space of 10 minutes, without which unexpected and sudden fall, every house and inhabitant in this town, must, in all probability, have perished: And before 3 o'clock the hurricane was entirely over.—Many people were drowned and others much hurt by the fall of houses.

At Sullivant's Island, the pest-house was carried away, and of 15 people that were there 9 are lost, the rest saved themselves by adhering strongly to some of the rafters of the house when it fell, upon which they were driven some miles beyond the island, to Hobcaw: At Fort-Johnson, the barracks were beat down, most of the guns dismounted, and their carriages carried away: At Craven's and Granville's bastions, and the batteries around this town, the cannon were likewise dismounted. . . .

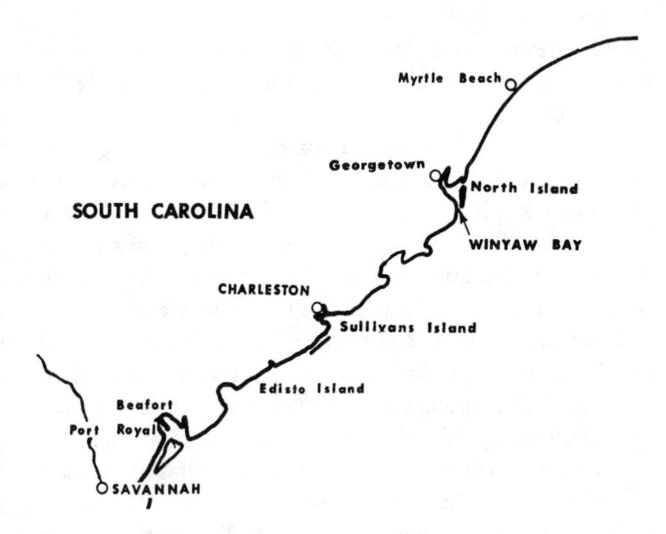

For about 30 miles around Charles-Town, there is hardly a plantation that has not lost every out-house upon it.—All our roads are so filled with trees blown and broke down, that traveling is rendered extremely difficult; and hardly a fence was left standing in the town or country.—Our loss in fine timber-trees, is almost incredible; and we have suffered greatly also, in the loss of cattle, sheep, hogs, and all kinds of provision.

From Winyaw and Port-Royal, our accounts are much more favorable than were expected, no damage having been done to the shipping in those harbors, and very little to the houses, as the hurricane was hardly felt at either place.[1]

Little doubt existed among the early writers on the subject that the hurricane of 1752 was the most severe in the Charleston area in colonial times. Ramsay (1809) declared: "This was the greatest and most destructive hurricane that has ever taken place in Carolina." [2] Dr. Prioleau, who made a study for the Medical Society about 1805, thought "the hurricane in the year 1752 far, very far exceeded, both in violence and devastation, the one in 1804. . . ." [3] Dr. Thomas Logan, writing in the *Southern Literary Journal* (1836), also repeated the above, saying that a "partial inundation" occurred in 1752, but that no complete inundation had ever taken place.[4] In modern times at Charleston only the hurricane of 1893 could be placed in the same class with that of 1752.

Practically all subsequent accounts of the Great Hurricane of 1752 have relied on the original *Gazette* story. In fact, the copies of the *Gazette* for this period in the files of the Charleston Library Society are so thumbworn that they are not considered in suitable condition for general use. Milligen-Johnston (1763), Hewatt (1779), and Ramsay (1809) repeat the facts cited in the original write-up and largely paraphase the newspaper account. Ramsay does add some new information, mainly gleaned from Dr. Prioleau's report to the Medical Society comparing the 1752 storm with that of 1804.

It is of interest to follow the speed with which the *South Carolina Gazette* account circulated through the colonies. In 1728, as we noted, there were only three American localities with a local press, but now there were seven. The Boston press first carried news of the September 15th hurricane on 16 October; at New York it was not until 23 October. The *Pennsylvania Journal* had the story on 2 November; the *Virginia Gazette* at Williamsburg on 10 November; and the *Maryland Gazette* at Annapolis on 16 November. Across the seas the *Gentleman's Magazine* for December 1752 gave the story wide circulation by repeating a verbatim account taken from the Charleston paper.

To learn the extent of the gale, the editor of the *Gazette* followed up his original story by gathering accounts from ships arriving after the end of the blow. One from Providence in the Bahamas encountered the storm when off St. Augustine, Florida, but received no damage. Another met gales on Wednesday, the 13th, seven leagues to the eastward of Charleston. With the wind continuing until the next afternoon, the ship lost mainmast and sails, had one side beat in, five seamen and one negro drowned, and all boats battered. The remaining crew were rescued off Edisto Island on Friday. Another vessel with masts gone put into St. Helena Sound south of Charleston where it rode out the hurricane without further damage.

From Georgia the editor of the *Gazette* learned that the hurricane was felt there, but the tide ran four feet lower than usual, though "incredible damage was done to the trees, but no house or vessel was hurt." It appears that the wind must have been northwest or west at Savannah, coming from a land quarter, to have driven the water seaward out of the river and estuary.

The Great Hurricane at Charleston was probably of rather small areal extent, but, nonetheless a most powerful storm. The veering of the wind at Charleston would indicate that the center passed inland just south and west of that place. This would give a northwesterly flow at Savannah, while a southeast hurricane was raging at Charleston. The coincidence of high tide and an easterly hurricane in the early part of the storm would account for the great destruction around the shores of Charleston harbor prior to 1100. Dr. Prioleau, relying on the word of Dr. John Moultrie, stated that if the wind had not shifted while the tide still had two hours to run, it is probable that the whole town would have been laid in ruins "since a foot more of water would have inundated the highest spot." Ramsay gives a vivid picture of the state of mind at this moment: "The consternation which seized the inhabitants exceeds all description. Finding themselves in the midst of a tempestuous sea, and expecting the tide to flow till one o'clock, they retired at eleven to the upper stories of their houses, and contemplated a speedy termination of their lives. At this critical time providence mercifully interposed and surprised them with a sudden and unexpected deliverance. Soon after eleven the wind shifted, in consequence of which the waters fell five feet in the space of ten minutes."

Though observations were currently being taken by at least two local persons, Drs. Lining and Chalmers, we do not have satisfactory meteorological data on all aspects of the storm. Lining does tell us that the rainfall during the storm "was only 3.74 inches, and the greatest part of that was the spray of the sea." One wonders how he was able to maintain a rain gage upright during the hurricane, and where it was located so that it was not inundated by the flood tide. No mention was made of the height of the barometer though this instrument was part of his equipment.

The origin of the storm is lost in historical silence. There are no reports of gales in any part of the West Indies or adjoining waters at this time. Our first report comes from the area between Florida and the Bahamas.

The loss of life also remains uncertain. A contemporary Boston press report, based on a Winyaw skipper's account, estimated that 95 persons were drowned. The Carolina *Gazette* merely stated that "many people were drowned and others much hurt by the fall of houses." Alexander in his summary (1896) gave the loss of life as 20.

The destruction of trees proved one of the most disastrous longtime effects of the hurricane. Many plantations lost their largest and best stands as the terrific gale mowed them down like ten pins. One plantation owner's loss from trees destroyed was estimated at the equivalent of $50,000. Most of the country roads were impassable for many weeks afterwards, halting normal business and social activities.[14]

An unfortunate loss involved the destruction of most of the papers in the Surveyor General's office which was five feet under water at the height of the storm tide. Many papers were washed away, and the remainder reported in the most perilous condition from the soaking. The utmost care was taken to preserve them.

A full listing of the destruction in Charleston cannot be given here. The *Gazette* summarized this contemporaneously, and Dr. Prioleau correlated the damage of 1752 with the map of the city as it existed in 1804, a half century during which great changes had been made in the city's physical aspect. He summarized the 1752 damage:

> In general all wooden fences and brick walls which were much exposed, and high stacks of chimnies were blown down. All wooden houses above one story in height, were either beaten down or shattered. Many gable ends of houses blown out. All tiled and slated houses, without exception, were more or less striped

of their covering; those on the Bay, in a manner quite uncovered. When the front was stripped, the wind blowing under the roof, burst the back part out in bulk . . . All the southwest point of the town comprehended between Tradd and King streets was inundated. White Point and South Bay were under water; it was two feet deep in Meeting Street . . . It flowed up the creek to Meeting Street through said street, round St. Michael's church into Broad street, as far down as the corner of Church street, where the South Carolina Bank now stands, where it met the water which flowed up from East Bay through Queen street into Church street.[15]

An idea of the force of the gale and storm tide can be gained from Ramsay's account of the fate of three vessels lying in the Charleston harbor when the storm came on:

When the gale came on there was a large ship at anchor at Sullivan's Island road. When it was over, that ship, no longer visible, was supposed to be foundered, but was shortly after found in Clouter's creek, about six miles north of Charlestown. During the gale she had drifted, with her anchors ahead, through the marsh opposite the city, called Shute's Folly, and also passed over another piece of marsh land three miles higher up, called Drum Island, without the loss of any of her crew, masts, or yards. After taking out two schooner loads of her cargo she was hove down at Hockbaw careening place. On examination it appeared that she had sustained no other damage than the loss of some of her sheathing plank, torn off by oyster shells. She was afterwards reloaded and safely arrived at London, after she had there been given over as lost.

Another vessel was driven with her anchors ahead from off White Point through the mouth of Vanderhorst's creek. In passing she carried away the southwest corner of the Baptist new church, and afterwards safely grounded on the west side of Meeting street. Her draft of water was from nine to ten feet.

A ship with a cargo of palatines had anchored in Ashley river a day or two before the gale. She, with her anchors, was driven into the marsh near to James Island where, by continual rolling the passengers were tumbled from side to side. About twenty of them, by bruises and other injuries, lost their lives. The Hornet sloop-of-war, with seven anchors ahead, drifted almost on shore near to the place where Gadsden's wharf now stands. Her bowsprit and foremast were cut away to prevent her foundering. She was the only vessel in the harbor that rode out the storm. All others were wrecked, damaged or driven on the wharves.[16]

[1] *The South-Carolina Gazette* (Charleston), 19 Sept 1752.
[2] Ramsay, II, 179n.
[3] *Idem.*
[4] *Southern Literary Journal* (Charleston), July 1836, 349.
[5] George Milligen. *A Short Description of the Province of South Carolina, 1770* (written 1763); Alexander Hewatt. *An Historical Account* (1779) David Ramsay, *Hist. of S. C.*, 1809.
[6] *S. C. Gaz.*, 19 Sept 1752.
[7] *S. C. Gaz.*, 26 Sept 1752.
[8] Ramsay, II, 180n.
[9] Robert Aldredge. *Weather Observers and Observations at Charleston, S. C. 1670–1871,* 188. Ms. U. S. Weather Bureau Library.
[10] *Boston Gaz.*, 16 Oct 1752.
[11] *S. C. Gaz.*, 19 Sept 1752.
[12] E. P. Alexander. *Monthly Weather Review,* 1896, 156.
[13] *S. C. Gaz.*, 10 Oct 1752.
[14] *Idem.*
[15] Ramsay, II, 179–182n.
[16] *Idem.*, 178.

THE SECOND CAROLINA HURRICANE OF 1752

By a vessel from North Carolina we hear that on the 1st instant they had there a most violent storm of wind, by which great damage was done in diverse places. At Johnston the Court House blown to pieces, and all the public records lost. Several other houses were also blown down, many trees tore up by the roots, and corn thrown down, much of it destroyed, and seven or eight persons killed. Several vessels were drove ashore upon Ocracoke Bars but all got off except two Virginia schooners, all of which bilged.[1]

This newspaper dispatch, dated Boston, 23 October 1752, is the only contemporary description from North Carolina of what appears to have been one of the most violent hurricanes ever to strike the Cape Fear country and the Outer Banks of North Carolina. Not only was the Onslow County Court House at Johnston (near Mittam's Point on the New River) destroyed and all the records lost, but the settlement was so materially damaged that it was abandoned and the county seat moved well inland.[2] The *Colonial Records of North Carolina* reveal that Beacon Island, a two-mile stretch of sand, completely disappeared under the fury of the pounding waves as the hurricane came ashore probably between Cape Fear and Cape Lookout.[3] Very heavy rains fell throughout the Cape Fear country causing floods.[4]

Tracing the antecedents of this storm offers an interesting problem. A severe gale struck the eastern side of Turks Island in the Bahamas on 23 September, continuing at full force from 0100 in the morning until 1700 in the afternoon. A large bank on the island was entirely washed away.[5] It is possible that the storm in the extreme southeastern Bahamas on the 23rd could have been the same that struck the Carolina coast on the 30th. Northwestward progress of hurricanes is frequently very slow and erratic until a steering current aloft permits a northward acceleration. But another severe hurricane struck the Isles of Pines, off the southern coast of Cuba, on the 26th, and apparently moved across the island as 16 vessels were reported lost off the north shore near Havana at that time.[6] Either of these storms, and it is more logical to believe that there were two separate hurricanes, could have sideswiped Charleston and driven ashore near Wilmington, North Carolina, on the 30th.

The press account from Charleston, dated 3 October, is of interest:

> On Saturday last we had here another terrible hurricane, which began with wind and rain at 4 o'clock in the afternoon, but cleared soon after 7 in the evening. For 2 or 3 hours before, the violence of the wind (which blew NE and at last settled at SE and the great quantity of rain that had fallen kept the tides from ebbing their due course and time, so that when this hurricane began to abate, tho' the water should have been low, it was higher than at common spring tides; and had the wind rose as expected, when the flood should have come in, our situation would have been most deplorable indeed! In the town it did little other damage, than to the goods of those people, who removed from the most exposed places, and the tops of some houses: But at Winyaw we are informed, it was severely felt than the former. About midnight the wind veered around to the northwest when another violent storm broke up the bad weather.[7]

The information that the storm was felt more severely in the Georgetown area agrees with the previously mentioned account of its violence in southeastern North Carolina. The course of the center northward from the Cape Fear country cannot be traced. Both Williamsburg and Annapolis papers carried accounts of extremely heavy rains, but gave no details as to the wind behavior. In the upper country of Virginia the greatest freshet ever known poured down the streams, some were reported to have risen 30 feet vertically. Many tobacco houses on low lands were swept away.[8] In Maryland the heavy two-day rain carried away dams and bridges, and a strong northwest gale on 1 October upset several vessels on the Bay.[9]

There are no storm reports at this time from New Jersey or New England. Perhaps the disturbance recurved over the Carolinas and lower Virginia and moved seaward. Contemporary papers, however, do list a violent storm in Nova Scotia and Cape Breton on October 1st, the same day that the Second Carolina Hurricane of 1752 was devastating the settlements in North Carolina. At Louisbourg harbor fifty vessels were driven ashore and only three men-of-war rode out the 16-hour gale.[10] This, possibly, could have been the gale noted at Turks Island on the 23rd which moved almost directly northward in a trough on September 30th at the same time that the Cuba storm was also drawn into the southern end of the low pressure trough. 1 October 1752 was a fateful day in the hurricane annals of North America with two major storms making spectacular landfalls on the same day.

[1] *Boston Eve. Post,* 23 & 30 Oct 1752.
[2] F. X. Martin. *History of North Carolina,* 1829, 2, 61; *The Historical Records of North Carolina, The County Records,* 3, 62.
[3] *Colonial Records of North Carolina,* 5, 596.
[4] *Va. Gaz.,* 12 Oct 1752.
[5] *Penna. Gaz.,* 6 Dec 1752.
[6] *Penna. Jour.,* 2 Jan 1753.
[7] *S. C. Gaz.,* 3 Oct 1752.
[8] *Va. Gaz.,* 12 Oct 1752.
[9] *Md. Gaz.,* 5 & 12 Oct 1752.
[10] *Penna. Jour.,* 2 Nov 1752.

THE EASTERN CAROLINA HURRICANE OF SEPTEMBER 1769—I

A hurricane of great power struck the Atlantic coastal areas from the Carolinas northward to Maine early in September 1769. Contemporary press accounts of the destruction along its trail would tend to place the storm among the severest of the century. Certainly, with the growth of the settlements, there was more to damage than in earlier days, and the track of the disturbance carried over practically all of the major settlements of that time.

Though neither Poey or Tannehill-Tingley mention a storm in the West Indies at this time, we have ship reports of hurricane winds 150 miles northeast of Nassau, Bahamas, on September 5th. Another report indicates the occurrence of a southeast gale 250 miles east of Norfolk on the 8th.[1] These with contemporary newspaper accounts would indicate a storm of great size and intensity.

The Charleston press, in taking note of the "late blowing weather" at sea, listed a number of ships which had been dismasted by the gales, but observed that the winds ashore had not been high enough to damage the rice crop which was approaching the harvest stage.[2] The increase of damage reports to the northeast would indicate a probable landfall of the hurricane near Cape Lookout and a traverse across the bays and peninsulas that make up eastern North Carolina.

The prominent colonial towns of Brunswick, New Bern, and Edenton, all on estuaries leading into Pamlico and Albemarle Sounds, appear to have been very close to the path of the great storm and suffered immense damage. Fortunately, we have excellent contemporary reports which are presented here in full:

Letter from Governor Tryon to Lord Hillsborough.

BRUNSWICK 15th Sept 1769.

> On Thursday the 7th instant we had a tremendous gale of wind here. It began about 10 in the morning

at North East and blew and rained hard till the close of the evening when both wind and rain increased. The wind shifted before midnight to the North West. The gale became a perfect hurricane between twelve and two o'clock on Friday morning the 8th instant. The fury of its influence was so violent as to throw down thousands and I believe from report hundreds of thousands of the most vigorous trees in the country, tearing some up by the roots, others snapping short in the middle. Many houses blown down with the Court House of Brunswick County. All the indian corn and rice leveled to the ground and the fences blown down, add to this upwards of twenty saw mill dams carried away with many of the timber works of the mills, and lastly scarce a ship in the river that was not drove from her anchor and many received damage. This my Lord is but the relation of what happened within fifty miles of this town. We are therefore in hourly expectation of receiving as melancholy accounts from other parts of the province. It is imagined that as the corn was within six weeks of its maturity, the planters may save about half a crop, but they have no hopes of recovering the rice lying at this period under water from the freshets that this gust occasioned. The country will I fear be greatly distressed this winter for provisions as far as this gale has extended, for the people will not only be short of corn, but the hogs which are the support of many families will lose the acorns and nuts in the woods which used to fat them for market, the wind having stripped every acorn from the trees before they were ripe. In short, my Lord, the inhabitants never knew so violent a storm; every herbage in the gardens had their leaves cut off. This hurricane is attributed to the effect of a blazing planet or star that was seen both from Newbern and here rising in the east for several nights between the 26th & 31st of August, its stream was very long & stretched upward towards the S° West.[3]

Letter from Mr. Stewart to the Secretary.

NEWBERN N. Carolina, Decr the 6th 1769.

REVEREND SIR,

My last to you was forwarded by Mr Blinn a Gentleman that went home in July last with Letters recommendatory from Governor Tryon to the Bishop of London for orders since which time very little could be done in our way in the Eastern parts of this Province, for on the 7th of Septr at night we had the most violent Gale of wind and the highest tide that has ever been known since this country has been inhabited. The tide rose in a few hours at my house 12 feet higher than I ever before knew it, and the wind blew so violent nothing could stand before it: Every Vessel, Boat or Craft were drove up in the woods and all the large Oaks, Pines &ca, broke either off or torn up by the roots, Our Indian Corn (which was not quite ripe, and which is the common Bread of the country) was mostly destroyed and in many places together with the Cattle, Sheep, Hogs &ca washed quite away. But no place has suffered so much as this Town of Newbern, one entire Street, Houses, Store Houses, wharves &c., to the amount of near £20,000 pounds were destroyed and swept off together with several of the Inhabitants in a few hours time. The roads were impassable for several weeks by reason of the trees fallen and the Bridges carried away and so great is the scarcity of small Boats at the Ferries &c that the people cannot travel nor attend the places of public Worship as usual. The damages have been great in many other provinces. But no parts that we have heard of have suffered any thing equal to the country on Pamlico and Neuse Rivers being in Mr Reed's parish and mine.

I had the misfortune to have one of my Legs much hurt the night of the storm in endeavoring to save some of my Houses. By neglect and by the rheumatic humour in that Leg, I am once more here under the Doctors hands but hope it will not be of long continuance as I have been obliged to have had my Foot laid open which has relieved me in some measure, and put me in a good way of recovery. My private losses in the Hurricane in Houses and Stores in this Town, and at my plantation is upwards of £600.—this currency, and I question whether these lower Inhabitants will ever get over it these seven years. I am Reverend Sir—Yours &c

ALEXr STEWART.[4]

[1] *S. C. Gaz.*, 14 Sept 1769.
[2] *Ibid.*, 21 Sept 1769
[3] "From Tryon's Letter Book," *The Colonial Records of North Carolina*, 8, 71.
[4] "From N. C. Letter Book S.P.G.," *ibid.*, 159. See also detailed damage account at New Bern, *ibid.*, 73–75.

THE ORDERING OF PROVIDENCE: THE HURRICANE OF AUGUST 1778—I

President Ezra Stiles of Yale College suggested that the August hurricane of 1778 might have been sent by Providence to prevent an impending sea battle between the French and the English off Rhode Island.[1] A previous section has discussed the storm's meteorological characteristics while in northern waters and its effect on the immediate course of military operations.[2] Since this disturbance from the tropics caused damage in the Southern Colonies on its way northward, its behavior there will be briefly noted in this section.

This history-making hurricane first appeared at Charleston. On the morning of the 10th the South Carolina coast was raked by a northeast storm which continued until late in the afternoon. Then the wind went to the north and northwest. Several vessels were damaged in the harbor, trees and fences bowled over, and the charred walls of many buildings, still standing after the disastrous fire in January, blown down.[3]

It appears that the storm made a landfall somewhere between Georgetown and Cape Fear since the wind on the evening of the 10th at New Bern, North Carolina, was blowing a gale out of the southeast at the same time that a northwest gale prevailed at Charleston, South Carolina. The blow continued all night at New Bern, but the tide did not rise as high as it had done in the castastrophe attending the hurricane of September 1769. The main damage came from the heavy rain which ruined the crops and fodder.[4]

In west-central North Carolina at the Moravian settlement near Winston-Salem there was an all-night "hard storm from the northeast," followed by much rain the next morning of the 11th. "The storm has beaten down the corn, which this year is very tall and heavily loaded with ears, and has broken down some of it. It cleared up in the afternoon," noted the Moravian diary.[5] All through central and eastern North Carolina came reports of the gale and of great losses of crops. At Hillsborough "from dusk yesterday to noon today there was a wind almost like a huricane [sic], which did much damage to the corn, tobacco, and in the woods." Later in the season the effects of the storm were felt especially in the southern counties where corn advanced to an inflationary price of 50 shillings a bushel.[6]

[1] Ezra Stiles. *Diary,* II, 295.
[2] A. T. Mahan. *The Influence of Sea Power upon History, 1660–1783,* 364; C. Laughton & V. Heddon. *Great Storms.* London, 1927. 251.
[3] *S. C. Gaz.,* 13 August 1778.
[4] *N. C. Gaz.,* 14 August 1778.
[5] *Records of the Moravians in N. C.,* III, 1244.
[6] *State Records of N. C.,* 13, 490.

THE OCCUPATION STORM AT CHARLESTON IN 1781

The harbor and city of Charleston surrendered to the British forces on 12 May 1780 after a month's investment; with a 5,400 man garrison and four ships, its capture comprised the greatest single loss for the Americans during the entire war. The Charleston *Royal Gazette* provides our main source of weather information during the occupation, and its columns show that the principal meteorological event during the period came with the tropical storm of 10 August 1781. It is interesting to note that the account of this storm in the *Royal Gazette* on 11 August appeared in the Philaphia press as soon as 5 September:

> The long period of dry weather we lately had, has been followed by a severe storm. On Thursday evening (9 Aug.) about 7 o'clock the wind began to blow from the NE. Between 8 & 9 it blew fresh, accompanied by rain. It continued to rain and blow from the same quarter with increasing violence during the whole night and yesterday, till one o'clock, when the wind shifted to ENE. The gale continued to increase till 4 P.M. when the wind having got around to SE, it blew a mere hurricane. The rain increasing at eight o'clock the wind began to fall; at high water, three o'clock this morning, the storm abated. The wind has now got around to SSW but still blows fresh.[1]

The ships at anchor in the harbor fared better than those tied up at wharves. The British *Thetis* sank at its dock as did the small ship *London.* Most vessels in the Bay dragged their anchors and went ashore, but were all gotten off without serious damage; even the *London* was raised within four days.[2]

The only other report on this storm was marked by William Redfield in his notebook, Record of Storms. He wrote that on 10 August 1781 at Wilmington, North Carolina, a gale prevailed for 40 hours, commencing in the NE, shifting to SE, and finally veering to SSW—the same wind behavior as experienced at Charleston.[3] The course of the storm must have been close, but to the west of both places. A track north-northeastward from coastal Georgia over the east-central portions of the Carolinas is indicated.

[1] *Royal Gazette* (Charleston), 11 Aug 1781; *Penna. Jour.,* 5 Sept 1781.
[2] *Royal Gaz.,* 15 Aug 1781.
[3] W. C. Redfield. Record of Storms. Ms., Yale Univ. Library.

THE CHARLESTON HURRICANE OF 1783—I

During the half century between the Great Hurricane of 1752 and the Gale of 1804, only two tropical storms seem to have rated any special mention at Charleston. These occurred in 1783 and 1797. Dr. Samuel Latham Mitchill referred back to "the great storm of 1783" when writing his account of the 1804 hurricane, indicating that the storm under study was considered of some stature by Carolinians.[1]

The tropical storm of 7–8 October 1783 is best described by the local press:

> On Monday last the wind blew very fresh the whole day from the northeast: toward the evening it increased considerably, and from midnight to Wednesday morning, we had a most violent storm of wind, accompanied with incessant rain, which has done considerable damage to the wharves and shipping in the harbor. Happily we hear of very few lives being lost. Many people were apprehensive of a hurricane: and the tide rising very high on Tuesday evening, increased their fears so much, as to cause several families which lived near the water, to remove from their houses. Fortunately, the wind backed from

north east to north west during the height of the storm, which of course kept the tide down: had it shifted to the east or south east, in all probability the city would have experienced all the horrors and destruction of September 1752.

From the country we learn, that although many fences were blown down, and other damage sustained, yet the crops in general have suffered little or nothing, the harvest being mostly over, and the grain secured.

The morning after the storm, several vessels arrived from sea, who were close upon the coast during the whole of it, but the wind being offshore, they have suffered little damage.[2]

Another press report detailed the damage to the wharves, but added that the shift of the wind from northeast to northwest at a critical juncture saved most of the vessels in the harbor from major damage:

> Every wharf (except Eveleigh's) has more or less suffered, the one on the Fish-market in particular (and so they will annually, for ever, on the present principle of constructing them). On the South-bay several families moved from their houses, the water having risen to the first floor—the tide flowed near a foot and a half higher than has been remembered for several years.[3]

The behavior of the wind at Charleston would indicate that the storm moved to the east and north of that place, yet close enough to give the city a good sweep of possible hurricane force winds. Great damage to valuable stores of salt were reported at Wilmington, North Carolina, probably as a result of the attending storm tide.[4] The high winds had a rather large areal extent as the Moravians inland at Winston-Salem made mention of the blow. At Wachovia the evening meetings were omitted "because of a severe storm, which during the night assumed the proportions of a hurricane, damaging buildings, fences, and buildings, and blowing down many trees in the woods." The wind lulled there about 0900 the morning of the 8th, and floods that day damaged goods in low-lying storehouses.[5] At nearby Bethabara the storm became so violent on the night of the 7–8th that some people could not sleep; at Bethania people thought they felt the earth quake during the height of the blow. The strength of the wind is attested by the many trees in the vicinity which were pulled up by the roots.[6] Farther north in the Richmond and Chesapeake Bay area a northeast gale raged on the 8th as the storm moved along the seaboard toward New England.[7]

[1] *Medical Repository* (New York), 8, 1805, 362.
[2] *S. C. Weekly Gaz.*, 11 Oct 1783.
[3] *S. C. Gaz. and Gen. Adv.*, 8 Oct 1783.
[4] *Penna. Gaz.* (Phila.), 29 Oct 1783.
[5] *Records of the Moravians in N. C.*, IV, 1843.
[6] *Ibid.*, 1865, 1869.
[7] Richmond, 18 Oct., *Penna. Gaz.*, 29 Oct 1783.

THE END OF OCTOBER STORM OF 1792 AT CHARLESTON

A tropical storm of considerable intensity passed very close to and just east of Charleston Harbor on 30–31 October 1792 after approaching from the south or southwest. Some meteorological details are supplied by an account in the *Charleston Daily Advertiser* and then reprinted and given wide circulation in the *Columbian Magazine* of Philadelphia which served as a national news periodical in those days:

> On Tuesday last (30 Oct.) a great quantity of rain fell in this city, attended by very high wind. In the evening the wind increased, and continued to blow with great violence the greatest part of the night. A quarter before 12 o'clock P.M. the wind was due N. —at three quarters after 12, it had shifted to ENE—at half past one, to SE—at quarter past two, to due E— and at three quarter after three, to SE at which time it blew a severe gale: but at four it veered, and got around again to due N. Considerable damage was done to the wharves and shipping in the harbor. . . . Several small buildings were unroofed and thrown down, and a number of fences leveled with the ground. Some of the inhabitants were apprehensive of an inundation, but it being ebb tide, and the wind shifting to the NW, prevented that calamity.[1]

Another press account of this storm, probably taken from the *City Gazette*, was also widely copied in the Northern press:

> On Tuesday night (30 Oct.) this town and harbor were visited by one of the heaviest gales it has felt this many years—one or two buildings were blown down and several unroofed and a number of fences destroyed, and most of the vessels in the harbor more or less injured; and two small ones sunk—several wharves were considerably damaged—Fortunately the tide being down, the injury to the vessels in the harbor was far less than might have been apprehended had it been high water.[2]

[1] *The Charleston Daily Adv.*, 2 Nov 1792.
[2] Charleston paper, 1 Nov 1792, in *U. S. Chronicle* (Providence, R. I.), 28 Nov 1792.

THE STORM OF OCTOBER 1797 AT CHARLESTON

TREMENDOUS STORM

For several days the weather had been dirty, with strong gales of wind from the N.E. which gave reason to expect a storm which came on, and has not been surpassed since that experienced in October 1783. About 3 o'clock yesterday morning the wind came in from the SSE, raised to such a dreadful degree of violence, accompanied with rain, as to tear away a vast number of vessels from their moorings at the wharves, & did great damage among the shipping and and on shore. . . .

The violence of the storm continued for about two hours, the wind changing at the latter part of it, to SSW and with increased violence. The tide was uncommonly high and covered all the wharves.[1]

The above account appeared in the *South Carolina Gazette and Daily Advertiser* on Saturday, 21 October 1797. This and other evidence in the same issue definitely date the major storm of the season as occurring on 19–20 October. Previous accounts have been in error on the dating of the event. John Drayton in his *View of South Carolina,* published in 1802, stated that the storm took place in September, and the *City Year Book, 1880* gave the date as 5 September 1797, and this has been repeated by others.[2] An examination of Charleston newspaper files revealed no storm on that day; in fact, mention was made of arrivals and departures on the 5th, as well as on the days previous and following. Further, on the 16th the *City Gazette* noticed that the planet Jupiter had been especially clear "from the serenity of our atmosphere for some days past." [3] A hard blow did occur off Cape Hatteras on the 6th to the 8th, but this was not felt in Charleston harbor.[4]

The above quoted account in the *South Carolina Gazette* of 21 October went on to describe the scene along the wharves when several ships broke loose from their moorings at the southeast point of the town and were impelled along by the rushing tide into many other vessels tied up at nearby wharves, with widespread damage resulting to rigging, sails, and superstructure. Some smaller boats were driven over the marshes and left high and dry when the storm tide receded. Several buildings under construction were demolished by the force of the gale; and many tiles, as usual, were torn from roofs to be hurled as deadly missiles into the streets. One seaman who was washed off a wharf comprised the only casualty known to the editor. An interesting aspect of the storm involved the destruction of a "aerial chair" or ferris wheel which had recently been finished for the amusement of Charlestonians.[5] The above evidence would seem to place this disturbance in the tropical storm category when in the latitude of Charleston, though it may have attained full hurricane strength when wreaking havoc among the dozen or so ships it drove ashore along the Florida coast.

There had been minor tropical storms in August and again in late September 1797 in the Bahamas and Atlantic shipping lanes, but the mid-October blow seemed to have been the major news of the maritime season. It was especially severe in the Bahama Gulf where on 15–16 October a number of American vessels were cast away on the east coast of Florida and nearby Bahama Islands. The crews and cargoes were taken into custody by the "wreckers of New Providence" and gathered at Nassau where they were well treated despite the confiscation of their ships.[6] One vessel enroute to Havana reported meeting the hurricane when only 12 leagues (42 miles?) from that port and was blown back through the Straits to the Bahamas.

One can surmise that the storm center moved northward from the Caribbean across Cuba and then along the Florida and Georgia shores. The behavior of the wind at Charleston in veering from southeast to south-southwest would indicate a track just to the west of that city. The only additional evidence of storm activity at this time was a notice in the *New York Gazette* to the effect that a ship was seen ashore on the south end of Long Island on 24 October with the comment that "she must have gone ashore in the late heavy gale." [7] Evidently the track of the storm took it northward to affect shipping along the Middle Atlantic coast.

[1] *S. C. Gaz. & Daily Adv.,* 21 Oct 1797.
[2] John Drayton. *View of South Carolina.* Charleston, 1802, 202. *City Year Book, 1880.* Charleston, 1881, 280.
[3] *City Gaz.,* 16 Sept 1797.
[4] *Ibid.,* 19 Sept 1797.
[5] *S. C. Gaz. & Daily Adv.,* 21 Oct 1797.
[6] *N. Y. Gaz.,* 22 Nov 1797.
[7] *Ibid.,* 28 Oct 1797.

THE OCTOBER STORM OF 1800 AT CHARLESTON

In early October a "tremendous and destructive" storm, probably of tropical origin and possibly reaching hurricane force, struck the Charleston area. The only meteorological details available are supplied by the *Charleston Times:*

On Saturday night [4 Oct.], from eleven to twelve o'clock as tremendous and destructive a storm was experienced in this city and harbor, as has happened for nearly 20 years [probably a reference back to 7–8 October 1783].

For several days past the wind has been excessively violent from the N.E. On Friday last about noon it shifted to N without any abatement. On Saturday it returned to NE and blew a very heavy gale until noon accompanied with rain. It then shifted to SE and continued from that quarter until after twelve o'clock at night, when it suddenly chopped around to SW and blew with dreadful violence, accompanied with repeated and heavy claps of thunder, and sharp lightning. At this time the tide was two feet high on the wharves." [1]

The *Times* account mentioned much damage to the wharves, but stated that losses to shipping were relatively light. Further details as to damage were supplied by the *City Gazette and Daily Advertiser*. New Street along the Bay was badly damaged, and the wharves along South Bay Street torn up by the surging water. The greatest wind damage occurred in the northwestern part of the city where trees and chimneys were blown down and fences laid low. One person was reported killed in bed in this sector of the city. Out on Sullivans Island, at the harbor entrance, three homes were washed away by the waves at high water.[2]

[1] *Charleston Times,* 6 Oct 1800.
[2] *City Gaz. & Daily Adv.*, 6 Oct 1800.

THE GREAT GALE OF 1804 IN GEORGIA AND CAROLINA

Of the three major hurricanes in 1804 which generated in the West Indies after September 1st, two aimed severe blows at the American mainland. At St. Kitts "the worst hurricane since 1772" struck on 3 September, and then next day moved west-northwestward to rake the western end of Puerto Rico where many vessels were sunk.[1] The storm was reported in the southeastern Bahamas on the 4th when Turks Island, always a check-point for American-bound hurricanes, had damaging winds.[2]

In approaching the American coastline, the hurricane had already commenced to recurve to the northward while striking a glancing blow at the north Florida coast. St. Augustine experienced a very severe northeast gale on the 6–7th, the tide rose to an uncommon height, and only one of nine vessels in the harbor rode it out, though "the town received no very serious damage." [3] The hurricane also lashed at the Nassau area in the Bahamas on the 6th.[4]

The low-lying coastal plantations of Georgia and the thriving seaport of Savannah, which had not been hit by a *great* hurricane since 1752, lay in the direct path of the whirling winds as the whole mass turned more and more to the north. It so happened that Aaron Burr, that controversial figure in American history at the turn of the century, was at St. Simon Island on the central Georgia coast on the day the hurricane struck. In his celebrated correspondence with his daughter Theodosia, Burr has left a graphic description of the impact of the wind and tide on his retreat, and has also preserved some essential meteorological details which demonstrate that the eye of the storm passed right over St. Simon:

St. Simon's, Wednesday, September 12, 1804.

In the morning the wind was still higher. It continued to rise, and by noon blew a gale from the north, which together with the swelling of the water became alarming. From twelve to three, several of the outhouses had been destroyed; most of the trees about the house were blown down. The house in which we were shook and rocked so much that Mr. C. began to express his apprehensions for our safety. Before three, part of the piazza was carried away; two or three of the windows burst in. The house was inundated with water, and presently one of the chimneys fell. Mr. C. then commanded a retreat to a storehouse about fifty yards off, and we decamped, men, women, and children. You may imagine, in this scene of confusion and dismay, a good many incidents to amuse one if one had dared to be amused in a moment of such anxiety. The house, however, did not blow down. The storm continued till four, and then very suddenly abated, and in ten minutes it was almost calm. I seized the moment to return home. Before I had got quite over, the gale rose from the southeast and threatened new destruction. It lasted great part of the night, but did not attain the violence of that from the north; yet it contributed to raise still higher the water, which was the principal instrument of devastation. The flood was about seven feet above the height of an ordinary high tide. This has been sufficient to inundate great part of the coast; to destroy all the rice; to carry off most of the buildings which were on low lands, and to destroy the lives of many blacks. The roads are rendered impassable, and scarsely a boat has been preserved. Thus all intercourse is suspended.[5]

The wind at Savannah, coming from the northeast at first, but later varying occasionally to the north, raged for 17 hours. The tide surged over the sand bars, into the bays, up the rivers, and over the wharves—over everything that was less than 10 feet above sea level. So powerful was the gale, the *Georgia Republican* reported, that the rain had a saline taste from being mixed with sea spray, and sand picked up by the gale was blown into the upper stories of houses as high as 30 feet above ground level.[6]

The story, as carried in the *Savannah Museum* of the 12th, gives pertinent meteorological details:

Storm commenced on Friday night, with a degree of violence by no means unusual or alarming. On Saturday morning it had abated; but about nine o'clock its violence increased until 4 or 5 o'clock, when it appeared to have gotten to its height, and continued to rage with dreadful fury until about ten o'clock when it began to subside; during the early part of the day the wind was from the north: but about the middle and during the latter part of the day was shifting almost continually from N to NE.[7]

Further details are found in an extract of a letter written from Savannah on the 11th:

I have barely survived the most tremendous hurricane that ever blew. The house in which I lodge was blown almost to pieces. The storm commenced on Friday night, and continued to increase until Monday morning, by which time every vessel in the harbor was either sunk, driven over the docks or dismasted. There is not a house in the lower part of the town but what has suffered very materially from the high tide which rose to the astonishing height of from twenty to thirty feet above the highest spring tides. Gun Boat No. 1 was driven from anchorage 7 miles over marshes, through woods, and finally came to in a field.[8]

Ramsay has given a vivid description of the effect of the hurricane and storm tide at Fort Green at the entrance to the Savannah River and along the coastal plain:

> On Cockspur Island Fort Green was leveled, all the buildings destroyed, and thirteen lives lost. Muskets were scattered all over the island. Cases of canister shot were carried from one hundred to two hundred feet, and a bar of lead of 300 pounds was likewise removed to a considerable distance. A cannon weighing 4,800 pounds is said to have been carried thirty or forty feet from its position. Broughton Island was covered with water, and upwards of seventy negroes, the property of William Brailsford, were drowned by the oversetting of a boat in which they attempted to escape from the island to the main. The barn on the island being raised on made high land stood the storm, and in it the negroes would have been safe. At St. Simon's Island great damage was done. The crops were generally covered with water, and several negroes were drowned. The like happened on St. Catherines, and on the other islands on the coast. At Sunbury the bluff was reduced to a plane, and almost every chimney leveled to the ground.
>
> The rice swamps and low lands within the reach of the tides were generally overflowed. The crops of rice and provisions were greatly injured, and in some places totally destroyed or washed away. The fields of cotton along the seashore which previously promised an abundant crop, were blasted and nearly destroyed by the violence of the wind and the spray of the sea.[9]

At Beaufort, the seaport in the complex of islands and bays between Port Royal Sound and St. Helena Sound (about half way from Savannah to Charleston), the damage was immense. The Port Republic Bridge Company saw the efforts of the past seven years swept away in a few minutes when the rising tide, capped by storm-tossed waves, smashed all the causeway close to the mainland and destroyed about half the structure on the island side. All dwellings on Bay Point were swept away as the eye of the savage hurricane roared in from the Atlantic. Cotton fields around Beaufort were overflown to a depth of 4 to 5 feet by the immense water surge.[10]

There are no reports of a complete calm anywhere except at St. Simon, though a lull was mentioned at both Savannah and Charleston, but this does not necessarily seem to have been associated with the passage of the center. At Savannah the wind is always mentioned at northeast or east—while at Charleston 85 miles to the northeast, the wind went around from northeast to east and finally to southeast, all-the-while blowing a hurricane. The wind still farther north at Georgetown, behaving in the same manner as at Charleston, commenced to blow at 0300 from the northeast and continued through the day and evening until a shift to south-southeast at midnight.[11] It is my surmise that the center of the Gale of 1804, skirted the coastline of Georgia, passing over St. Simon Island, just to the east of Savannah, right over Beaufort, South Carolina, where it came inland, and then to the west of Charleston and Georgetown.

We have a good press report at Charleston from the *City Gazette:*

> GALE OF WIND IN THE HARBOR OF CHARLESTON
>
> On Friday night last, about 11 o'clock, a dreadful gale of wind came on this harbor, and continued to blow with the most extreme violence until Saturday morning, one o'clock: the wind at first was at northeast, in the course of Saturday morning it changed to east, and in the afternoon it changed to southeast. It is impossible for us at this time to describe accurately the destruction caused by this gale; the whole of the wharves from Gadsden's on Cooper river, to the extent of South-Bay, have received very considerable damage, the heads and sides of most of them are washed away. Of the vessels in the harbor, but three or four have escaped without injury, several are totally lost and many more are much damaged.
>
> At seven o'clock on Saturday morning, the period of low water, the tide was as high as it generally is at spring tides: it appeared that at the preceding ebb but little water had left the river; at twelve o'clock it was from two to three higher than it had been seen for many years, and made a complete breach over the wharves and drove many vessels on them, where they now lie. On General Gadsden's wharf several stores were washed or blown down, and their contents of rice and cotton much damaged and some lost. On South-Bay, the whole of the bulwark made against the water is in ruins and the house of Mr. William Veitch, built on made ground was washed down—the new street made to continue East-Bay to White Point is greatly damaged, the sea made breaches through it in many places. On Blake's wharf, a brick building occupied as a scale and counting-house, was beat down by the bowsprit of the ship Lydia.
>
> In the city no other damage is done than many houses which were covered with slate, are in part unroofed, and most of the trees in the streets and many fences are blown down.
>
> Great apprehensions were entertained for the safety of the families on Sullivan's Island, but accounts received from thence yesterday were very favorable, not a life was lost there except a black boy. From fifteen to twenty houses were undermined by the water and washed away; the inhabitants of which lost almost everything that was in them.
>
> [Eleven ships severely damaged, five sunk and lost].[12]

David Ramsay, who made a particular study of this storm, states that the gale was not felt north of Wilmington, but he appears to have been in error on this point. The extent of the gale inland may be judged from reports that trees were blown down as far as 100 miles from the sea coast. Near the entrance to Cape Fear River a brig, the *Wilmington Packet,* was cast away on Bald Point during the height of the blow after striking Frying Pan Shoals in "the gale of the 8th." [13]

Little information is at hand about the storm from North Carolina to New England. At Norfolk the wind was east-northeast on the 7–8th with a shift to east on the afternoon of the 8th, and to east-southeast early on the 9th.[14] If these winds are an accurate portrayal of conditions at Norfolk, it would appear that the storm center moved west of that place.

Southeastern New England had a severe blow on the 11–12th of September. Several ships were sunk in Boston harbor, while at Salem the steeple of the new South Church of Mr. Hopkins was demolished by the high wind on the afternoon of the 11th.[15] Whether this was a separate tropical storm or part of the Georgia-Carolina Gale of the 7th is an open question. The absence of any damage reports from the Middle Atlantic coast or the Long Island Sound area, along with the relatively long time from the 8th to the 11th to move from Norfolk to Boston, would lead us to believe that we are dealing with two separate storms. Possibly the 11 September 1804 blow in Massachusetts may have been a small tropical storm which followed tracks similar to those of 1869 and Edna in 1954.

The hurricane and flood tide of 1804, being the first major disaster from a tropical storm to hit the United States in almost 15 years, aroused a great amount of interest. Accounts were published in most of the more important papers throughout the country. The public interest in the scientific causes of the storm led Dr. Samuel Latham Mitchill to make a study in accordance with the generally held theories of the day. Dr. Mitchill, from his post at Columbia College in New York City, served the country as the leading authority on all matters scientific. He was serving a term as a United States Senator at this time, and he used his contacts with senators and representatives from all parts of the country to gather data on various storms. His account of the Hurricane of 1804 was issued in the form of a letter to Baron von Humboldt, dated George-Town, Maryland, 19 January 1805, and was subsequently published in the New York *Medical Repository,* of which Mitchill was editor and the leading spirit.[16]

The title on the front-page of the article is practically an abstract: *Some particulars of a terrible hurricane which began to the Windward of the Caribee Islands on the third of September 1804, and preceeded Northwestwardly over the Virgin Islands, and Bahamas, on the fourth, fifth, and sixth, until it reached Florida, Georgia, and South Carolina on the seventh, eighth, and ninth; and of a furious Gale from the northeast which prevailed at the same time, and preceeded southwesterly until it met the former: showing that storms of the most destructive violence sometimes arise to Windward, and bear down everything before them in their passage to Leeward.* In his treatise Dr. Mitchill upheld the rectilinear theory of straight-line wind storms. He accounted for the wind shift at Charleston by the meeting of two currents; first one from the northeast, to be succeeded by one from the southeast. He saw no suggestion of a vortex or a circular pattern of winds. The correct surmise had been made by William Dunbar from his experiences in the New Orleans hurricane of 1779 and also by Colonel James Capper in 1801 from his earlier observations of an Indian Ocean cyclone on the Coromandel Coast, but neither of these pioneer thinkers secured any following for their vortex theories.[17] Even when William Redfield took up his work in the early 1830's, it was some time before his attention was called to these earlier original studies of the hurricane circulation problem.

[1] Poulson's *Amer. Adv.* (Phila.), 13 Oct 1804; *N. Y. Post,* 22 Oct 1804.
[2] S. L. Mitchill. Some Particulars of a Terrible Hurricane. *Medical Repository* (New York). 8, 354–365.
[3] Charleston, 17 Sept, *Amer. Adv.,* 1 Oct 1804.
[4] Redfield Ms., Yale Univ. Library.
[5] *Memoirs of Aaron Burr.* (Matthew L. Davis, ed.). 2, 339.
[6] *Georgia Republican* in *Amer. Adv.,* 13 Oct 1804.
[7] *Savannah Museum,* 12 Sept, in *Aurora* (Phila.), 2, Oct 1804.
[8] Extract of letter, Savannah, 11 Sept, in *Aurora,* 2 Oct 1804.
[9] Ramsay, 2, 184.
[10] *Amer. Adv.,* 1 Oct 1804.
[11] *Wilmington Gaz.,* 12 Sept, in *Public Ledger* (Norfolk), 27 Sept 1804.
[12] *City Gaz.* (Charleston), 10 Sept 1804.
[13] *Wilmington Gaz.* in *Amer. Adv.,* 6 Oct 1804.
[14] *Public Ledger* (Norfolk), 18 Sept 1804.
[15] *Boston Weekly Magazine,* 15 Sept 1804; William Bentley. *Diary.* 3, 110.
[16] S. L. Mitchill. *Medical Repository.* 8, 354–365.
[17] William Reid. *An attempt to develop the Law of Storms.* 3d ed., 1850, 2.

THE GREAT COASTAL HURRICANE OF 1806—I

The tropical storm season of 1806 rates top ranking as a producer of major hurricanes in the shipping lanes along the eastern coast of North America. From 19 August to the end of October we have a six-week period packed with almost constant accounts of shipwrecks and disasters. The Bahama Islands served as the focal point for the storm activity. The first blow of the season on 19–22 August passed to the north of the Bahamas before striking the North Carolina coast.[1] The next followed soon afterwards on 30 August. Alex. K. Johnston states: "This storm commenced at Eleuthera at 8 p.m., it entirely destroyed vegetation. On September 13 a more severe gale threw down the houses and tore up the trees by the roots. These were followed by a severe hurricane on the 27th September and another on the 5th of October of the same year." [2]

The first storm of the 1806 season packed the greatest wallop for the American seaboard. Though not listed by Poey, Johnston, nor Tannehill, this storm received great attention in the press. Both Noah Webster in his diary and Henry Laight in his meteorological log gave special mention to the damage to shipping and installations along the coast.[3] The most spectacular event during the blow occurred with the upsetting of the *Rose In Bloom* off the central New Jersey coast on Sunday morning, 24 August, with the loss of 21 lives.[4] This tragedy was played up in the press in much the same way that such events are treated in modern papers.

The whole stretch of the Atlantic Ocean from Bermuda to the Carolinas and south to Eleuthera in the Bahamas appeared to have been the scene of very extensive gales commencing on 19 August. A ship 150 miles east of Eleuthera reported a three-day gale from the 19–21st, and others were in trouble to the northeast of that point.[5] The circling winds caught both a British and a French war fleet at sea. The squadron of Jerome Bonaparte was scattered and several ships dismasted, necessitating individual ships to put into Norfolk or New York for refitting. The same blow dismasted the mighty British man-of-war *L'Impeteaux* of 74 guns which then drifted under jury masts for 23 days until beached near Cape Henry.[6]

The storm center appears to have churned around the area north of the Bahamas awaiting a steering current to carry it northward. We can watch the scene as the storm approached Charleston. The press there, as usual, took special note of things meteorological. After a long period of west to southwest winds, an easterly flow varying from northeast to southeast set in on the 17th. By Wednesday, the 20th, it was blowing fresh out of the north-northeast. By 0200 on Friday, the 22d, it had increased to a complete gale with rain. Some ships were driven ashore and trees blown down, but the press stated there had been no major damage as the harbor was well shielded from northerly gales. The storm center must have passed well to the east of Charleston.[7]

It was a different story from Georgetown northeastward along the coast. The lighthouse on North Island at the entrance to Winyaw Bay blew down. The wind damage in the Georgetown area was considered more severe than in 1804, but since the wind never got into the east, the tide surge was much less than might be expected from a consideration of wind speed alone.[8] At a nearby plantation at St. James on the Santee River, the following excerpt of a letter appeared in the press:

> I never expected to have seen another gale equal to that of September 1804; however, in this neighborhood, we have witnessed a much severer one. On Friday night last it blew, if possible, twice as hard as in the last hurricane—you may imagine how hard, when I state there is scarsely a tree left in our yard; such as are not blown up by the roots are broken off, and the roads covered with trees. My rice being planted late has not suffered much, as it was just getting into ear. Had it been three weeks later, I should not have made a barrel. My cotton is completely ruined, twisted and smashed all to pieces, and I do not expect to make ten bales out of 94 acres.[9]

The main force of the hurricane struck in the Cape Fear area of southeastern North Carolina where the towns of Wilmington and Smithville stand exposed to tropical storms as they are about to recurve to the north and northeast. The editor of the *Wilmington Gazette* thought it the "most violent and destructive storm of wind and rain ever known here." On Thursday evening, the 21st, the wind had gone around to northeast and gradually increased until Friday at 1000 "it then became a hurricane and blew with utmost violence." On Saturday, the 23rd, about daylight the wind began to veer and about 0700 it settled at southwest and continued at full blast until 1200–1300 when it seemed to abate, but did not subside entirely until 1700–1800 that evening. The tide rose to a height hitherto unknown and "when the wind shifted to southwest it seemed to threaten universal destruction." [10]

At Smithville, on the immediate coast, great damage occurred. The wind pattern was reported in the press: northeast on Thursday; between 1000–1200 on Friday, a sudden shift to southwest and a hurricane until 2200, then a shift to west. There is some discrepancy between the Wilmington and Smithville reports as to the time of the southwest wind shift. The two places are 25 miles apart. It is possible that the storm center remained stationary in the area for a number of hours with only a local oscillating movement. The long continuance of hurricane winds would suggest this. The tide at Smithville rose higher than ever known, or at least since 1762 or 1763 when the New Inlet was broken through.[11]

The center of the hurricane must have passed very close to Wilmington and Cape Fear. No lull was reported, but the quick shift of wind from northeast to southwest would suggest the proximity of the center. The course of the storm took it across extreme eastern North Carolina and out to sea again south of Norfolk where a sound wind lashing caused the press to describe the blow as "the tail of a West Indian hurricane." A press report follows:

The appearance of the weather, had for some days past, indicated a storm, which came on Saturday last attended with heavy rain. The wind thro' the day was from South to East. In the night it got more to Northward; and from midnight to about three in the morning, the wind blew with great violence, from North to Northwest, after which it abated. It was fortunate that the violence of the gale was of such short duration, for had it continued it must have done great damage.[12]

[1] *Public Ledger* (Norfolk), 1 & 8 Sept 1806; *New England Palladium* (Boston), 9 Sept 1806.

[2] Alexander K. Johnston. *The Physical Atlas of Natural Phenomena,* 56.

[3] Noah Webster. *Diary,* I, 567; Henry Laight. Ms. Met. Register.

[4] Webster. *Diary,* I, 566; *N. Y. Adv.* in *U. S. Gaz.* (Phila.), 1 Sept 1806.

[5] *Palladium,* 9 Sept 1806; *Public Ledger,* 8 Sept 1806.

[6] *Nat. Int.* (Wash.), 19 Sept 1806.

[7] Charleston, 23 Aug., *U. S. Gaz.,* 8 Sept 1806.

[8] Georgetown, 25 Aug., *Nat. Int.,* 12 Sept 1806.

[9] *Idem.*

[10] *Wilmington Gaz.,* 26 Aug., in *Courant* (Hartford), 17 Sept 1806.

[11] *Idem.*

[12] *Public Ledger,* 25 Aug 1806.

THE TROPICAL STORM OF SEPTEMBER 1810

A tropical storm passed inland to the south and west of Charleston on 11–12 September 1810. Apparently its visit did not coincide with an unusually high tide nor did the winds attain full hurricane force. Thus, damage remained relatively light. The *Charleston Courier,* as usual concerning weather events, gave good coverage:

> A storm of wind and rain has been experienced in this city since Tuesday evening last [11 Sept.]. The wind, which had been fresh at NE all day on Tuesday increased toward the evening of that day, and at dark some rain fell; there was but little however in the course of the night. On Wednesday the wind was very high through the day, with an almost incessant fall of rain; the wind increased in the evening to an almost complete gale. In the course of the night it hauled around from NE to SE and blew with increasing violence; but toward morning it lulled considerably, the rain ceased to fall, and in the course of yesterday, the wind having gotten to the southward and westward, ceased to blow with violence. In the course of Wednesday and Wednesday night some damage was done by the sea to several of the wharves, but we believe that all vessels in port rode out the gale with little or no damage, excepting some small wood sloops, two or three of which sank at the wharves. The street on South-Bay, between Meeting and King Streets, was completely washed away by the waves. Several trees were blown down in the streets.[1]

[1] *Charleston Courier,* 14 Sept 1810.

THE CHARLESTON HURRICANE AND TORNADO OF 1811

The occurrence of a tornado within a hurricane circulation is not an infrequent phenomenon. It is quite likely that many local whirls, some of tornado stature, develop in these tropical masses of turbulent winds and dense clouds without being observed. But to have one traverse a major settlement such as Charleston during the presence of an easterly gale is quite a meteorological event, indeed.

A tropical storm is listed by Tannehill in the Leeward Islands on 7–8 September 1811, but as usual he cites no reference nor gives any details.[1] Tingley in his notes makes no mention of any such storm. It is possible that the Leeward Islands disturbance could have been the same one which struck Charleston on the morning of the 10th, though the rate of travel would have been unusually rapid. Whether this storm should rate as a tropical storm or as a true hurricane is a nice question, but it deserves a high place in Charleston's disaster list if only for the presence of the tornado.

The *Charleston Times* presented a full account of the storm but we shall leave the details of the tornado itself to a monograph on the subject of tornadoes:

> On Sunday evening last, the wind which had been for some days light and variable, shifted to the northeast, and blowing very fresh through the night, it continued in the same quarter on Monday and Monday night; on Tuesday morning it blew with increased violence, and during the whole time from Sunday evening, there was an almost uninterrupted fall of rain. About ten o'clock in the forenoon of Tuesday, the wind shifted to SE and at half past twelve, A Tornado, unprecedented here in its extent and effects, crossed a section of our city. It first took effect at Ft. Mechanic, situated on the southeast point of the city, and passed from thence in a northwest direction, it crossed the town in a direct line to the pond on the north side of Cannon's bridge. . . . Path about 100 yards wide. . . . Last evening the wind shifted to southwest and although it continued to blow with some violence, we trust that the storm has spent its force.[2]

The only other check point to give notice of this meteor was Columbia, the capital of South Carolina, lying inland about 100 miles northwest of Charleston. The press notice there follows: "On Sunday and Monday last the weather appeared cloudy and unsettled, with a smart wind from the NE and early on Tuesday morning the 10th instant it increased to a perfect hurricane, and continued without the least interruption for upwards of 14 hours, accompanied by an unusual fall of rain." [3] There was no material damage, but the planters suffered great loss from the combination of wind and rain. It was said to have been the worst ever experienced at that inland point.

Such evidence as we have would indicate that this was only of tropical storm intensity over South Carolina. It appears that the center came inland north of Charleston. Unfortunately, no report from the Georgetown area has been located which might supply more light on the northern side of this interesting storm.

[1] Tannehill, 249; Tingley Ms. U. S. Weather Bureau Library.
[2] *Charleston Times,* 11 Sept 1811.
[3] Columbia dispatch, n.d., in *Raleigh Register,* 20 Sept 1811.

THE DREADFUL STORM ON 27–28 AUGUST 1813

CHARLESTON COURIER—30 AUGUST 1813

Again we have been visited with one of those disasters, which have of late years so frequently desolated our city and seaboard. On Friday night last, we experienced one of the most tremendous gales of wind, that ever was felt upon our coast. . . . For some days previous to Friday, last, the unsettled state of the weather was such as to indicate a gale; the uncommon roaring of the sea upon the bar, an unerring indication of such an event, was noticed by many. . . . About 3 o'clock P.M. the wind began to blow very fresh at N.E. by E., between 6 and 7 o'clock it had increased to a strong gale, and at 9 o'clock it was a complete hurricane, prostrating in its course houses, chimnies, fences and trees. It continued to blow with equal violence, until about one o'clock in the morning; when the wind having shifted to westward, it lulled considerably, but still blew with much force until daylight, when it became moderate. Torrents of rain accompanied the gale; and the tide, which should have been high before 10 o'clock, continued rising until nearly 12, at which time it was about 18 inches higher than in the great gale of 1804.* The rising sun, notwithstanding it disclosed to us the ruins produced by the storm, was cheering to the eye; after such an awful night of uncertainty, the return of day was hailed with joy.[1]

The Hurricane of 1813 rates a position close to the top of Charleston's meteorological list for its combination of severe winds, height of flood tide, and general destruction. Probably those of 1752 and 1893 were of greater physical stature and more lives were lost in each, but 1813 must be considered among Charleston's major disasters. This storm has never received its deserved attention. Very likely the explanation lies in the fact that Ramsay published his study of hurricanes in 1809, playing up the storms of 1752 and 1804 as classic examples.[2] Subsequent historians have in the main followed Ramsay without doing any original investigation, and without bringing his data up-to-date.

The hurricane season of 1813 was an active one. There were many warships at sea that year and this may account for the surprising number of 13 storms listed by Tannehill and Tingley. During the last week of August tropical storms were reported at Dominica on 25 August, Jamaica on the 28th, in addition to the Charleston occurrence on the 27–28th. It is doubtful that the Dominica storm, if the dating is correct, could have been the same which struck Charleston two-and-a-half days later. Since the airline distance between the two points is approximately 1400 miles, the rate of progress would have been 24 miles per hour, a rather fast clip for the tropics. The extreme violence of the wind at Charleston and its relatively short duration, six hours, might indicate a small areal storm of great intensity and speed.

* "At 10 o'clock a great part of Charleston was under water for two squares from the wharves, by the tide water; in some places breast deep."

A vivid eyewitness account of the ravages of the storm at Charleston, written on the day following, appeared in the Philadelphia press:

The very foundations of the houses shook. My bedstead had so much motion, that it was not till morn and with incessant fatigue that I got to sleep. In the morning I made a sortie from my lodgings to survey the ravages the storm had made, and the sight was distressing. The trees which afforded a shelter from the scorching rays of the sun were blown down, and lay in every direction in the streets. Slates, tiles, windowshutters, signs, &c. were scattered over the pavements in quantities.

When I reached the wharves the sight was distressing beyond expression. Every wharf was tore up and at least, one half destroyed. Some were entirely washed away, the logs and timbers washed up into the streets and against the stores; where shallops, boats, shingles, hogsheads, spars, and every individual article that was in or near the wharves was washed promiscuously one over the other, and crushed to pieces.

Every vessel appeared to be more or less damaged, some were thrown partly in the wharves, some sunk in the river, others at the wharves, some were up in the very streets; one large ship was driven against the market and stove part of it in, the smaller vessels were all either upon top of the wharves, or sunk along side.

The guard ship and prison ship drove ashore in James Island, high and dry. I have not heard yet whether any prisoners escaped. Nearly half a mile of Charleston bridge washed away, and drifted on James Island. For want of boats there was no information from Sullivan's Island till the afternoon. I learned that 12 dwelling houses have been blown down and washed away, & 15 to 20 persons drowned.—The whole Island was completely inundated four to five feet *—the furniture was washed out of many houses. The soldiers from the fort were sent out to afford assistance to some families who were in danger of being washed away. A great many ladies were rescued and taken to the fort, which was the only place of security which was not under water.

Fort Johnston is very much injured; considerable part of it is destroyed. The men's barracks were washed to sea. Many stands of arms, clothing, provisions, &c. also lost. One of the officers told me last evening that they were up to their breasts in water several hours during the night, saving men, women, and children, who were in the barracks, and it was only at the point of the bayonet that they were forced to seek other shelter, although the barracks were then washing away by the sea. To repair the damages and losses sustained at the Fort, will require $150,000.

The damage sustained by this city and shipping is in rough calculation estimated at near two millions.[3]

The damage about the city was summarized in the local press:

- More than half of the New Bridge over Ashley River, was swept away by the violence of the storm—water rose and floated the top from the piers.
- Cannon's bridge carried away and the causeway much injured—also Peyton's bridge.
- Property stored warehouses on the wharves damaged especially in Vendue Range on Prioleau's wharf.
- Nearly every slate- or tile-covered house received damage.
- Second Presbyterian Church suffered much—lead and slates thrown off.
- Several small houses on South Bay undermined.
- The top of Mr. Ross's windmill at the head of Tradd Street blown off.
- About 20 feet of the stone wall under the wooden battery lately erected on East Bay, has fallen down, occasioned by the weight of the guns.
- A blacksmith's shop leveled.

In the harbor, John H. Dent, commander of the U. S. Navy flotilla, reported that only three vessels of all those anchored there rode out the gale—the *Nonesuch, Carolina,* and a hospital ship which was dismasted.[4]

There is little doubt that the eye of the cyclone moved inland north of Charleston as the wind there and at Edisto Island, just to the south, backed constantly during the blow from northeast to north, and finally to west. At Edisto, exposed to the full fury of the open sea, "it began to blow with the greatest violence from the northeast, accompanied with torrents of rain, and continued, with unabated fury for six hours. . . . It happened very fortunately that the wind soon changed to north, which drove the waves back from the land, otherwise it is supposed the whole Bay would have been overflowed. Those who have long resided here, say that it was the most severe gale which they ever experienced." [5]

The violence of the wind at Georgetown, S. C., was emphasized in a dispatch, but the details of its compass behavior were omitted. "The houses on the adjacent islands have received no injury of any consequence. This place has fared much better than could be expected from the violence of the gale—some of the wharves are a little injured; two schooners, a sloop, and the hull of a brig, were driven onshore, and expect there will be no difficulty in getting them off. About one-third of the rice crop lost. Roads in dreadful condition." [6]

Lacking accurate wind data from any point north of Charleston, we are limited to stating that the landfall of the eye was somewhere on the South Carolina coast, probably close to Charleston. At Camden, S. C., 100 miles northwest of Georgetown, James Kershaw in his weather diary noted a "hurricane" on the 27th with "hard rain and high winds in the night" of the 27–28th. His single daily wind notation was southeast on the 27th and north on the 28th.[7] Based on this scanty evidence, one might surmise that the center moved northwestward from Charleston over eastern South Carolina and central North Carolina.

Our only wind observation to the north comes from Point Lookout, Maryland, at the junction of the Potomac River and Chesapeake Bay. An all-day easterly gale there on the 28th greeted the fleets maneuvering in the Bay.[8] Heavy rain fell inland throughout the Carolinas on the 27–28th causing a fresh on the Cooper and Santee Rivers.

* "The tide was 2½ feet high in the officers' quarters, and about 4 feet on the parade." Charleston paper in *U. S. Gaz.,* 10 Sept 1813.

[1] *Courier* (Charleston), 30 Aug 1813.
[2] Ramsay, II, 175–184.
[3] Letter: "Dear John, Charleston August 29, 1813" in *U. S. Gaz.,* 19 Sept 1813.
[4] *Courier,* 30 Aug 1813.
[5] *City Gaz.,* 31 Aug 1813.
[6] *U. S. Gaz.,* 14 Sept 1813.
[7] James Kershaw. Ms. Weather Diary. National Archives.
[8] *U. S. Gaz.,* 8 Sept 1813.

THE GULF COAST: 1722–1814

THE GREAT HURRICANE OF 1722 IN LOUISIANA

The first great hurricane in Louisiana history of which we have an adequate first-hand account occurred on 12–13 (23–24) September 1722. The middle Gulf coast from Mobile to the mouth of the Mississippi, laying in the dangerous eastern sector of the storm, suffered severe damage. The waters of the Gulf were reported to have risen 7 or 8 feet above normal at Bay St. Louis on the present Mississippi coast, and the wind at the newly-established capital of New Orleans blew down private houses and public buildings alike. The plantation crops along the Mississippi River north and south of the Louisiana trading post were laid low by the combined onslaught of wind and rain.[1] No doubt it was a storm of major proportions and should be listed among the *great* hurricanes of the area.

The year of occurrence of the storm has been variously reported as 1721, 1722, and 1723, but the days of the month are usually given as the 12th and 13th, old style. Two contemporary documents now at hand definitely fix the year as 1722. De L'Orme in a letter of 30 October 1722 to French officials at home mentioned the storm, and the very informative *Journal of Diron D'Artaguiette* described its effects in the New Orleans and surrounding country in detail.[2] We quote in full all references to the event in this diary, with dates corrected to new style:

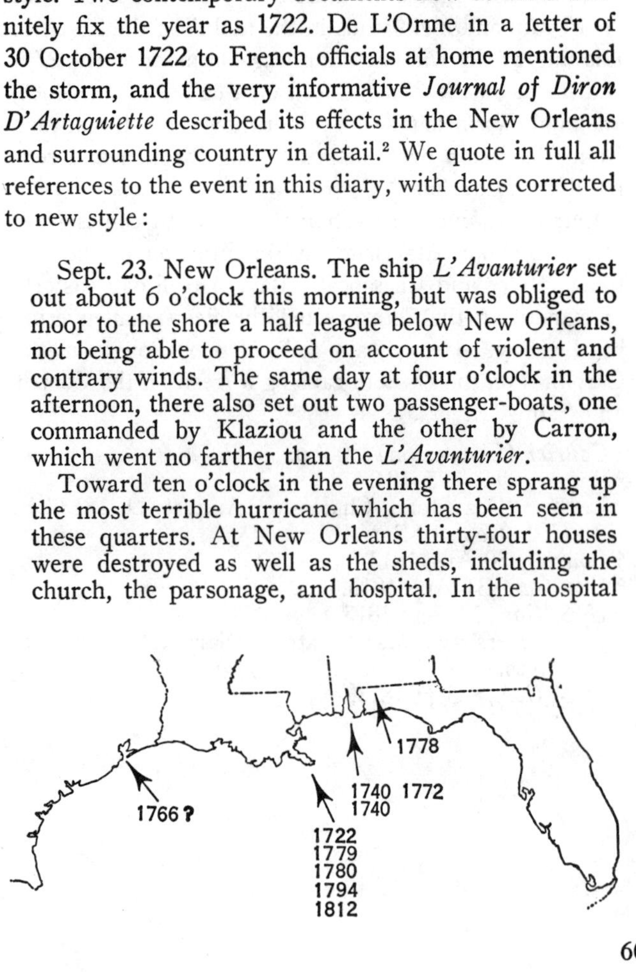

> Sept. 23. New Orleans. The ship *L'Avanturier* set out about 6 o'clock this morning, but was obliged to moor to the shore a half league below New Orleans, not being able to proceed on account of violent and contrary winds. The same day at four o'clock in the afternoon, there also set out two passenger-boats, one commanded by Klaziou and the other by Carron, which went no farther than the *L'Avanturier*.
>
> Toward ten o'clock in the evening there sprang up the most terrible hurricane which has been seen in these quarters. At New Orleans thirty-four houses were destroyed as well as the sheds, including the church, the parsonage, and hospital. In the hospital were some people sick with wounds. All the other houses were damaged about the roofs or the walls.
>
> It is remarkable that seven years ago there was a similar calamity which caused a terrible destruction at Massacre or Dauphine Island, where I was at the time. [This paragraph was in the margin of the original manuscript.]
>
> There were ten flat-boats broken up and sunk together with launches, canoes, and pirogues, and in fact everything in port was lost. The wind came chiefly from the southeast. The ships, the *Santo Christe* and the *Neptune,* and two passenger-boats, one of which was used as a powder magazine, were damaged and grounded far ashore.
>
> It was remarkable that if the Mississippi had been high this hurricane would have put both banks of the river more than 15 feet under water, the Mississippi, although low, having risen 8 feet.
>
> Sept. 24. The hurricane, continuing until mid-day, has not ceased to rage, but at noon, it having become much calmer, we learned from some people who had just come from the settlements of Srs. Trudeau and Coustillas, that the houses there were blown down and their crops lost. The same day, in the evening, we learned that three pirogues had been lost up toward the Tensas, five leagues above New Orleans. These pirogues were loaded with fowl, Indian corn, and other goods which a man called Poussin, living at the Thonniquas, was coming down here to sell.
>
> Sept. 27. We have been informed that a boat was lost the night of the hurricane toward Bay St. Louis. The people aboard escaped to shore.
>
> Sept. 30. The Bayou which is situated a league from here and by which one goes and comes from Biloxi to this place, has overflowed by two or three feet, by reason of the hurricane. The waters having subsided, they found upon the surface of the water many dead fish, which caused a great stench.
>
> Oct. 5. We have learned by means of a canoe which arrived from the lower part of the river that the two ships, the *Loire* and the *Deux Freres,* which had set out from the roads of the Isle aux Vaisseaux a few days before the hurricane to come to New Orleans, had returned there without having suffered any damage. They also say that all the houses and storehouses which are in Fort Louis were either blown down or damaged and that the sea rose 7 or 8 feet more than is ordinary, and that it had partially inundated the place.
>
> Oct. 6. I set out from New Orleans to go to Cannes Bruslees. The same day I spent the night at the house of Sr. Massy where I noticed that the greatest injury which the hurricane had done there was to their houses which it had completely destroyed. The crops are not in such a hopeless condition; the rice, which was only in flower, having straightened up again.

> The loss to the colony, however, will be very considerable, because there was a great deal of rice ready to cut but which is entirely lost.
>
> Oct 7. I arrived at Cannes Bruslees where I found all the houses and barns blown down and not fit to be used. We set to work to repair the loss and to build sheds for the workmen and the grain.
>
> Oct. 8. The continual rains which fall here prevent us from getting in our crops and cause a considerable loss.
>
> Oct. 9. The report that the ships had returned to the road of Fort Louis without having been damaged and that four passenger-boats belonging to the company had been stranded on the beach, has been confirmed.[3]

Additional wind data came from two old German settlements located about 30 miles north of New Orleans. J. H. Deiler in a history of these villages stated that the wind set in from the southeast and shifted around to southwest, indicating that the center of the storm moved inland west of the mouth of the Mississippi River, passed through central Louisiana, and probably curved northeastward when well inland.[4] This is corroborated by D'Artaguiette's statement that the main wind at New Orleans was southeast. Deiler added that the storm continued for five days, probably a reference to the rainfall and subsequent flood which inundated the two German villages. Eight thousand quarts of rice ready for harvest were lost, certainly a grevious blow for the struggling colony.

The destruction at New Orleans was not as serious as first represented since all buildings at the time when the hurricane struck were of a temporary nature, having been put up in 1717–18 in a hurry when the site was selected for the future capital of the Louisiana Company. Much political bickering had been involved in the choice of New Orleans as a site upon which to build the capital. The older settlements resented this and continually pointed to the disastrous New Orleans flood of 1719 as another example of nature's frowning on the location as a place for the major French settlement of the colony.

The origin of the great hurricane of 1722 is not readily ascertainable. Port Royal in Jamaica was overwhelmed by a storm tide on 28 August (8 September) with a great loss of life and damage to shipping, and Poey lists another storm "in the Antilles" on 31 August (September 11), perhaps referring to the same one.[5] It seems unlikely that either of these was a direct antecedent of the Louisiana storm of 10 days later, but does indicate that the season of 1722 as an active tropical storm period in the Caribbean and Gulf of Mexico.

D'Artaguiette stated later in his diary that the hurricane was not felt on the Wachita River to the northwest in the interior, but that along the coast to the east of New Orleans a ship had been sunk at Bay St. Louis by the blow.

Another possible reference to the widespread effect of this storm came from South Carolina. Bartram in his list of hurricanes related the following account of a tremendous tropical downpour at Charleston:

> In the year 1722 in September was the most violent rain for 3 days and 3 night without intermission that the English inhabitants ever knew. It so flowed in the country as to cause great destruction of grain and other necessaries of life. It seemed as if that collection of vapours that used periodically to operate in currents of wind was discarded in rain.[6]

Though no date is given for the flooding rains, they might very well have been the aftermath of the great hurricane after it had slashed ashore on the middle Gulf coast and then spread a canopy of very heavy downpours over all the Southeastern area. This frequently occurs when tropical storms from the Gulf of Mexico lose their forward momentum after crossing the coastline. The Carolina rains provide additional evidence that the season of 1722 was a memorable one from the meteorological point of view in Gulf and South Atlantic colonies.

> About the middle of September here fell the greatest floud attended with a Hurricane that has been known since the country was settled. Great numbers of Cattle, Horses, Hogs, and some people were drowned. The Deer were found frequently lodged on high trees. The wind was so violent that it tore up by the roots great numbers of Trees, and disrobed most others of their leaves, Cones, and seed: so that had I been well, the collection would have fallen short of other Years. Particularly it dispersed all the Laurels, Umbrella, and many other things I sent out for, but none to be found.[7]

[1] Le Baron Marc de Villiers. *Histoire de la fondation de la Novelle-Orleans, 1717–1722.* (M. Gabriel Hanotaux, ed., 1917), 105–107.

[2] Charles Gayarre. *History of Louisiana. The French Domination.* New York, 1855, I, 282.

[3] Journal of Diron D'Artaguiette in *Travels in the American Colonies.* (Newton D. Mereness, ed., 1916), 24–27.

[4] J. Hanno Deiler. *The Settlement of the German Coast of Louisiana.* Phila., 1909, 51.

[5] Poey, 9–10.

[6] Bartram, 20.

[7] Catesby to Sherard, Charles City, Dec 10, 1722, Sherard Letters, II, No. 166, quoted by G. R. Frick and R. P. Stearns. *Mark Catesby. Colonial Audubon.* Urbana, 1961, 24.

THE TWIN MOBILE HURRICANES OF 1740

Charles Gayarre in his massive *History of Louisiana* (1846) featured the two hurricanes which struck the Mobile area within a week of each other in mid-September 1740, a year "which produced very extensive disasters" for the colony. He draws his interesting material from a letter of Commandant Beauchamp, dated Mobile, 25 February 1741:

> This hurricane was so violent, that, here, it blew down several houses, and among others, the edifice which M. Bizoton had constructed, not only as a store, but as a house of refuge for sailors. Unfortunately, it contained all the flour and other provisions destined for the subsistence of the garrison. I was obliged to send the garrison a fishing along the coast for the barrels which had been blown into the water, and part of which was staved off. Without this barrel fishing, we should have run the risk of dying of hunger, as our resources were limited to six or eight barrels of flour, which were in the fort.
>
> The wind was so furious that, if it had continued forty-eight hours, as all hurricanes generally do, we should have been inundated. Fortunately, it blew only during twelve hours, but with such force, that half of Dauphine Island was carried away, and more than three hundred head of cattle were drowned on the island. We have lost a great number of them on this coast, and at Pascagoulas. This loss is severely felt by the poor population of this section of the country.
>
> The effect produced by the force of the wind is almost incredible. There was lying before the guard house of Dauphine Island, a cannon of four pound caliber. The wind transported it eighteen feet from where it was. This fact is sworn to by all the inhabitants of the island.
>
> This hurricane, which lasted twelve hours, began in the night of the 11th of September (22nd), and ceased on that day at noon. But although its duration was not long, it caused much damage. . . . To cap the climax of our misfortunes, there came another hurricane on the 18th of September (29th), which destroyed the rest of our resources. This wind, which blew from N.N.E. and which was accompanied by heavy rains, caused an overflowing of all the rivers, by which were laid waste all the plantations of the Indians from Carolina to this place. The first hurricane was from E.S.E.—luckily these hurricanes did not pass over New Orleans and the adjacent country, where the crops have turned out to be pretty abundant. Otherwise, the whole colony would have been in a frightful state from the scarsity of provisions, and it would have been utterly impossible to make presents to the Choctaws, in whose debt, on this score, we have been for two years. . . .[1]

Poey makes no mention of any hurricanes in September 1740. That the damage from the storms was widespread is indicated by the mention of "Carolina to this place," and also from another contemporary document written by Louisbois on 7 March 1741 in which he stated that the twin hurricanes had so damaged the battery at Belize, covering the mouth of the Mississippi River, that the fortifications could easily be carried by an attacking force of only four gunboats. Louisbois added that the settlements were experiencing the lowest degree of misery: "There are many families reduced to such a state of destitution, that fathers, when they rise in the morning, do not know where they will get the food required by their children." [2]

[1] Gayarre. *Hist. La.,* I, 514.
[2] *Ibid.,* 516.

THE GULF HURRICANES OF 1766

The *South Carolina Gazette* in December 1766 carried a dispatch relating the occurrence in the Gulf of Mexico of two severe storms, both presumably of hurricane force, which struck at two spots especially susceptible to storm tides. One swept into the Galveston Bay area in the northwestern Gulf on 4 September, the first of a long line of storm disasters to hit there.[1] Other storms would strike in 1818, 1837, 1867, 1900, and 1915 and overflow the low-lying Galveston Island. The storm on 4 September 1766 received prominence since a richly-laden Spanish treasure fleet of five galleons enroute from Vera Cruz to Havana was driven ashore during the gale and had to wait many weeks for assistance to come.

A second storm lashed at the northeastern Gulf region during October coming ashore somewhere near Pensacola on the 22nd. The dispatch to Charleston follows:

> By Captain Henderson, from Pensacola, we learn that a most terrible hurricane was felt there on the 22nd October, which has done much damage as well onshore as in the harbor. That the gale began at E.S.E., was extremely violent from about 10 o'clock at night until 6 in the morning, when the wind shifted to the W.S.W. it abated. That only four vessels rode it out, not without receiving some damage; and all the rest were driven ashore. . . . And that about five very rich Spanish galleons, from Vera Cruz, bound for Havannah, were drove ashore in the Bay of St. Bernard, W. and S. from Pensacola, sometime before the hurricane of the 22 October, to whose relief a vessel had been sent from New Orleans, but was supposed to be lost.[2]

The historian of colonial Pensacola adds the further information that two brigs, four schooners and sloops were "driven from their anchors and wrecked in the storm of 22nd last month, although the height of the gale was off the nighest shore." [3] This would presumably refer to a wind from a northerly quadrant. Dunn and Miller supply the fact that the tide in 1766 at St. Marks rose 12 feet above normal; they consider this storm as one of major proportions.[4]

The season of 1766 produced at least seven severe storms in the Antilles and along the American shore. Major hurricanes struck the Leeward and Windward Islands in August and October, and one in September swept the Atlantic Coast from Florida to Virginia.

[1] R. D. Frazier. Early Records of Tropical Hurricanes on the Texas Coast in the Vicinity of Galveston. *Monthly Weather Review,* 49–8, Aug 1921, 455.
[2] *S. C. Gaz.,* 12 Dec 1766.
[3] C. N. Howard. Colonial Pensacola. *Fla. Hist. Quart.,* 19–1, July 1940, 113–114.
[4] G. Dunn and B. Miller. *Atlantic Hurricanes.* Baton Rouge, 1960, 297.

BERNARD ROMANS' GULF COAST HURRICANE OF 1772

The fatal hurricane of August 30, 31, September 1, 2, 3, anno 1772, was severely felt in West Florida. It destroyed the woods for about 30 miles from the sea coast in a terrible manner. What were its effects in the unsettled countries to the eastward, we cannot learn.

In Pensacola it did little or no mischief except the breaking down of all the wharves but one; but farther westward, it was terrible.

At Mobile everything was in confusion—vessels, boats, and loggs were drove up into the streets a great distance. The gullies and hollows as well as all the lower grounds of this town were so filled with loggs, that many of the inhabitants got the greatest part of their yearly provision of firewood there. All the vegetables were burned up by the salt water, which was by the violence of the wind, carried over the town, so as at the distance of half a mile, it was seen to fall like rain.

All the lower floors of the houses were covered with water, but no houses were hurt except one, which stood at the water side, in which lived a joiner. A schooner drove upon it, and they alternately destroyed each other.

But the greatest fury of it was spent in the neighborhood of the Pasca Oocolo river. The plantation of Mr. Krebs there was almost totally destroyed, of a fine crop of rice, and a large one of corn was scarsely left any remains. The houses were left uncovered, his smith's shop was almost washed away, all his works and out houses blown down. And for thirty miles up a branch of this river which (on account of the abundance of that species of cypress, vulgarly called white cedar) is called cedar river, there was scarsely a tree left standing. The pines were blown down and broke; and those which had not entirely yielded to this violence were so twisted, that they might be compared to ropes.

At Botereaux's cow pen, the people were about six weeks consulting on a method of finding and bringing home their cattle. Twelve miles up the river live some Germans who, seeing the water rise with so incredible a rapidity, were almost embarked, fearing an universal flood, but the waters not rising over their land, they did not proceed on their intended journey, to the Chactaw nation.

In Yoani, in this nation, I am told the effects were perceivable. In all this tract of coast and country, the wind had ranged between south south east and east, but farther west its fury was between the north north east and east. A schooner belonging to the government having a detachment of the sixteenth regiment on board, was drove by accident to the westward as far as Cat Island, where she lay at an anchor under the west point. The water rose so high, that when she parted her cables, she floated over the island, the wind north by east, or thereabout she was forced upon the Free masons island, and lay about six weeks before she was got off; and if they had not accidentally been discovered by a hunting boat, the people might have remained there and died for want, particularly as water failed them already when discovered.

The effect of this different direction of the current of the air or wind was here surprising, the south easterly wind having drove the waters in immense quantities up all the rivers, bays, and sounds to the westward, being here counteracted by the northerly wind, this body of water was violently forced into the bay of Spirito Santo at the back of the Chandeleurs, grand Gozier, and Breton isles; and not finding sufficient vent up the rogolets, nor down the outlets of the bay, it forced a number of very deep channels through these islands, cutting them into a great number of small islands. The high island of the Chandeleur had all the surface of its ground washed off; and I really think, had not the clay been held fast by the roots of the black mangrove, and in some places the myrtle (Mryica) there would have been scarse a vestige of the island left.

At the mouth of the Mississippi all the shipping was drove into the marshes. A Spanish brig foundered and parted, and a large crew was lost; some of the people were taken from a piece of her at sea, by a sloop from Pensacola a few days after. In the lakes at Chef Menteur, and in the passes of the rigolets, the water rose prodigiously and covered the low islands there two feet. At St. John's Creek, and New Orleans, the tide was thought extraordinarily high, but at all these last places there was no wind felt, being a fine serene day with a small air from the eastward.

The most extraordinary effect of this hurricane was the production of a second crop of leaves and fruit on all the mulberry trees in this country, a circumstance into which I very carefully enquired, but could not learn from the oldest and most curious observers that this had ever happened before. This tardy tree budded, foliated, blossomed, and bore ripe fruit with the amazing rapidity of only four weeks time immediately after the gust, and no other trees were thus affected.[1]

The above account appeared in Bernard Romans' intriguing study, *A Concise Natural History of East and West Florida,* which was published in New York City in 1776. This is the only first-hand account of the 1772 hurricane which has been uncovered. Fortunately, it relates the essential meteorological details so that we know that the storm made a landfall a short distance to the west of Mobile since the waters of the Bay were carried into the city, presumably by a southeast wind. At the same time at Cat Island, 55 miles to the west of Mobile Bay, the wind came from north by east. The force of the combined wind and waves is well attested by the huge damage done.

At the same time that the Middle Gulf Coast was being raked by this severe hurricane, another blow of equal or greater strength struck the Island of St. Croix in the Virgin Islands close to Puerto Rico, and was felt in Hispaniola on the first of September.[2] Though these occurrences have often been connected with the Middle Gulf hurricane of the same days, it is obviously impossible that they were one and the same. Poey mentions a tropical storm at Jamaica on 28 August, and this seems a much more logical antecedent. Romans' dating of the storm as lasting five days is puzzling for a tropical type unless he referred to the continuance of the floods which followed the main part of the gales. The track of the storm over the Southeast is unknown except for the occurrence of a hurricane in eastern North Carolina on 1 September 1772. The *South Carolina Gazette* reported: "a violent gale of wind or hurricane was felt on the coast of North Carolina the 1st ulto. which raised the tide to a height beyond the memory of man and did a great deal of damage among the shipping." [3] A report from Edenton, North Carolina, spoke of 14 of 15 ships at Ocracoke Bar being driven ashore and near 50 persons perishing. The hardest part of the gale was from the north, the report indicated. This would place the storm center offshore in that latitude. It is possible that the Mobile-Pensacola storm described by Romans could have swung northeastward across Alabama, Georgia, and South Carolina to strike at the Outer Banks of North Carolina. But the Charleston press made no mention of any unusual weather activity there at this time. In the absence of positive evidence connecting the two, the North Carolina storm must be considered as a separate cyclone, one of several which occurred in this active season.

[1] Bernard Romans. *A Concise Natural History of East and West Florida.* New York, 1776, 4–8.

[2] Hugh Knox. *Discourse . . . on the Occasion of the Hurricane which happened on the 31st day of August.* St. Croix, 1772, 16–18.

[3] *S. C. Gaz.,* 1 Oct 1772.

THE OCTOBER 1778 STORM

During the mid-years of the American Revolution, hurricanes struck the Gulf Coast from Louisiana to Pensacola in three successive years. New Orleans at this time was held by the Spanish as a result of the Treaty of Paris in 1763, while the British remained in possession of both East and West Florida, a territory stretching all the way from the Atlantic Ocean to the east bank of the Mississippi River. An uneasy neutrality existed between the two nations in 1778.

The first of the triple hurricanes came in late 1778 before Spain had declared war on Great Britain. The Spanish Governor, Bernardo de Galvez, mentioned the storm of 7–10 October 1778 in his dispatches: "The sea rose as it had never been seen to do before, and the establishment at the Belize, Bayou St. John, and Tigouyou were destroyed." [1] These were Louisiana Delta locations.

The hurricane also struck farther east at Pensacola were the local British governor wrote to the Colonial Office:

"N.B. The reason of no business being done in the House of Assembly from the 9th to the 15th of October was owing to a very meloncholy cause—the severest hurricane ever felt or known in this part of the world since West Florida has belonged to the Crown of Great Britain happened on the night of the 9th with such irresistable fury and violence as entirely to sweep away all the wharfs, stores and houses contiguous to the Water Side, with part of the front batteries of the Garrison, besides destroying several houses and making general havock of the ferries in the Town of Pensacola. All of the ships and vessels in the harbor were either lost or driven ashore, except His Majesty's Sloop of War, the Sylph, which with difficulty rode out the gale. . . ." [2]

[1] Spanish Ms. in La. State Hist. Doc. quoted by Alcee Fortier. *A History of Louisiana.* New York, 1904, II, 61.

[2] Public Record Office, C. O., 5: 628; 246, quoted by C. N. Howard. Colonial Pensacola. *Fla. Hist. Quart.,* 19–4, April 1941, 393.

DUNBAR'S HURRICANE AT NEW ORLEANS IN 1779

By 1779 Spain had declared war on Great Britain and the reduction of the British holdings along the Gulf Coast and on the east bank of the Mississippi River became a prime Spanish military objective. But for two successive years hurricanes were to provide unforeseen obstacles. Governor Bernardo de Galvez was preparing in secret an expedition to seize the British post at Baton Rouge. On 18 August, however, a severe storm lashed the New Orleans countryside to disrupt the plans. The Governor described the scene on the day following the storm:

> Although the wind and the rain began on the night of the 17th, it was not until three o'clock in the morning that it attained its full violence, keeping its strength continually until ten o'clock in the morning, then it began to lose its force a little, but not until all the houses, barges, boats and pirogues were demolished, some with many people from these settlements, among the frigate *America La Reseda,* and one belonging to Don Bernardo Ogaban. Others have gone aground, half destroyed and useless, stranded in the woods and finally, there are others of whose fate we are still ignorant. To aggravate the situation, the schooners and gunboats have also sunk.
>
> The village presents the most pitiful sight. There are but few houses which have not been destroyed, and there are so many wrecked to pieces; the fields have been leveled; the houses of the near villages, which are the only ones from which I have heard to this time, are all on the ground, in one word, crops, stock, provisions, are all lost.[1]

A contemporary dispatch to the British press describing the hurricane appeared in *The London Magazine:*

> We have had here on the 18th of August the most dreadful hurricane that ever was remembered; all the vessels that were in the river were either sunk or blown on shore; among the number of those that were sunk was the Morris, an American frigate, commanded by Capt. Pickles, and some of her crew drowned, as were several other persons who were on board of vessels in the river; great numbers of houses in the town, though very low, were entirely blown down; and all the others suffered very considerably; all the plantations from the bottom of the river, to six or seven leagues above the town, were entirely laid waste; trees in the forest were torn up for several miles together; the disaster that the hurricane occasioned is so great, that it will require two or three years labour to put the country in the state it was before.[2]

All of the vessels which Galvez was preparing for the expedition were sunk except the *El Volante*. It appears, though, that the damage to the small craft was quickly repaired as four of the boats were raised and the attacking force moved northward on the 27th only a week behind schedule.[3]

William Dunbar, a resident of the lower Mississippi country from 1773 to 1810 who had previously studied mathematics and astronomy in London, chanced to be in New Orleans at this time and has left an important document in the history of storm study.[4] Though his remarks were not published until a quarter-century later, his observations in 1779 perceived the true nature of the tropical storm: that it had a progressive forward movement and that the winds revolved around a vortex at the center. It is interesting to note that Colonel James Capper of the East India Company described Bay of Bengal typhoons as whirlwinds with vortices as a result of his experiences with two storms on the Coromandel Coast in 1760 and 1773.[5] Again, one of those historical coincidences appeared when Capper published his *On the Winds and Monsoons* in 1801, the same year that Dunbar's paper on the storm of 1779 was read before the American Philosophical Society.

Dunbar's reputation as an observer of natural phenomena lead Thomas Jefferson to assign him the task of exploring the sources of the Arkansas River and to make meteorological and astronomical observations there as part of the extensive surveys of the Louisiana Territory that the President instigated soon after its purchase in 1803.

Dunbar's "Remarks of the Climate of Mississippi" was published in 1804 by the American Philosophical Society. He noted that "August and September are called the hurricane months, and I believe that there never happens a hurricane of great extent and duration at any other season, and this seldom reaches much higher than New Orleans, sweeping along the seacoast."[6] His description of the 1779 hurricane follows:

> Since I have resided in this country, two or three hurricanes only, of great magnitude, have ravaged New Orleans and its vicinity. Two of them burst forth in the months of August in the years 1779 and 1780; I was at New Orleans during the first of those two. More than half of the town was stript of its covering, many houses thrown down in town and country, no ship or vessel of any kind was seen on the river next morning. The river which at this season is low was forced over its banks, and the crops which were not yet collected, disappeared from the face of the earth. The forests for some leagues above and below New Orleans assumed the dreary appearance of winter, the woods over large tracks were laid flat with the ground, and it became impossible to traverse the forests, but with immense labor on account of the multitude of logs, limbs and branches with which the earth was every were (sic) strewed within the extent of the hurricane; which might be estimated at about 12 miles due North and South, New Orleans being in the center of its passage. The partial hurricanes [tornadoes?] which frequently traverse this territory do not merit the name and ought rather to be called whirlwinds, which seldom last above 5 to

10 minutes, occupying a narrow vein from 50 to 500 feet in width; whereas that which I witnessed at New Orleans was of some hours duration; it continued blowing from the East or S.E. for two or three hours with undescribable impetuosity, after which succeeded all at once a most profound and awful calm, so inconceivably terrific that the stoutest heart stood appaled and could not look upon it without feeling a secret horror, as if nature were preparing to resolve herself again into chaos. The body became totally inelastic and a disposition was felt to abandon one's self prostrate upon the ground as if despair alone at that moment, could find abode in the human mind, entirely divested of all energy. How is this extraordinary effect to be counted for? It is generally believed by philosophers, that hurricanes and perhaps the gentlest zephyrs are connected with electrical phaenomina, may we then be permitted to suppose that by the violent operation of natural agents (of which we can form no conception) the electric fluid has been in a manner abstracted from the central parts of the hurricane (which we may consider as a vortex) and a species of vacuum formed with respect to the electric fluid—hence that otherwise unaccountable relaxation and dejection of spirits, similar to (though infinitely exceeding) what has been observed of the influence of the sirocco wind in Naples and Sicily upon the human body and mind, no perceptible sign of electricity being discoverable in the amosphere during the time of its blowing. Happily the wind was arrested but for a short time, by this horrid state of suspense, for in 5 or 6 minutes, perhaps less, the hurricane began to blow from the opposite point of the compass and very speedily regained a degree of fury and impetuosity equal if not superior to what it had before possessed. Floating bodies, which had been driven up the stream with vast rapidity against the natural course of the river, now descended with a velocity of which the astonished eye could form no estimate, it rather resembled the passage of winged inhabitant of the air than that of a body born upon the more sluggish element of water. Vessels were left upon dry land or dashed in pieces against the shores. An American armed ship being overset was precipitated into the ocean and never more heard of; the officers and men were chiefly saved by leaping ashore, sometimes by the assistance of rafts and logs of timber, watching the opportunity of the vessel impinging against the bank as she darted from side to side of the river, hurled along by the ungovernable fury of the torrent.

From every information I can procure, I believe the center of the hurricane passed over the city of New Orleans, the general progress of its course being at that time from about N.E. to S.W. and as its first fury was felt in the direction of S.E. nearly, and ended about N. W. It is evident that the circular course of the vortex followed that of the sun's apparent diurnal motion.—It is probable that as similar observations are made upon all hurricanes, tornadoes and whirlwinds they will be found universally to consist of a vortex with a central spot in a state of profound calm, which spot will probably be greater or less according to the magnitude of the vortex.[7]

[1] Bernardo de Galvez to Senor Don Diego Joseph Navarro. New Orleans, Aug 19, 1779. Ms. Cabildo Records. Book 2, Vol. 9, 59; also John W. Caughey. *Bernardo de Galvez in Louisiana, 1776–1783*. Berkeley, 1934. 152.

[2] *The London Magazine for the Year 1780*. 49–2, 93–94.

[3] Alcee Fortier. *A History of Louisiana*. New York, 1904, II, 63.

[4] *Life, Letters and Papers of William Dunbar*. (Mrs. Dunbar Rowland, ed.). Jackson, 1930, 69.

[5] James Capper. *Observations on the winds and monsoons; illustrated with a chart, and accompanied with notes, geographical and meteorological*. London, 1801.

[6] William Dunbar. Meteorological Observations, Made by William Dunbar, Esq. at the Forest, four miles east of the River Mississippi . . . for the year 1800; with remarks on the state of winds, weather, vegetation, &c. calculated to give some idea of the climate of that country. Natchez, Aug 22, 1801. Read Dec 18, 1801. *Trans. Amer. Phil. Soc.*, VI–1, 1804, 43–55.

[7] *Ibid.*, 53–55.

THE GREAT HURRICANE SEASON OF 1780

Of all the hurricane years in our period that of 1780 proved outstanding and deserves special consideration. Not only did eight damaging storms batter West Indian and American waters, but in the first three weeks of October three successive storms swept the area and each deserved the rating of *great* in its own right, not only for its tremendous physical energy, but also for the economic and military consequences attending its passage.

The season opened early with a severe storm striking St. Lucia in the Windward group on 13 June and later battering Puerto Rico. In August, almost simultaneously, separate storms struck Louisiana on the 24th and St. Kitts in the Leewards the next day. No disturbances have been reported in September, but this served only as a deceptive interlude for the great outbreak of activity in October when five hurricanes occurred.[1]

By 1780 all the major seafaring nations of Europe had been drawn into the world-wide conflict which originated with the stand of the embattled farmers at Lexington and Concord. At this time the harbors and sea lanes of the West Indies were thick with warships and transports, as well as with merchantmen of many nations. Choosing this critical time, three fierce hurricanes struck blows in quick succession at the assembled fleets and their stores.

Since the fortunes of the rebellious American colonies

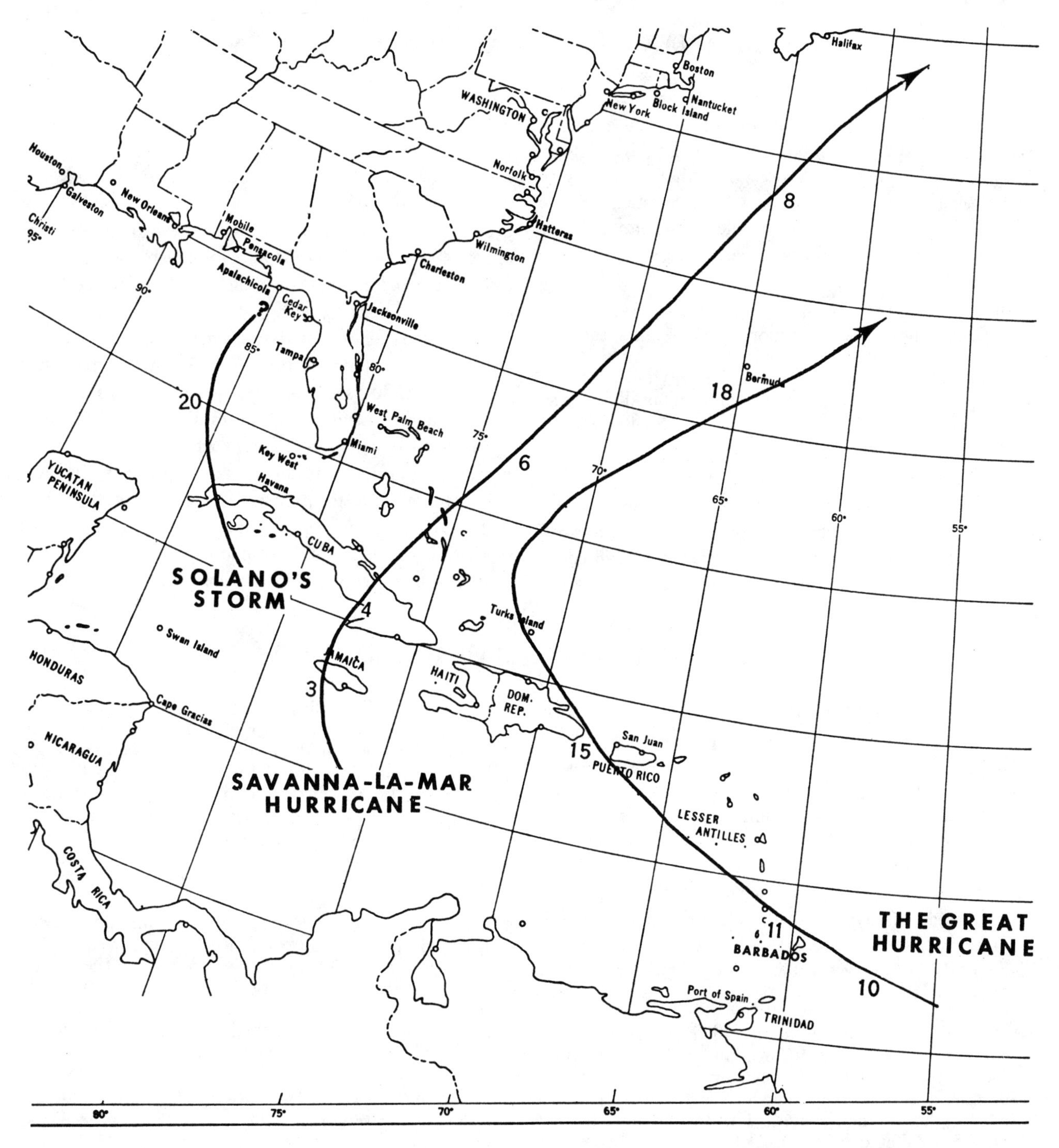

were so intimately tied with the success or failure of naval operations in West Indian and Atlantic waters, we shall shift our attention momentarily from the American mainland to chronicle the meteorological events in West Indian waters. Fortunately for this purpose, Colonel William Reid of the Royal Engineers took up the study of the hurricane problem in the 1830s. He had been sent to Barbados to direct the rehabilitation of the government buildings after the severe hurricane disaster of August 1831, and there his scientific curiosity lead him to attempt to determine the physical laws governing hurricane structure and movement. In this work he followed the acknowledged lead of William C. Redfield of New York City with whom he corresponded frequently during a twenty-year period. Reid's major contribution here lay in his transcription of the 1780 logs of British, French, Spanish, and Dutch ships involved in the three major hurricanes. He assembled this material in orderly, scholarly fashion to prove without doubt that hurricanes were large atmospheric whirls with a definite systematic structure of circulatory winds surrounding a vortex, and that they showed a somewhat regular progressive motion.[2] We shall return to Col. Reid's documents shortly.

THE LOUISIANA HURRICANE OF AUGUST 1780

Just a year and a week after the great hurricane of August 1779, another tropical storm struck the troubled settlements around the mouth of the Mississippi River. We have no contemporary document that supplies any meteorological data about the characteristics of the storm. Charles Gayarre, the detailed historian of early Louisiana, relates the following account:

> On the 24th of August 1780, as if intended to be the last pound of weight wanting to break the camel's back, a hurricane much more furious than the one which had prevailed the year preceding, on the 18th of the same month, swept over the province, destroying all the crops, tearing down the buildings, and sinking every vessel or boat, which was afloat on the Mississippi or the lakes.[3]

The distress of the inhabitants of New Orleans became acute and caused the Intendant, Don Martin Navarro, to issue a public pronouncement of his sympathy. He reminded the people, however, that "all countries have their inconveniences; some suffer from extreme cold or heat of climate, and others are convulsed with earthquakes; this one is infested with wild animals and insects; that one is exposed to inundations; and I know of none which are not occasionally devastated by the fury of storms and hurricanes." He urged the people "to put their faith in the divine providence" and appealed to their loyalty to the Spanish sovereign not to abandon the country regardless of the adverse blows of nature. The people appreciated the sympathy of Navarro, but again listed their grievances against the adverse circumstances with which they were struggling in August 1780 "such as war, two hurricanes, inundation, contagion, a summer more rainy and a winter more rigorous than had been known before."[4] There was a prevailing opinion that the climate there was changing and definitely for the worse. The events of the 1780s in the lower Mississippi Valley gave them little cause to change this belief as more floods and record-breaking cold continued to plague the communities.

THE SAVANNA-LA-MAR HURRICANE

The first of the October series aimed a savage blow at western Jamaica on the afternoon and evening of the 3rd. The settlement of Savanna-la-Mar near the southwestern corner of the island is said to have been completely destroyed by the combined force of wind and a storm tide which flowed one-half mile beyond the shoreline to engulf and destroy all the buildings of the erstwhile thriving seaport. Several hundred persons were thought to have perished.[5] The principal first-hand account appeared in the *Royal Jamaica Gazette* of 12 October 1780 and was copied widely in the American and British press:

> About one P.M. the gale began from the S.E., and continued increasing with accumulated violence until four in the afternoon, when it veered to the south, and became a perfect tempest, which lasted in force until near eight; it then abated. The sea during the last period exhibited a most awful scene; the waves swelled to an amazing height, rushed with an impetuosity not to be described on the land, and in a few minutes determined the fate of all the houses in the Bay.
>
> Those whose strength or presence of mind enabled them to seek their safety in the Savannah, took refuge in the miserable remains of the habitations there, most of which were blown down, or so damaged by the storm, as to be hardly capable of affording a comfortable shelter to the wretched sufferers. In the Court-House, 40 persons, whites and of colour, sought an asylum, but miserably perished by the pressure of the roof and side which fell upon them. Numbers were saved in that part of the house of Mr. Finlayson, that luckily withstood the violence of the tempest. Himself and another gentleman had left it, when the wind had forced open the door, and carried away the whole lee-side of it, and sought their safety under the wall of an old kitchen; but finding they must inevitably perish in that situation, they returned to the house determined to submit to their fate.
>
> About ten the water began to abate, and at that time a smart shock of an earthquake was felt. All the small vessels were driven ashore, and dashed to pieces. The ships, *Princess Royal,* Captain Ruthven; *Henry,* Richardson; and *Austin Hall,* Austin; were forced from their anchors, and carried so far into the morass that they will never be got off. The earthquake lifted the *Princess Royal* from her beams-ends, righted her, and fixed her on a firm bed. This circumstance has been of great use to the surviving inhabitants, for whose accomodation she now serves as a house.
>
> The morning ushered in a scene too shocking for description—Bodies of the dead and dying scattered about the watery plains, where the town stood, presented themselves to the agonizing view of the son of humanity, whose charity lead him in quest of the remains of his unhappy fellow-creatures! The number who have perished, is not yet precisely ascertained, but it is imagined 50 whites, and 150 persons of colour are lost.—Amongst them are numbered Dr. King, his wife, and four children, his partner, an assistant, Mr. Nesbit, a carpenter, and 24 negroes, all in one house. Dr. Lightfoot and Mr. Antrobus were found dead in the streets. In the whole parish, it is said, there are not five dwelling-houses, and not one set of works remaining; the plantain walks all destroyed; every cane piece levelled; several white people and some hundreds of negroes killed.

In the adjoining parish of St. Elizabeth, although the face of the country wore a less horrible aspect than at Westmoreland, much damage was done, and several lives lost.

Our account from Lucea, though not particular, are terrible indeed. The town, except two houses, those of Messers A. and D. Campbell, and adjoining tenement of Mr. Lyons, levelled to the ground; many lives lost;—and in the whole parish of Hanover, but three houses standing;—not a tree, bush, or cane to be seen—universal desolation prevails! [6]

The damage track across Jamaica was confined to the western tip of the island, west of a line drawn from Bluefields to near Montego Bay. On the northwest coast the same disasters from wind and waves were repeated with Lucea and Montego Bay suffering. The *Cornwall Gazette* described the scene at Montego Bay:

On Monday night the 2nd instant about 12 o'clock there came on a storm of wind and rain which continued with unremitted perseverance and violence from the S.E. until 12 o'clock on Tuesday; the weather then appeared to be a little moderated, and continued so much abated until between 3 and 4 o'clock in the evening, as to furnish no immediate indication of an approaching storm. About 4 o'clock the wind seemed to be quite southerly, but increased (accompanied by incessant rain) to such an amazing degree, at about dark, to threaten general ruin and destruction. The darkness of the night now added horrors to the general apprehensions. . . . About 12 o'clock, from the best of our information and our own recollection, the storm began to abate.[7]

From Jamaica the storm center, as plotted by Col. Reid, raced across east-central Cuba, through the north-central Bahamas east of Nassau, and then north-northeastward into the shipping lanes between Hatteras and Bermuda. On the morning of the 5th the hurricane overtook a British fleet under Admiral Rowley when in the latitude of Daytona Beach, but about 500 miles to the east. His ships suffered greatly from the hurricane winds which came out of the east-northeast at the peak of the blow. The center must have been just to the east of *H.M.S. Bristol* at 0100/6th when "a sudden shift of wind" from east-northeast to northwest took place. Most of Admiral Rowley's ships suffered split sails, broken booms, and several lost their mainmasts. Their struggle with the raging elements makes exciting reading in Reid's cool, factual recital from their logbooks.[8] This same fleet had the misfortune a few days later, while in such a damaged condition, to encounter the whip of the Great Hurricane when in the vicinity of Bermuda.

As the Jamaica hurricane moved northward, it struck another British fleet patroling off the Virginia Capes. On the 6th and 7th *H.M.S. Terrible, Cyclops, Triumph,* and *Guadaloupe* suffered such damage that the entire squadron had to put into port for repairs. Admiral Rodney gave a valuable clue as to the extent of this hurricane by stating that his flagship off Sandy Hook experienced fine weather throughout, but that another group off Rhode Island had a severe blow at the same time, the wind veering from east-northeast to west-northwest "with strong gales and squalls, with rain." [9]

THE GREAT HURRICANE

The second storm of October 1780 is still referred to as *The Great Hurricane*. From the enormous toll of human lives exacted and the tremendous total destruction done to property, it certainly rated as the severest hurricane of the Eighteenth Century to strike a land area and should be placed at the top among the most destructive of all-time in the West Indies. Further, the historical importance of the storm was heightened by the presence of the powerful fleets of Britain and France, both maneuvering on nearby islands to strike blows at each other's rich possessions in the Antilles.

The area of development of this hurricane is lost in historical silence. The first land to feel its effects was the sugar island of Barbados, the easternmost of the Windward Islands, at that time a bastion of British economic, political, and military power.

The first news of the disaster was carried to England in a letter from Major General Vaughan, commander-in-chief of His Majesty's forces in the Leeward Island, dated 30 October 1780:

I am much concerned to inform your Lordship, that this island was almost entirely destroyed by a most violent hurricane, which began on Tuesday the 10th instant, and continued almost without intermission for nearly forty-eight hours. It is impossible for me to attempt a description of the storm; suffice it to say, that few families have escaped the general ruin, and I do not believe that 10 houses are saved in the whole island: scarse a house is standing in Bridgetown; whole families were buried in the ruins of their habitations; and many, in attempting to escape, were maimed and disabled: a general convulsion of nature seemed to take place, and an universal destruction ensued. The strongest colours could not paint to your Lordship the miseries of the inhabitants: on the one hand, the ground covered with the mangled bodies of their friends and relations, and on the other, reputable families, wandering through the ruins, seeking for food and shelter: in short, imagination can form but a faint idea of the horrors of this dreadful scene.[10]

The totality of the destruction impressed all observers. As expressed by Admiral Rodney: "The whole face of the country appears an entire ruin, and the most beautiful island in the world has the appearance of a country laid waste by fire, and sword, and appears to the imagination more dreadful than it is possible for me to find words to express."[11] In September Rodney had split his fleet into two groups, unwisely according to his military critics, and at the time of the hurricane was cruising off New York with a strong squadron, thereby luckily escaping the losses which befel the ships which remained near their bases at Barbados and St. Lucia. It was Rodney's belief, expressed in a widely-quoted letter, than an earthquake must have accompanied the hurricane to achieve such total destruction. This seems to be a common surmise when a particularly destructive hurricane strikes. Col. Reid, from his study of the evidence, doubted the occurrence of an earthquake, and we now know, from many recent instances, the gigantic force that can be generated by wind and tide acting together.

The only first-hand account of the storm's passage over Barbados that has come to light is a journal of the event contained in a letter by Major General Cunninghame, governor of Barbados, to General Vaughan, then British commander in the Leeward Islands, for forwarding to Lord George Germaine in London. It was published in the *London Gazette* in late December along with other accounts from the West Indies and subsequently reprinted in many of the leading periodicals of the day. We present the original account of the storm in Barbados, excising only the section dealing with some personal experiences of the governor's family:

> The evening preceding the hurricane, the 9th of October, was remarkably calm, but the sky surprisingly red and fiery; during the night much rain fell. On the morning of the 10th, much rain and wind from the N.W. By ten o'clock it increased very much; by one the ships in the bay drove; by four o'clock the *Albemarle* frigate (the only man of war then there) parted her anchors and went to sea, as did all the other vessels, about 25 in number. Soon after, by six o'clock, the wind had torn up and blown down many trees, and foreboded a most violent tempest.
>
> At the Government House every precaution was taken to guard against what might happen; the doors and windows were barricaded up, but it availed little. By ten o'clock the wind forced itself a passage through the house from the N.N.W. and the tempest increasing every minute, the family took to the center of the building, imagining from the prodigous strength of the walls, they being three feet thick, and from its circular form, it would have withstood the wind's utmost rage: however, by half after eleven o'clock, they were obliged to retreat to the cellar, the wind having forced its way into every part, and torn off most of the roof.
>
> * * *
>
> Anxiously did they wait the break of day, flattering themselves, that with the light they would see a cessation of the storm; yet when it appeared, little was the tempest abated, and the day served but to exhibit the most melancholy prospect imaginable; nothing can compare with the terrible devastation that presented itself on all sides; not a building standing; the trees, if not torn up by their roots, deprived of their leaves and branches; and the most luxuriant spring changed in this one night to the dreariest winter. In vain was it to look round for shelter; houses, that from their situation it was to have been imagined would have been in a degree protected, were all flat with the earth, and the miserable owners, if they were so fortunate as to escape with their lives, were left without a covering for themselves and family.
>
> General Vaughan was early obliged to evacuate his house; in escaping he was very much bruised; his secretary was so unfortunate as to break his thigh.
>
> Nothing has ever happened that has caused such universal desolation. No one house in the island is exempt from damage. Very few buildings are left standing on the estates. The depopulation of the negroes, and cattle, particularly of the horned kind, is very great, which must, more especially in these times, be a cause of great distress to the planters.
>
> It is as yet impossible to make any accurate calculation of the number of souls that have perished in this dreadful calamity; whites and blacks together, it is imagined to exceed some thousands, but fortunately few people of consequence are among the number. Many were buried in the ruins of the houses and buildings. Many fell victims to the violence of the storm and the inclemency of the weather, and great numbers were driven into the sea, and there perished. The troops have suffered inconsiderably, though both the barracks and the hospital were early blown down.
>
> Alarming consequences were dreaded from the number of dead bodies that lay interred, and from the quantity the sea threw up; which however are happily subsided. What few public buildings there were, are fallen in the general wreck; the fortifications have suffered very considerably. The buildings were all demolished; for so violent was the storm here, when assisted by the sea, that a twelve-pound cannon was carried from the south to the north battery, a distance of 140 yards.
>
> The loss to this country is immense, many years will be required to retrieve it.[12]

The best source of meteorological facts at Barbados came from *H.M.S. Albemarle* which put to sea from Carlisle Bay about 1400/10th in the face of "strong gales" from the north-northeast and tremendous seas from the southwest that were breaking over the fort ashore. The later log read: "at 2200, cut away the foretop mast to save the foremast; lowered down the foreyard. At 0030/11th still blowing very hard; a hurricane with rain; wind shifting round to westward." By 0500 the wind had got around to southerly "still blowing very hard of wind, with constant rain." And by noon: "still blowing a hurricane of wind with constant, heavy rain" now from southeast-by-south at which time the ship had drifted to a point only 5 miles off the northwest end of the island.[13] The above would indicate that winds of whole gale force commenced about 1300/10th, certainly reaching hurricane force in early evening, and that the center passed shortly after midnight, though hurricane winds continued well into the morning, possibly until noon of the 11th—a full 24-hour duration of damaging winds.

The loss of life on the island of Barbados, alone, was estimated at 4,326 by Bryan Edwards, a contemporary historian of the West Indies writing in 1793.[14] Another author, writing in 1844, placed the death toll from this single hurricane disaster at 22,000 souls for the entire West Indies.[15] Edwards also wrote that the property losses amounted to over 1.3 million pounds. To assist in the relief and rehabilitation of Barbados, the British Parliament appropriated 80,000 pounds soon after the news of the catastrophe reached London.

Next in the path of the storm lay St. Vincent about 100 miles almost directly west. Here the wind exhibited the same pattern as at Barbados and the destruction appeared equally great according to a press dispatch from St. Vincent:

> The north-east winds began in the night of the 10th and 11th, and the hurricane continued 24 hours, without intermission. It overset, from its foundations, almost all the town of Kings-Town; of 600 houses, of which it was composed, there remained only 14 standing, and these very much damaged. . . . The sea carried away, overset, and destroyed everything that was within 30 yards of the sea.[16]

Farther south, still within the left semi-circle, an account from Granada told of "great devastation on shore; nineteen sail of loaded Dutch ships stranded and beat to pieces." Even reports from Tobago, 250 miles south-southwest of Barbados, spoke of damage to shipping but gave no details.[17]

To turn to the dangerous right-hand side of the advancing hurricane, we find the island of St. Lucia, the largest of the Windward group, about 90 miles to the northwest of Barbados, directly in the path of the advancing whirl. From Col. Reid's analysis of ship logs, the center appeared to have cut right across the center of the island in the late afternoon of the 11th. The ships which put to sea from St. Lucia in a southerly direction got into the left semi-circle of the storm and had backing winds such as experienced at Barbados; those remaining in the harbor of Castries at the northwest point of the island stayed in the right semi-circle with winds constantly veering during the height of the blow.[18] The storm appeared to have had a forward speed of about 7 mph at this stage, a not unusual rate for a large developing hurricane in this latitude.

The *St. Lucia Gazette* of 21 October 1780 carried an informative account of the gale:

> On the 10th there was a heavy sea at the entrance of the harbour and along the coast, attended with heavy rains, and at times severe gusts of wind, insomuch that his majesty's ships, which lay far out, were in danger of parting from their anchors.
>
> In the evening the wind was felt much severer in the harbour; but on shore nothing was particularly observed that seemed to indicate such a severe a change of weather. The gale cannot be said to have fairly commenced till about one o'clock the next morning, when it began to blow from about N.E. in the most sudden and severe gusts.
>
> At day-light there was scarse a vessel left in the road but what had drifted, and many, though so early, were on shore. In short, the whole were driven in a cluster, on the Morne Fortune side. . . .
>
> It would seem as if the former part of the gale was intended to prepare the mind for the redoubled violence with which, in the afternoon, it increased. From three to five, which was esteemed the height of it, those vessels which had rode it out were torn from their anchors, and carried with astonishing force against the other shipping. . . .
>
> The gale continued till one in the morning of the 12th (twenty-four hours) during which time the wind came very little, if any to the southward of E. or farther to the northward than N.N.E. At the end of the gale the wind went to the southward, and great quantities of rain fell. It is unknown the mischief that would have been done, had the wind during the gale come to the westward or southward.[19]

Directly north of St. Lucia, the French island of Martinique, so often subject to cruel blows of nature, underwent a major disaster when the storm-driven sea invaded the low-lying streets surrounding the harbor area of St. Pierre. A report to the French Minister of Marine stated that in an instant 150 homes and buildings were gobbled up by the gigantic swell and that many more were moved from their foundations and their contents destroyed. The wind was reported to have commenced blowing from the east-northeast, with a shift to east-southeast from which the major blow was experienced; it then passed through south to west.[20] The report in the British press follows:

> At Martinique, all the ships were blown off the island that were bringing troops and provisions, and the lives of more than 3000 soldiers and seamen they have on board, much dreaded.
>
> On the 12th four ships foundered in Fort Royal Bay, and could not save a soul; every other ship was blown out of the Roads, and many must of course be lost.
>
> In the noble town of St. Pierre every house is down, and more than 1000 people perished; at Fort Royal town the cathedral, the seven churches, and other noble and religious edifices, the governor's house, the record-office, senate-house, prisons, hospitals, barracks, shore-houses of government and merchants, and upwards of 1400 other houses were blown down, and an incredible number of persons lost their lives; the new hospital of Notre Dame, the most convenient and elegant in the West-Indies, in which were 1600 sick and wounded patients, was blown down, and the greatest part of them, with the matrons, nurses, and attendants, &c. buried in the ruins. Every store-house in the deck-yard is blown down, and filled with ruins; the sick-house of the ship-wrights, &c belonging to the yard, shared the fate of that of Notre Dame and about 100 perished.
>
> By the reports of the day, the number susposed to have perished upon the island, including negroes, is

computed at upwards of 9000, and the damage at upwards of 700,000 louis d'or.[21]

From Dominica, the next island northward along the curving Windward chain, we have a local press report:

> The 11th from day break until noon, the atmosphere was very cloudy; there fell a large quantity of rain, and it thundered all the morning. The wind, which had varied much without being violent, suddenly shifted to S.E. At one P.M. it was a perfect hurricane and its sad effects began to be felt on the most exposed plantations. At 11 o'clock at night the sea became so high, that part of the houses on the shore, at Roseau, were carried away by the waves;—this terrible element rose with redoubled fury until the next day at noon, and finished the destruction of everything that resisted it, casting into the streets of Roseau the small vessels that were at anchor in the road, and in fine everything that it found in its passage; it has not left any remains of the magazines and storehouses that were not 22 feet perpendicular above high-water mark.[22]

The continued northwestward advance of the Great Hurricane put Guadaloupe along with the westermost of the Leeward Islands in the dangerous right side of the storm. It appeared that the important isle of Antigua lay far enough east of the track to escape severe land damage; all ships in the harbor rode out the gale with minor damage. Very heavy rainfall hit Antigua.[23] A thanksgiving proclamation was issued there for the good fortune in being spared the fate of their sister islands to the southwest. But tiny St. Eustatius, an important entrepot for American contraband trade, suffered severely from a harbor storm wave which inundated the shoreline. The same tidal conditions wrought great damage on nearby St. Christopher Island (St. Kitts).[24]

As the whirl crossed the eastern Caribbean, it took a slightly more northerly turn bringing the central track very close to the western end of Puerto Rico. Reid believed that the center passed over or very close to Mona Island, between Hispaniola and Puerto Rico, on the morning of the 15th, as the logs of the *Venus* and *Convert* showed them to have been in the right side of the storm, while the *Diamond* and *Pelican,* of the same squadron and cruising nearby, were in the left side. The storm center apparently bisected the Mona Passage.[25]

Upon reaching a point just east of Turks Island, in the extreme southeastern Bahamas about 70°W 22°N, the Great Hurricane came under the influence of a westerly steering current and commenced its recurve to north and then northeast sometime on the 16th. Again Admiral Rowley's squadron, already severely mauled by the Jamaica storm ten days earlier, felt the whip of the western side of the hurricane on the 16th and 17th. Now with a greatly accelerated forward motion to the northeast, the hurricane passed about 50 miles south of Bermuda on the 17–18th. Fifty vessels were driven ashore on Bermuda by the still powerful disturbance. Reid tells us that living witnesses at Bermuda in 1839 still referred to the storm of 1780 as the greatest ever experienced in that latitude.[26]

SOLANO'S HURRICANE

Of more immediate concern to the actual mainland was the third storm of the month which lashed the central and eastern Gulf of Mexico soon after mid-month, catching a powerful Spanish war fleet of 64 vessels under Admiral Solano enroute from Havana to strike a decisive blow at Pensacola. This small armada had 4,000 men aboard under the military command of Bernardo de Galvez, the New Orleans commander in 1779 who had previous experiences with hurricanes. The immediate task was to reduce the main base left to the British in the Gulf of Mexico area.

Though the location of the birthplace of Solano's Hurricane remains in doubt, it most probably had a rapid development in the western Caribbean, as many others have done at this season, and then passed quickly northward very close to the extreme western tip of Cuba. Thanks to Col. Reid and his researches in the 1830's, we have a good account of the disturbance when in the Gulf of Mexico area for he secured through diplomatic channels in Madrid a transcript of the logs of Solano's flagship and other vessels accompanying. Reid believed that the circulating winds on the eastern side of the storm, having a southerly and easterly component, carried the *San Juan* and others through the eastern Gulf to a position directly south of the mouth of the Mississippi River before the five-day blow subsided.

The hurricane scattered the entire fleet over the central Gulf with some of the ships getting into the western side of the vortex. Here again Reid found confirmation of his thesis of progressive, rotary hurricane circulation. Damage to the fleet was extensive. Those warships which tried to hold to course were dismasted, while the transports scattered to the four winds, some fetching up in Mobile, some in New Orleans, and some in Campeche. Galvez after fruitless attempts to reunite his fleet returned to Havana a month and a day after he had set sail, and the attack was called off. For a third time in as many years the elements had spared Pensacola.

Whether the storm ever made a landfall on the Gulf coast as a full hurricane we do not know. Reid plotted the track as leading into the northern Gulf only, and we have no land station reports indicating an unusual wind activity.

The log of the *San Juan* as she whirled helplessly around the Gulf follows:

> Oct. 17th. At dawn it was calm in shore [Cuba]. By eight o'clock a breeze sprang up from the N.E. . . . By noon the wind freshened at N.E., scud and heavy cloud closing upon us. . . . At 9 p.m. the breeze freshened. . . . By 10 at night, the wind increased, and was then at N.E. ½ E., with torrents of rain and some hard squalls, shifting as far as E.N.E.
>
> Oct. 18th. At daybreak, heavy clouds, rain, wind, and sea. At 9 a.m. the wind was E.N.E. . . . From the 18th to the 20th, continued lying-to in the fourth quadrant; the weather still dark and increasing; the wind at N.E.; continued rain, with a heavy sea . . . at 10 p.m. on the 20th, our tiller broke; secured the rudder; the ship sustaining heavy and repeated squalls, whilst she came up from the E.N.E. as far as E., as the wind veered around from the S.E. to the E.S.E.
>
> Oct. 21st. By half-past four in the morning, the wind changing, made the ship come up head to sea. The ship then pitched away all her masts as well as her bowsprit, and with it lost the greater part of the cutwater. By the exertion of the officers and crew the wreck was cleared by six o'clock; at this hour it began to clear up from the S.S.E. Lightened the decks of everything we could. The sea ran so high that we were still unable to ship another tiller.
>
> At 11 a.m., set top-gallant-sails on the stumps on the main and fore masts, and the sail of the launch on the stump of the mizen, keeping her head to the N.E. At noon, latitude 26° 42′ N., longitude, 290° 9′ E. of Teneriffe; longitude 86° 11′ W. of Greenwich.
>
> Oct. 22nd. Commenced with less sea and wind.[27]

Soon after Solano's Hurricane churned up the Gulf of Mexico, another storm struck St. Lucia on the 23rd and apparently moved northward into the broad Atlantic. The story of *H.M.S. Elizabeth* perhaps best portrays the tribulations of navigating the seas north of the West Indies during the tempestuous month of October 1780. This ship had sailed from Jamaica for England on September 5th, but had run into contrary winds. On October 8th she met a "hard gale of wind and mountainous seas" which necessitated cutting away the maintop after the mizzen mast had snapped off. Then on the 18th the ship encountered the lash of the Great Hurricane described as "a second more severe gale and heavy sea" which sprang the foremast in three places. And finally, on the 25th "a third and more fierce gale seemed to unite the fury of the former two" to split the sails of the British ship to rags. This must have been the severe disturbance which ravaged St. Lucia on the 23rd and then passed directly northward toward Bermuda. The *Elizabeth* proceeded to England under jury masts, arriving a month overdue with the crew not only physically battered, but in a semi-starvation state due to the unanticipated length of the storm-tossed voyage.[28]

Even October did not end the storm session along the Atlantic seaboard, for Admiral Rodney's fleet guarding the ports of the rebellious colonies was scattered by "a violent gale of wind which continued forty-eight hours" on 17 November when only one day out of Sandy Hook, New Jersey. Every ship of the sizable squadron was separated from the *Sandwich,* Rodney's flagship, and it was not until two weeks later, and then south of the Tropic of Cancer, that the Admiral again sighted one of his command.[29] We have no other information about this storm and cannot determine whether it was of tropical origin or not. Its occurrence again demonstrated the outstanding severity and long continuance of the hurricane season of 1780.

[1] Tannehill, 246; Alan Burns. *History of the British West Indies.* London, 1954, 507.
[2] William Reid. *An Attempt to Develop the Law of Storms.* 3d ed., London, 1850, 289–402.
[3] Charles Gayarre. *History of Louisiana.* III, 151.
[4] *Ibid.,* 152.
[5] Reid, 292–335.
[6] *Royal Jamaica Gazette* (St. Jago de la Vega), 12 Oct 1780 quoted in John Fowler. *A General Account of the Calamities occasioned by the late tremendous Hurricanes and Earthquakes in the West-India Islands.* London, 1781, 27–29. Also *Gentleman's Magazine* (London), 50, 1780, 620–621; and *The Annual Register for the Year 1780.* London, 177, 292–295.
[7] *Cornwall Gazette* in *Penna. Jour.,* 13 Dec 1780.
[8] Reid, 329.
[9] *Ibid.,* 332.
[10] *The Political Magazine* (London), Jan 1781, 48.
[11] Letter-Books and Order-Book of George, Lord Rodney. 1780–1782. *Coll. N. Y. Hist. Soc.,* 65, 1932, I, 91–92.
[12] *The Gentleman's Magazine,* 1780, 621–623; *Annual Register 1780,* 295–298; Fowler, 33–36.
[13] Reid, 349.
[14] Bryan Edwards. *The History Civil and Commercial of the British West Indies.* 5th ed., London, 1819, I, 347.
[15] Henry H. Breen. *St. Lucia: Historical, Statistical, and Descriptive.* London, 1844, 136.
[16] *Penna. Jour.,* 13 Dec 1780.
[17] *The Scots Magazine* (Edinburgh), 42, 1780, 658.
[18] Reid, 342.
[19] *St. Lucia Gazette* (Carenage), 21 Oct 1780 in Fowler, 43–46.
[20] Reid, 340.
[21] *Annual Register,* 1780, 297–298.
[22] Martinico, 26 Oct 1780, in *Penna. Jour.,* 13 Dec 1780.
[23] *The Political Magazine,* Dec 1780, 772.
[24] *Idem.*
[25] Reid, 342.
[26] *Idem.*
[27] *Ibid.,* 397–399.
[28] *The Political Magazine,* Dec 1780, 772.
[29] Rodney, I, 90–91.

THE STORM TIDE OF 1794 IN LOUISIANA

A hurricane struck the Louisiana coast sometime in August 1794 to cause the first tropical storm inundation of the lower Mississippi Delta region since 1780. We have no eyewitnesses who have left firsthand accounts of the event. New Orlean's first newspaper, the *Moniteur de la Louisiane,* commenced publication in 1794, but only the copy of 25 August 1794 has survived for use by historians of all the issues published prior to 1800.[1] That issue was fortunately reproduced by photostat in 1896 prior to its destruction by fire.

Louisiana at this time was under Spanish domination with the channels of communication to Madrid rather than through the United States. Thus, the American press did not have ready access to New Orleans exchange papers; our search in Charleston, Philadelphia, and New York papers for this period has proved fruitless. Perhaps the Havana or Madrid press would reveal a more satisfactory meteorological account than we now have at hand.

A contemporary reference in 1797 to "hurricanes" was made in a letter of Betram Gravier, the Attorney General of New Orleans. He referred in a letter to "the three hurricanes we had in 1793 and 1794, experienced during the month of August, the time for harvesting the crops, have successively devastated the rural sections of the country and also ruined the farmers of lower Louisiana." [2] In the absence of additional data about two of the three "hurricanes," it is assumed that they were tropical storms with damage limited to crops and that only one in 1794 achieved full hurricane stature.

Our main source of information about the 1794 hurricane and storm tide came from a United States Government report of 1803, popularly known as Jefferson's *Account of Louisiana.* In discussing some aspects of the storm, it stated that the calamity occurred on an unspecified day in August 1794. In the surviving issue of the *Moniteur de la Louisiane* of 25 August, though it does mention a wind storm at Natchez upsetting a boat on the 11th, there is no hint that a general storm had recently taken place in the region, so we are inclined to place the date later than the 25th.

Desiderio Herrera in his history of Cuban hurricanes listed a major storm sweeping that island on the 27–28 of August.[3] Aligning with this, the recent historian of Avoyelles Parish, some 130 miles northwest of New Orleans, unearthed "an account of high wind lasting 9 hours on September 1, 1794" in her search for local material.[4] It is well within meteorological probability that these two could have been aspects of the same hurricane. In the absence of contrary arguments (and we would certainly welcome any), we shall tentatively mark the date as 31 August in lower Louisiana and 1 September in the interior.

As soon as the Louisiana Territory was purchased from France in 1803, Thomas Jefferson directed the State Department and Treasury Department to assemble all the known facts about the vast area. The section devoted to the climate of Louisiana gave due space to the hurricane problem, listing 1780 and 1794 as years of major storm inundations of the lower Mississippi Delta. The calamity in 1794 put all the plantations under two to ten feet of water from the post of Belize at the mouth up to the English Turn at Plaquemines, a distance of some 80 miles. Many lives were lost along the river together with most of the horses and cattle. The surging storm tide, engulfing the site of Fort St. Philip, drowned the chief engineer along with some of the men who were building the bastion at the strategic English Turn. All subsequent writers on the subject—Warden, Monette, Drake, and Darby—appear to have drawn their meager material about the storm from this portion of the *Account of Louisiana:*

> These hurricanes have generally been felt in the month of August. Their greatest fury lasts about 12 hours—They commence in the south east, veer about to all points of the compass, are felt more severely below and seldom extend more than a few leagues above New Orleans. In their whole course they are marked with ruin and desolation. Until that of 1793 (sic), there had been none felt from the year 1780.[5]

For additional material, we have a statement by Paul Alliot, writing in 1804, that a dozen plantations along with their animals were overflowed: "everything in that region was destroyed." [6] Warden in 1819 said the storm was accompanied by hailstones of uncommon size, certainly an unusual occurrence in a hurricane.[7] Dr. John Monette thought that in 1794 "nearly the whole province of lower Louisiana was desolated by a terrible storm or hurricane which caused immense damage to crops." [8] The above presents all the material so far uncovered about this seemingly very destructive hurricane. It would appear that 1794 should rank with the major inundations experienced by the lower Louisiana coast region.

[1] *Moniteur de la Louisiane,* 25 Aug 1794. Facsimile copy in *Publications of the La. Hist. Soc.* 1896, 1–4, 33.

[2] Ms. Betram Gravier, New Orleans, March 17, 1797. Cabildo Records. Vol. 4, No. 1. New Orleans City Archives, 1936. Typescript.

[3] Desiderio Herrera. *Memoria sobre los huracanes en la Isla de Cuba.* Havana, 1847. 47–49.

[4] Corinne Saucier. *History of Avoyelles Parish.* New Orleans, 1943. 169.

[5] *An Account of Louisiana, being an abstract of Documents in the offices of the Department of State and of the Treasury.* Philadelphia, 1803.

[6] "Paul Alliot's Reflections" in J. A. Robertson. *Louisiana, 1785–1807.* 1, 164.

[7] David Warden. *A Statistical, Political, and Historical Account of the United States of North America.* Philadelphia, 1819. 2, 501.

[8] Monette Ms., 21. Clements Library, Univ. of Michigan.

THE GREAT LOUISIANA HURRICANE OF 1812

Captain W. H. Gardner of Mobile in his *Record of the Weather from 1701 to 1885* mentioned a very destructive hurricane at New Orleans on 19 August 1812, but stated that "the details are meagre, but of which awful stories are told by old citizens." [1] The "awful" reputation of this storm remained for many years—in 1831 when a major storm struck, at the time of the Last Island disaster in 1856, and again after the severe 1860 blow, editors harked back to 1812 for a precedent of equal stature.

Dr. John Monette's unpublished study of the history and geography of the Mississippi Valley supplied some details of the storm: "In August, a constant south wind drove the waters of Lake Pontchartrain until it inundated all the marshes toward the city, as well as all the plantations & settlements for many miles below New Orleans." [2] Poey listed a storm at Jamaica on the 14th, and this could possibly have been the same as struck the Louisiana coast five days later.[3]

A search of published works on the early American period in Louisiana has failed to reveal any essential details about this major storm which rates with the *great* in the region's hurricane annals. One can readily understand why Captain Gardner could not find additional information. Perhaps the absorption of local historians in the exciting events attending the opening phases of the War of 1812 caused them to overlook a storm which in peacetime would have received special attention.

A check of the local press for the period, however, proved fruitful. There were at least four newspapers published at New Orleans in 1812, and all gave coverage to the hurricane. Under the head, "AWFUL AND DISTRESSING," the *Orleans Gazette* reported:

> On Wednesday night last (19th) about 10 o'clock, a gale commenced occasionally accompanied with rain and hail, and which continued with a most dreadful violence for upwards of four hours. As we have never witnessed anything to equal it, neither do we believe the imagination can picture to itself a scene more truly awful and distressing, than that which its consequences present.[4]

The account in the *Louisiana Gazette and New Orleans Advertiser* contributed a few meteorological details:

> It would be a vain expectation in any of our readers to suppose any pen capable of giving a faithful picture of the scene exhibited after the tremendous gale of Wednesday night—Tuesday evening was remarkably warm and sultry—Wednesday morning the wind was from the north, the weather very cool for the season, and the horizon covered with dark heavy clouds which indicated a storm;—before eleven o'clock it commenced raining, the wind still at north, and continued with short intervals during the day; at dark the wind (still from the north) began to increase, and the rain fell in torrents, the wind shifting a little to eastward; at half after eleven, wind at ESE, the storm raged with great fury,—and from that to one o'clock, the whole of the damage was sustained. At one o'clock, or soon after, the clouds began to break, and at three o'clock the storm had nearly subsided.[5]

There was some disagreement among editors as to the exact time of the onset of the hurricane force winds. The editor of *Moniteur de la Louisiane,* in a column headed, "Ouragan," declared: "Yesterday at five in the afternoon the wind commenced to blow from the northeast, but without much force. It increased from six to nine. At this hour one could say that the real storm commenced as it increased gradually until eleven when it blew with a violence that threatened the entire destruction of the city. It blew until two o'clock when it commenced to relent by degrees. Losses were estimated at three million piastres." [6] The editor of the *Courier* clocked the main force of the hurricane as commencing at 2200, commenting: "the storm lasted four hours but the havock was dreadful." [7] All agreed that the period of destructive winds was relatively short—no mention was made of any calm period—and no reference to any wind shift during the storm period has been found. It is apparent that this storm was moving at a fast pace for a large hurricane, and that it packed very destructive winds in its eastern sector: "the greatest part of the houses in the city as well as in the suburbs were uncovered and a great many of them entirely destroyed," noted the *Courier*.[8] The exact course of the hurricane cannot be definitely tracked in view of the lack of reports from the sparsely settled country in central and western Louisiana. The inundations in the Delta area and the wind behavior at New Orleans would place the center as moving a short distance to the west of the Mississippi River throughout its passage over Louisiana.

The editor of the *Orleans Gazette* summarized the damage in the center of the town: The Market House was blown down and its 24-inch diameter columns sent tumbling. The roof of the Convent Church was rolled up and blown away, and the trees in the garden were felled. Many brick houses were demolished or severely damaged. "The levee almost entirely destroyed; the beach covered with fragments of vessels, merchandize, trunks, and here and there the eye falling on a mangled corpse. In short, what a few hours before was life and property, presented to the astonished spectator only death and ruin." Fifty-three vessels were listed as being dismasted, ashore, or damaged in some way. All the small river craft such as barges and market boats were "crushed to atoms." [9]

Intelligence came from down-river at Plaquemines where St. Philip at the English Turn was anticipating an attack from the British fleet which had been hovering off the Balize at Northeast Pass:

> The Hurricane having spread desolation thro' this part of the country, you will be surprised at our miraculous escape in the old barracks.
>
> John Dennis and his wife, Sylvan Dennis and his wife and his five children, are all drowned and a number more. Oh!—the afflicting, heart-wounding sight to behold 32 persons committed to the earth in one common grave.
>
> The numbers between this and the Grand Prairie is 45; 16 of whom are negroes. The new Commandant's house, contractor's store, and in fact nearly all the buildings on the outside of the fort, have been hurled God knows where. Mr. Willis and his family have escaped by the interposition of Mr. Williams from the Balize, who fortunately had a valuable boat. The water was six or seven feet deep on the parade; the waves threw large pieces of timber over the front parapet. The U. States have suffered great damages—almost the whole of the medical and hospital stores are rendered unfit for use.[10]

The inundation of Ft. St. Philip caused many rumors to spread to the effect that the British had landed and seized the fort during the discomfiture of the American defenders. In early September, however, the *Louisiana Gazette* assured its readers that the post was still occupied by Americans, the Balize was free, and the river passes apparently not blockaded. The hurricane had struck the gathering British fleet a heavy blow and scattered them widely over the Gulf. Several American gunboats had been driven ashore in Lake Pontchartrain during the hurricane, but these events proved indecisive since the great military action in the area was still two and a half years away.[11]

[1] W. H. Gardner. *Record of the Weather from 1701 to 1885. Special Papers of the Alabama Weather Service, No. 1.* A & M College, Auburn, 1886?, 7.

[2] Monette Ms., No. 364. Clements Library, Univ. of Michigan.

[3] Poey, 20.

[4] *Orleans Gazette,* 21 Aug 1812, in *Nat. Int.* (Wash., D. C.), 22 Sept 1812.

[5] *The Louisiana Gazette and New Orleans Advertiser,* 22 Aug 1812.

[6] *Moniteur de la Louisiane* (New Orleans), 20 Aug 1812. Issue was delayed by damage to the press caused by storm.

[7] *Louisiana Courier,* 21 Aug 1812.

[8] *Idem.*

[9] *Orleans Gazette,* 21 Aug 1812.

[10] *Louisiana Gazette,* 25 Aug 1812, in *Courier,* 26 Aug 1812.

[11] *Louisiana Gazette,* 5 Sept 1812.

REFERENCES: 1815–1870

In the second section, 1815–1870, many footnotes refer to manuscript material now in the National Archives, Washington, D. C. The following standard abbreviations are employed: SI—Smithsonian Institution meteorological records; SG—Surgeon General, U. S. Army, meteorological records; and PR—records of private meteorological observers deposited in the National Archives.

HATTERAS NORTH: 1815–1870

THE GREAT SEPTEMBER GALE OF 1815

Of all the storms in New England's history the Great September Gale of 1815 was long accorded first place by local historians of the region. Noah Webster, a lifelong observer of the weather as well as of words, noted in his diary that "the storm was a proper hurricane" when it passed through his place of residence at Amherst in west-central Massachusetts.[1]

The path of the center of this gigantic tropical disturbance cut southern New England into two almost equal segments as it roared northward through eastern Connecticut, central Massachusetts, and New Hampshire on the morning of Saturday, 23 September 1815. Few local histories fail to mention some incident or reminiscence to add to the chronicle of devastation and horror left by the most powerful atmospheric disturbance to lash the area since the Great Colonial Hurricane of 1635. And 123 more years would pass until September 1938 when another great storm of like dimensions and intensity would create such widespread havoc.

The storm of 1815 was of great stature. It constituted one of a family of severe summer and fall oceanic storms that struck the North Atlantic shipping lanes in the first season of peaceful commerce following the conclusion of the War of 1812 and the Napoleonic struggles. Of definite tropical origin, this most destructive storm of a major hurricane season probably developed in the central Atlantic Ocean close to the Cape Verde Islands off West Africa, a region of warm seas, humid air, and frequent squalls where most of the *great* September hurricanes have their birth.

The first reported landfall in the West Indies came at little St. Barthelemy in the exposed Leeward Islands group. Great destruction to shipping and shore installations occurred there on September 18th. Spinning northwestward, the storm struck a blow two days later at Turks Island in the extreme southeastern Bahamas north of Hispaniola.[2] Its forward movement at this time was relatively slow as is characteristic of big hurricanes at this stage.

The next reported landfall came some thirteen hundred miles to the north-northwest on the south shore of Long Island, New York, on the morning of the 23rd. We have an ominous ship report, made at 0700 on that fateful morning when off Barnegat Inlet on the central New Jersey coast, indicating that a dead calm existed as an interim between "severe gales of great violence," first from the east-northeast and then from the west-northwest. This ship, no doubt, passed through the eye of the hurricane, when only 50 miles south of the Long Island coast.[3]

The exact location of its penetration of the Long Island shore cannot be pinpointed in the absence of adequate local reports of wind behavior. Noyes Darling of New Haven later assumed the historian's task of constructing a timetable of the Great September Gale by assembling all then available newspaper notices of wind and weather behavior on that exciting day.[4] The later course of the hurricane eye across Long Island

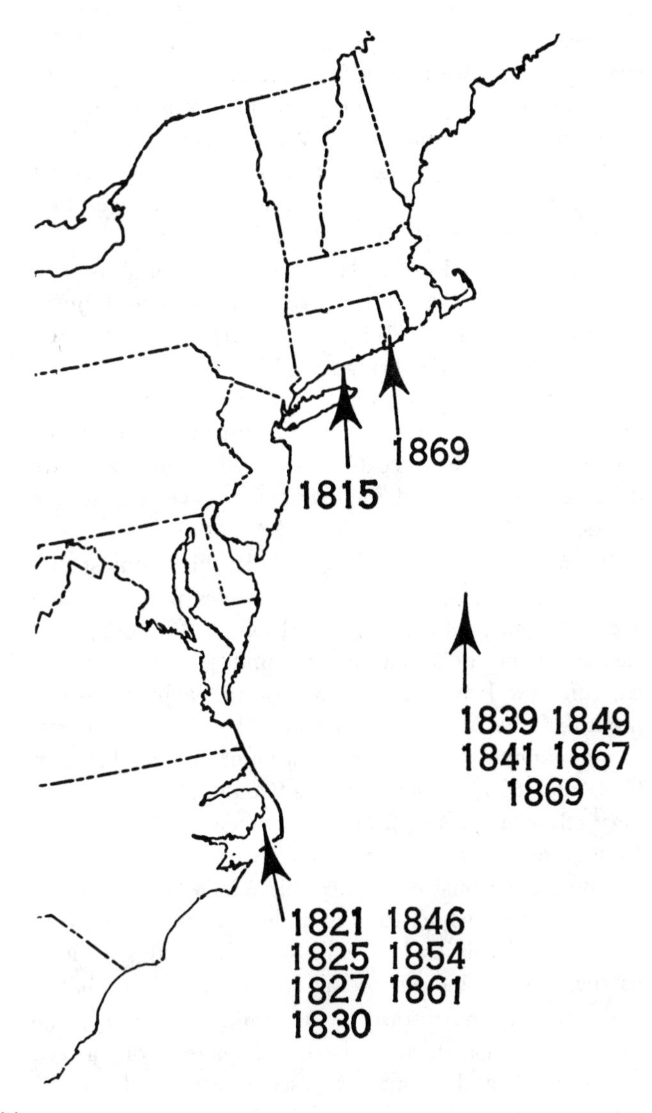

Sound and through southern New England would indicate that the center struck the southern shore of Long Island near Center Moriches, only a scant 5 or 10 miles east of the landfall of the Great Long Island-New England Hurricane of 1938.

Sag Harbor, to the east of the center's track, reported "a tide six feet higher than ever known before" and "trees strewn in every direction about our streets." [5] One house at Southampton was unroofed and two others were saved by lashing timbers across the roof and tying them to interior beams.[6] The lighthouse at Montauk Point was so injured that no beacon could be kept, and over at Gardiners Island the landing wharf was washed away, orchards stripped of their fruit, and much timber laid low. At Patchogue the vessel of Captain Spencer was cast up on the shore for a total loss.[7]

New York City, as usually experienced at points to the west of the track, was buffeted by "very heavy rain and gales from the northeast," but escaped with only moderate damage.[8] With a gale of increasing violence building up during the early morning, the river wharves were generally under water by 1000, three hours before flood tide; but the storm center was then well into Connecticut. With a shift around to southwest, the gale subsided, resulting in a gradual lowering of the water. Only minor damage resulted along the waterfront.

Shade trees and orchards in the city suffered as they had not since the storm of October 1804. The military telegraph on Signal Hill at the Narrows was blown down and the draw bridge at Flushing swept away. The *New York Spectator* concluded its damage summary with the statement: "considering the violence of the gales, we are agreeably surprised to find so little damage has been sustained among the vessels at our docks." [9]

It was not so in New England. Rushing almost due north now at a speed close to 50 miles per hour, the great cyclonically-spinning whirl churned across Long Island Sound in a few short minutes to roar inland east of New Haven and very close to Saybrook at the mouth of the Connecticut River. The time of landfall is not known exactly—one account stated between 0800 and 0900. Our analysis would place the time very close to 0900. Both the river ports of New London and Norwich lay close to the path of the center in the dangerous eastern semicircle where forward momentum of a hurricane is added to maximum wind speeds; both places had excessive river tides as long as the winds came out of the southeast and south.[10]

Continuing northward at undiminished speed, the eye of the vast storm crossed the plateau of eastern Connecticut and central Massachusetts, well to the east of Hartford and Springfield. The line of advance lay very close to an axis passing through Saybrook and Willimantic in Connecticut, through the Massachusetts settlements of Southbridge and Gardner, and into New Hampshire close to Jaffrey and Hillsboro. The peak of the storm passed between Amherst and Worcester in Massachusetts at approximately 1100 and thence into the hill country of New Hampshire.[11] The editor of the *Farmer's Cabinet,* published at Amherst in New Hampshire close to Nashua and Manchester, reported: "at 1130 the severest gale of wind from the Southeast ever known. The damage in this quarter is immense." [12] The *New Hampshire Patriot's* editor at Concord presented a vivid picture of the storm in that area:

> Last Saturday was experienced in this vicinity the most severe gale of wind, or rather hurricane, known by the oldest inhabitants. The wind commenced in the morning at N.E.—about noon it changed to S.E. and for two hours it seemed to threaten everything with ruin. The sturdy oak, the stately elm, and the pliant popular, were alike victims to its fury. The destruction of orchards and buildings has been great; there is scarcely an apple left on the standing trees. Many cattle have been killed by the falling trees. Had this violent wind taken place in the season of vegetation, there is no calculating its effects; it might have produced famine.[13]

Vermont and western Massachusetts were to the west of the storm track where wind forces are usually of less strength but rainfall more intense than to the east. The *Hampshire Gazette* at Northampton mentioned "a tornado of wind and rain," and the *Vermont Journal* to the north at Windsor referred to "the violent storm on Saturday last." [14] Both dispatches spoke of high waters following the storm rather than of wind damage during the blow. Over the mountains to the west neither the *Pittsfield Sun* nor the *Vermont Gazette* at Bennington made any mention of the hurricane in their issues following the storm, though later carried many columns about the destruction in eastern New England.[15]

To the west of the storm track in Connecticut came reports of very heavy rains, extremely high tides, heavy gales. At both New Haven and Bridgeport the strongest winds came from the northeast with a shift to northwest and west as the storm moved northward. President Day on the Yale campus measured a total three-day rainfall of 6.71 inches from the 21st to 23d (0.25″ on the 21st, 3.26″ on 22d, and 3.20″ on 23d). At his sunrise observation on the 23d the temperature stood at 48°, wind northeast, and rain was falling. By 1300 the clouds had broken and the temperature had risen to 65°. Unfortunately, the barometer readings for most of September were not entered.[16] In fact, no barometer readings close to the center of the storm have come to notice from any source.

As to destruction in the vicinity of New Haven, the *Columbian Register* commented:

> The damages sustained in this town by the late gale were trifling, considering the violence of the storm—The tide rose to an extraordinary height, and drifted several hogsheads of rum and mollasses from the wharf, besides other articles of some considerable value.
>
> It appears by accounts from the eastward that the storm raged more severely, and that more disasters were sustained by sea-coast towns between this and Boston than in any other part of the country.[17]

The greatest wind damage in the Great September Gale occurred in eastern Connecticut, over all of Rhode Island, in east-central Massachusetts, and in southeastern New Hampshire. Maine came off lightly with high winds and forest losses, but not much structural injury. The path of severe destruction did not extend more than 75 miles to the east of the track of the center. Samuel Rodman, New Bedford's weather man for some 50 years, reported a "tremendous gale," but 40 miles due east at Nantucket our weather-watching astronomer and congressman, Hon. William Mitchell, remembered "that the gale was not severe" there.[18]

The line of demarcation between light and heavy destruction in central Massachusetts was noticed by the Albany-to-Boston postrider in the vicinity of Brimfield, about equidistant between Springfield and Worcester on present Route 20: "From Brimfield to this place (Boston), there appeared to be one continued scene of devastation, in the unroofing of houses, upsetting of barns, sheds, and other buildings, and in the general prostration of fences, trees, grain, and every description of vegetation." [19]

Let us return to the landfall on the Connecticut shore to consider the firsthand account of Mr. V. Utley who witnessed the storm at River Head, Lyme, close to the point where the tropical intruder first entered New England:

> The morning was cloudy, with a cool wind and very heavy rain from N.E. which continued till 8 A.M. and then abated a little, when the wind veered to the S.E. and a much warmer wind was sensibly felt by most people who stood in the open air.
>
> The violence of the wind increased gradually till 9 o'clock, at which time it blew a perfect hurricane, and continued with the utmost fury until 11 A.M..
>
> In its course it tore down barns, unroofed dwelling-houses, upset cider mills, carried away carriage houses, &c. &c. It tore up the largest trees by the roots; some orchards are nearly destroyed, the trees lay level with the ground; the heart sickens at the sight.—The forest from New London to Connecticut River, (which is as far as I have heard) exhibits to the eye the most dreadful destruction ever made by a tornado in this part of the country; whole forests of trees are either broken down, or torn up by the roots and crossing each other; fences level with the ground. At River Head the tide rose 6 feet higher than ever it was known to rise before, carrying away bridges, &c. I have not heard of any lives being lost.
>
> P.S. The leaves and exterior parts of limbs of trees taste salt by water taken up from the ocean and probably carried into the country.[20]

Since the gale moved over a well-populated area, there are available many firsthand accounts of its characteristics and behavior. A study of the wind data indicated that a rather abrupt change in atmospheric conditions occurred during the period, and that there were three phases to the storm. As the wind strength commenced to pick up, the flow came from the northeast or east-northeast with cool temperatures for the season and a moderate rain. Then came an abrupt shift, as if a front had quickly passed. At Brooklyn, in northeastern Connecticut, the wind "changed in a moment from north to east, or a little southward of east. The air which had previously been cold and chilly became at once more than warm." [21]

Again at Worcester in central Massachusetts, "the wind increased in violence, and the rain descended in torrents, and continued with but short intermissions until about half past ten when the rain abated, and the wind suddenly shifted to the S.E. and blew a perfect hurricane." [22] The destructive portion lasted only one hour. Another Worcester press account mentioned a "suffocating current of air, as from a hot bath, which accompanied the middle stages of the tempest." [23]

From the above reports it is readily apparent that a weather front or temperature discontinuity moved across eastern New England during the forenoon, bringing an abrupt shift of wind from a northeasterly to a southeasterly quarter—a cool air stream of maritime polar origin was replaced by a warm air flow from a maritime tropical source. The rain diminished appreciably in the second phase, while the winds mounted to full hurricane force from the southeast as the warm core of the whirl arrived. It was at this time that the greatest damage occurred.[24]

The third phase of the storm, as the center came abreast and passed to the north, brought a slackening of wind force and a gradual veering of the wind through south to southwest, for points in eastern New England in the right semicircle of the circulation. At New London the shift to south and southwest came soon after 1100 and at Norwich, also in eastern Connecticut, the high winds had generally subsided by noon.[25] At Salem well to the east of the storm center, Dr. William Bentley reported the storm's duration as only two hours with the high winds subsiding soon after 1400 followed by "pleasant weather" in the evening.[26]

Practically all accounts from southeastern New England made mention of the salty flavor of the rain.

This was noticed as far inland as Worcester where grapes were said to have a salt taste and windows to have lost their transparency. At Providence the air seemed impregnated with saline particles, quite perceptible to the taste.[27] Houses took on a whitish appearance; leaves of trees turned white as if frosted and then became black as the froth and spray of the Atlantic Ocean was carried through the New England atmosphere. Streams emptying into the Thames River in eastern Connecticut and wells in the Hartford area were considered brackish for some days afterwards. Sea gulls were seen as far inland as Grafton and Worcester.[28]

The impact of the storm at Cambridge has been told by Prof. John Farrar of Harvard who was a qualified weather observer of many years' experience. His report, as published in the *Memoirs of the American Academy of Arts and Sciences,* was widely circulated in scientific circles both here and abroad:

X.

An account of the violent and destructive storm of the 23d of September 1815.

BY JOHN FARRAR,

PROFESSOR OF MATH. AND NAT. PHIL. IN THE UNIVERSITY AT CAMBRIDGE.

This storm was very severely felt throughout a greater part of New England. It was most violent on and near the coast, but does not appear to have extended far out at sea. It was preceded by rain, which continued to fall for about twenty four hours with a moderate wind from the N. E. Early in the morning of the 23d the wind shifted to the east, and began to blow in gusts accompanied with showers. It continued to change toward the south and to increase in violence while the rain abated. Between 9 and 10 o'clock A. M. it began to excite alarm. Chimneys and trees were blown over both to the west and north, but shingles and slates, that were torn from the roofs of buildings, were carried to the greatest distance in the direction of about three points west of north. The greatest destruction took place between half past 10 and half past 11. The rain ceased about the time the wind shifted from southeast to south; a clear sky was visible in many places during the utmost violence of the tempest, and clouds were seen flying with great rapidity in the direction of the wind. The air had an unusual appearance. It was considerably darkened by the excessive agitation and filled with the leaves of trees and other light substances, which were raised to a great height and whirled about in eddies, instead of being driven directly forward as in a common storm. Charles river raged and foamed like the sea in a storm, and the spray was raised to the height of 60 or 100 feet in the form of thin white clouds, which were drifted along in a kind of waves like snow in a violent snow storm. I attempted with several others to reach the river, but we were frequently driven back by the force of the wind, and were obliged to screen ourselves behind fences and trees or to advance obliquely. It was impossible to stand firm in a place exposed to the full force of the wind. While abroad, we found it necessary to keep moving about, and in passing from one place to another, we inclined our bodies toward the wind, as if we were ascending a steep hill. It was with great difficulty that we could hear each other speak at the distance of two or three yards. The pressure of the wind was like that of a rapid current of water, and we moved about almost as awkwardly as those do who attempt to wade in a strong tide.[29]

The low-lying islands off the southern shore and the southeastward-facing estuaries from the mouth of the Connecticut River eastward to Cape Cod took the worst beating as the hurricane winds, sweeping in from the ocean out of the southeast, pressured the waters into a tidal bore which rushed up the ever-narrowing valleys and mounted higher and higher. The action of wind and water over Fisher's Island off New London swept all the trees from the eight-square-mile piece of land rising only a few feet above the sea. At Buzzard's Bay, which almost separates Cape Cod from the mainland, the peak of the winds coincided with high tide, and the waters swelled within 15 inches of covering the narrow isthmus and creating a natural canal where the Cape Cod Ship Canal later was to be dug.[31]

As in 1938, Narragansett Bay, with its many inlets of northward-jutting fingers, bore the full brunt of the wind-driven storm tide. The area lay east of the center track of the storm and in the zone of maximum southeasterly wind speeds, and the period of peak winds here also came close to the time of high tide. Moses Brown, both a leading merchant and citizen of Providence as well as a weather-watcher for many decades, conducted a special survey of meteorological conditions attending the hurricane. His report was deposited in the library of the Rhode Island Historical Society.[32] According to his log, the wind set in from the east at 0900 after having been in the northeast with rain for the previous 24 hours. Soon it veered to east-southeast and by 1000 was howling straight out of the southeast. From 1000 to 1130 Brown described the storm as "tremendous." Another contemporary estimate placed Merchant Brown's losses from the hurricane at the sizable figure for the day of one million dollars. Fortunately, there was little loss of life reported—in direct contrast to the tragedies of 1938 in this area. Two persons were killed during the storm at nearby India Point, but we hear of no others.[33]

Sidney Perley is at his best in *Historic Storms of New England* when describing the rush of wind and water up Narragansett Bay and into the center of Providence:

The force of the gale was principally and most severely felt in Narragansett bay in Rhode Island. The wind swept the bay, and Providence suffered from its effects more than any other place. From ten to half-past eleven o'clock it blew a hurricane. About the wharves and lower part of the town generally confusion reigned. High water was about half-past eleven o'clock in the forenoon, and the wind brought in the tide ten or twelve feet above the height of the usual spring tides, and seven and a half feet higher than ever known before, overflowing and inundating streets and wharves. The vessels there were driven from their moorings in the stream and fastenings at the wharves, with terrible impetuosity, toward the great bridge that connected the two parts of the town. The gigantic structure was swept away without giving a moment's check to the vessel's progress, and they passed on to the head of the basin, not halting until they were high up on the bank. All the vessels were driven ashore, or totally destroyed. There were wrecked in the cove four ships, nine brigs, seven schooners and fifteen sloops. After the storm they lay high and dry, five or six feet above high-water mark, in the streets and gardens of the town. One sloop stood upright in Pleasant street before the door of a Mr. Webb, and a ship was in the garden of General Lippet. Nine of the vessels that were driven ashore were successfully launched again, but more than thirty were totally lost.

The owners of the stores, wharves, and other property in the inundated district exerted themselves to the utmost to save all they possibly could from destruction, but with little success on account of the terrible violence of the gale. The water was rising rapidly, trees were falling, chimneys were crashing through roofs of houses or into the street, tiles and railings from the tops of buildings, and other dangerous missiles were flying through the air, making it hazardous to be out of doors.

The storm raged with increasing violence, and the water was rapidly rising and deluging the lower parts of the town. Wharves were being washed away, stores and other buildings on them were about to leave their foundations, and the water surged around the houses of the people who resided in the lower sections. All considerations of property soon gave way to a more important concern. Every one in the more exposed parts of the town became solicitous for his own personal safety and that of his family and friends. Stores and dwelling houses were seen to reel and totter for a few moments, and then plunge into the deluge. A moment later their fragments were blended with the wrecks of vessels, some of which were on their sides, that were passing with great rapidity and irresistible impetuosity on the current to the head of the cove, to join the wrecks already on the land.[34]

[1] Noah Webster. *Diary*. II, 197.
[2] Noyes Darling. Notice of a Hurricane that passed over New England in September 1815. *Amer. Jour. Sci.*, 42, 1842, 243.
[3] *Ibid.*, 245.
[4] *Idem.*
[5] Sag Harbor, 24 Sept, in *N. Y. Spectator*, 30 Sept 1815.
[6] George R. Howell. *The Early History of Southampton*. 2d ed. Albany, 1887. 194.
[7] *N. Y. Spectator*, 30 Sept 1815.
[8] *N. Y. Post*, 25 Sept 1815.
[9] *N. Y. Spectator*, 23 Sept 1815.
[10] Norwich, 27 Sept, and New London, 26 Sept, in *Courant* (Hartford), 4 Oct 1815.
[11] *Mass. Spy* (Worcester), 27 Sept 1815.
[12] *Farmer's Cabinet* (Amherst, N. H.) *in N. Y. Spectator*, 30 Sept 1815.
[13] *N. H. Patriot* (Concord), in *North Star* (Danville, Vt.), 30 Sept 1815.
[14] *Hampshire Gazette* (Northampton), 27 Sept 1815; *Vermont Journal* (Windsor), 25 Sept 1815.
[15] *Pittsfield Sun*, 28 Sept 1815; *Vermont Gazette* (Bennington), 25 Sept 1815.
[16] Ms. Met. Reg. (Yale).
[17] *Columbia Register* (New Haven), 2 Oct 1815.
[18] Samuel Rodman, Ms. Met. Reg. (Harvard); Wm. Mitchell, Ms. note in Redfield Log (Yale), 1844–45, C-2.
[19] *Boston Gazette*, 25 Sept 1815.
[20] Lyme, River Head, 25 Sept, in *N. Y. Spectator*, 30 Sept 1815.
[21] Letter of Daniel Putnam, 24 Sept, in *Boston Herald*, 6 Nov. 1938.
[22] *Mass. Spy*, 27 Sept 1815.
[23] *Ibid.*, 4 Oct 1815.
[24] *Boston Gazette*, 25 Sept 1815.
[25] *Courant*, 4 Oct 1815.
[26] *Salem Gazette*, 26 Sept 1815.
[27] John B. Beck. Observations on Salt Storms. *Amer. Jour. Sci.*, 1, 1819. 389.
[28] *Mass. Spy*, 4 Oct 1815.
[29] John Farrar. An account of the violent and destructive storm of the 23d of September 1815. *Memoirs Amer. Acad. Arts & Sciences*, 4–1, 1821, 92–97.
[30] C. R. Stark. *Groton 1705–1905*. 418.
[31] E. S. Goodwin. Gale of September 1815. *Mass. Hist. Soc. Coll.*, 2nd Ser., 10, 1823, 45. Also *Boston Journal*, 1, May 1824, 370.
[32] W. R. Staples. *Annals of the Town of Providence*. 379–82.
[33] *The September Gale of 1869*. Providence, 1869. 23.
[34] Sidney Perley. *Historic Storms of New England*. Salem, 1891. 190.

THE NORFOLK AND LONG ISLAND HURRICANE OF 1821

On the late afternoon of 3 September 1821 a tropical storm of full hurricane intensity smashed across extreme western Long Island with the center moving onshore partly within the limits of present-day New York City. The landfall appears to have been near Jamaica Bay close to the present site of New York International Airport (Idlewild). Crossing through Nassau County, the center of lowest barometric pressure traversed Long Island Sound on approximately the Oyster Bay-Stamford axis, and then raced northward through the hill country astride and to the east of the New York-Connecticut border. Available

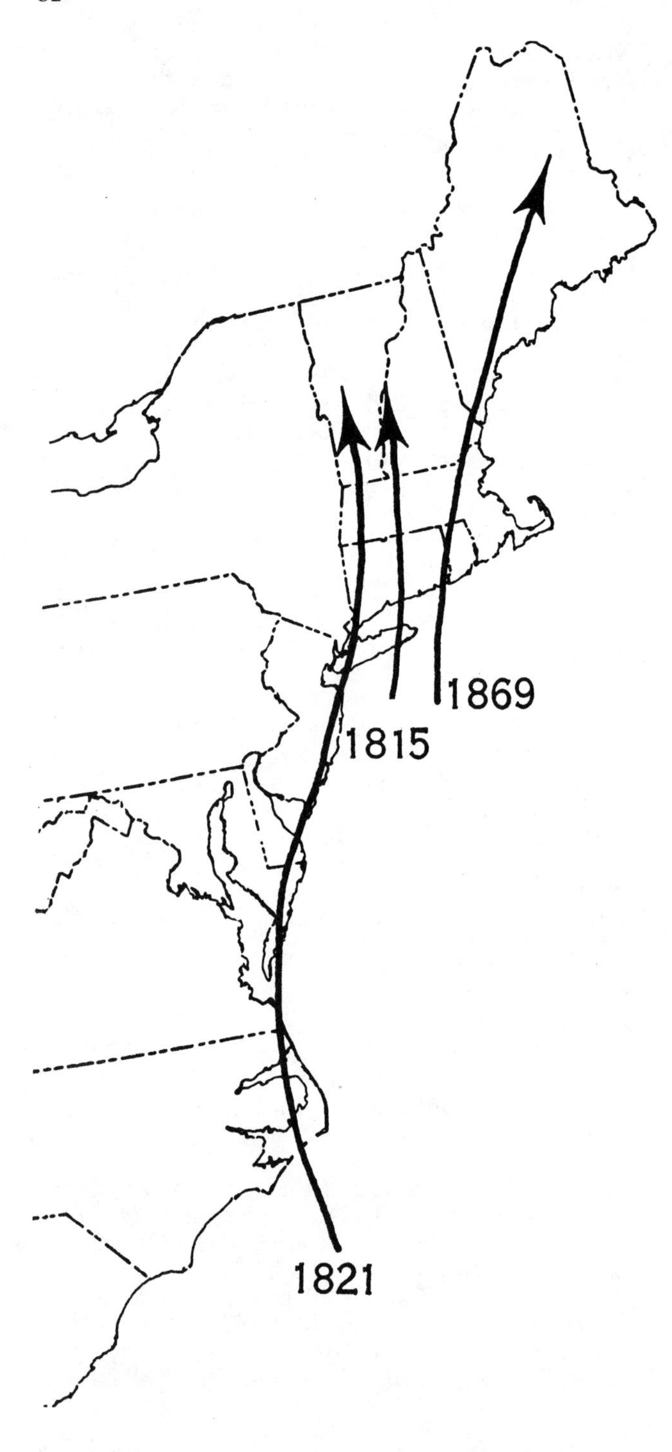

weather records indicate that this is the only *major* hurricane whose center actually passed over a part of New York City in the last 200 years.

The event provided a notable meteorological and scientific occurrence for the storm moved over a well-populated area where a number of observant people, mindful of the Great September Gale only six years before, were able to take careful notes of the storm's characteristics. Thus, the 1821 hurricane provided firsthand laboratory material for students of the atmosphere, material which ultimately initiated the first careful analysis of an American storm. Disagreements over the evidence were destined to spark controversies at scientific meetings and to enliven editorial pages for many years to come.[1]

In the weeks following the storm William C. Redfield, then the keeper of a small store at Cromwell near Middletown, took a trip that carried him across most of central and western Connecticut. He noted that the trees in the vicinity of his home in the central part of the Nutmeg State had fallen toward the northwest, while in the northern reaches of Litchfield County near Canaan the tree trunks were all pointing toward the southeast. Redfield concluded that the storm must have had a rotary wind system, contrary to the generally held belief that winds in a moving storm system were rectilinear.[2]

Several years later Redfield chanced to encounter Prof. Denison Olmsted of Yale College while on a steamboat returning from New York City, and they struck up a conversation on a popular topic of the day—the hurricane of 1821. The Yale scholar's keen interest in the unschooled Redfield's original observations and conclusions led in July 1831 to the publication in the *American Journal of Science* of the latter's classic paper: "On the Prevailing Storms of the Atlantic Coast." This marked the start of serious discussion in this country of the nature of storms. The celebrated Redfield-Espy feud of the 1830's and 1840's soon followed.

Our first indication of the existence of the September 1821 Hurricane came from the vicinity of Turks Island in the southeastern Bahamas where shipping reported its presence on September 1st.[3] Before dawn of the 3rd the storm raced inland through the Cape Lookout and Cape Fear region of the exposed southeastern North Carolina shoreline. Available wind reports indicate that the center passed west of Ocracoke Inlet at the entrance to Pamlico Sound where a severe east-southeast gale raged at daybreak of the 3rd. "It is supposed," said a report from there, "if the gale held to SE for 2 hours more, there would not have been a living thing on Portsmouth Island." [4]

The wind report from Edenton on Albemarle Sound, about fifty miles inland from the ocean, would indicate that the center passed to the east of that place:

> On Monday last a little before day it began to rain which was accompanied by a light breeze from the NE, but before 8 o'clock the wind increased to a gale, and blew with tremendous violence until about 9 o'clock, when it veered around to NW and blew with increased violence until about half after 11 o'clock. Before 1 o'clock the gale entirely ceased, and left only a few flying clouds.[5]

The above would indicate that the storm center was abreast of Edenton and Kitty Hawk about 0900 when the wind shifted from northeast to northwest. With the center passing over the bay and swamp country of eastern North Carolina just inside Cape Hatteras, the exposed island settlements suffered severely. At Currituck Island, guarding the entance to Albemarle Sound, all but a half-dozen houses were blown down and several persons killed. On Mattamuskeet a total of 70 houses fell victims to the hurricane force winds. "So tremendous a gale is not remembered by the oldest inhabitant," commented the editor of the *Edenton Gazette.*[6]

Next in the path of the "awful hurricane" lay the active Virginia seaport of Norfolk and the lower Chesapeake Bay complex, and fortunately we have good eye-witness accounts of both the storm's meteorological characteristics and of the great damage inflicted on the shipping and installations there. According to local press reports, rain commenced to fall at 0600 on the morning of the 3rd. By 0800 a gale raged from the northeast and gradually increased in intensity, reaching full hurricane force from 1130 to 1230. About noon a shift to northwest took place when the center was directly east of the city. The rain ceased at 1230, and by mid-afternoon the sun was shining and only a moderate breeze disturbed the atmosphere. The impact of the storm on the community was vividly captured by the editor of the *Norfolk Herald*:

> From half past 11 till half past 12, so great was the fury of the elements, that they seemed to threaten a general demolition of everything within their reach. During that period the scene was truly awful. The deafening roar of the storm, with the mingled crashing of windows and falling of chimnies—the rapid rise of the tide, threatening to inundate the town—the continuous cataracts of rain sweeping impetuously along, darkening the expanse of vision, and apparently confounding the 'heaven, earth, and sea' in a general chaos; together with now and then a glimpse, caught through the gloom, of shipping, forced from their moorings and driven with rapidity, as the mind might well conjecture in such a circumstance, to inevitable destruction. Even to those, if there were any, who could contemplate such a scene unappalled, it must have been painful to reflect on the widespread devastation which could not but be the result of this fearful 'war of the elements.'[7]

The editor of the local *American Commercial Beacon* compiled a quick survey of the damage which he estimated ran to $200,000:

> Very few house-keepers have escaped injury, either in their enclosures or houses, and nearly all of the most highly improved lots in the Borough have been despoiled of their attractions, by the prostration of their walls or fences, the uprooting of trees, whose sturdy trunks and luxurious foliage have heretofore defied the utmost fury of the tempest, or the partial destruction of gardens, many of which have been cultivated with uncommon care, and were about to reward their tenders with abundant supplies.
>
> Part of the front of the Episcopal Church blown in and the organ entirely demolished and much of the interior ruined—Court House partially unroofed—two chimnies of the Farmers Bank blown down—the back windows of the Bank of Virginia shattered—many private dwellings lost roofs, chimnies, and sustained major structural damage—several new houses demolished completely—ground stories of the warehouses on the wharves as high up as Wide Water Street were overflowed and their stores damaged considerably—drawbridge across the Elizabeth River entirely swept away—damage to shipping was immense—four new steamboats broke their moorings and carried up the Eastern Branch—U. S. Frigates Congress and Gurriere grounded—innumerable brigs, schooners, and smaller ships capsized, sunk, and grounded.

The low-lying communities surrounding Chesapeake Bay and on the Delmarva Peninsula suffered greatly. "The crops are laid low, and the country exhibits one scene of widespread desolation and ruin," observed the *Beacon*. At Pungoteague the water rose 10 feet and caused "unexampled destruction."[9] A packet schooner off Chincoteague experienced a southeast gale and was thrown on her beam-ends as the hurricane swept up the coast. The gale was reported to be at its height about 1600 at Poplar Island in the latitude of Annapolis; the wind direction then was given as north.[10]

The next check-point of the northward-rushing whirl came at Cape Henlopen, at the mouth of Delaware Bay about 140 miles north-northeast of Norfolk. The gale had commenced there at 1130 from the east-southeast, shifted twenty-minutes later to east-northeast, and blew hard for three hours. A calm of about thirty minutes followed. Then the wind picked up from the west-northwest and raged with still greater violence than before. At Cape May, 15 miles to the northeast across the Bay, the wind commenced from the northeast, but veered to the southeast at 1400 and blew with great violence. Cape May had a 15-minute calm, after which the gale recommenced and blew with great force for two hours.[11] It seems apparent from these reports that the center passed over these two points at the mouth of Delaware Bay—probably directly over Cape Henlopen and slightly west of Cape May where the calm was of shorter duration and the winds had previously been from the southeast.

The impact of wind and tide at the neck of Cape May in New Jersey has been related by Charles Ludlam of South Dennis who witnessed the hurricane passage at that exposed point:

> The morning of September 3d, 1821, commenced with a light wind from the west, there was nothing in the looks of the atmosphere that indicated bad weather. At about 9 o'clock the wind hauled round to the southeast, steadily increasing. At 11 o'clock it might be called a gale, at 12 it was blowing a hurri-

> cane with intermittent gusts that drove in doors and windows, blowing down outbuildings, trees, fences and overflowing the marshes between the beach and mainland several feet. At this time it was difficult to stand without some support; no clouds were to be seen, but in their place was a universal haze like a thick fog. The salt spray of the sea was driven inland some miles so as to kill vegetation. At about 10 o'clock [Ludlam in error here; 1430 approximately correct] it fell perfectly calm for about fifteen minutes, then the wind suddenly burst out from the northwest the directly opposite quarter, and blew with increasing violence for about three hours, then gradually subsided, and by six o'clock had nearly ceased and cleared off at sundown. The evening was as clear as the morning, but oh dire was the devastation it left in its progress. Vessels foundered, driven ashore, or dismasted, woodland nearly ruined by being broken down or blown up by the roots, the writer of this had a favorite weeping willow that made three-quarter of a cord of wood that was blown down by the southeast wind and when it came out, northwest blew it over to the opposite. Cape Island lost from 16 to 20 feet of its bank, and what is most singular, a ship anchored that evening immediately opposite the present breakwater that carried top gallant sails all day and knew nothing of the hurricane. The vortex or center of this cyclone as laid down in Blunt's Coast Pilot, struck our continent at or near the point of the Cape and passed over the center of the county and could not have exceeded 50 miles in width as it was but partially felt in Bridgeton and Salem. It was a providential circumstance that it was low water and a low run of tides, otherwise it would have been calamitous in the extreme in the loss of life and property.
>
> On our bay shore the tide was higher than on the seaside of the Cape by several feet; persons who witnessed the overflow said it came like a perpendicular wall some five feet high driven by the wind when it changed to the northwest and came in an overwhelming surgle. From the formation of the land in the cove in our bay, in the vicinity of Goshen and Dias Creek and Cedar Hammocks, the water was concentrated as a common center and the tide was higher there than anywhere along the shore; drift was lodged in the tree tops at the Cedar Hammocks nine feet high; in all probability the heave of the sea had something to do in this.[12]

The storm, now racing north-northeastward at about fifty miles an hour, roared through southern New Jersey only a short distance inland from the coast. At Tuckerton, 55 miles northeast of Cape May, the main storm struck at 1400 with a violent gale from the southeast which continued extreme for two hours. Extensive structural damage resulted at the small port on Little Egg Harbor, and other communities along the coastal strip must have suffered likewise.[13] The presence of southeast winds at Tuckerton would place the movement of the center as roughly parallel to the northeast-trending coastline of New Jersey and very close to the present route of the Garden State Parkway.

To the westward, the Philadelphia area also had a destructive gale, but the direction in that sector was steady northeast. Trees and chimneys were blown down in the Quaker City, and the roof of the covered bridge at the Upper Ferry blew away. All reports to the west mention very heavy rainfall—3.92 inches fell at Philadelphia. The temperature remained at an even 73° all afternoon.[14] Both Baltimore and Washington experienced heavy precipitation, but there was no mention of high winds at either place.[15]

The New York Metropolitan area lay directly in the path of the onrushing hurricane. The editor of the Newark *Centinel of Freedom* supplied the following meteorological summary: rain commenced at 0900—wind was fresh from the south to southeast all day—about 1700 a shift to northeast took place, and it "blew violently" until 1830 when the gale was at its peak—there was little diminution until about 1930 "at which time the wind suddenly veered to northwest and quickly swept away the clouds, and presented the stars brightly twinkling in the face of heaven." [16]

The arrival and behavior of the storm in the New York City area was described in the *New York Post* under the heading:

> TREMENDOUS GALE
>
> From Saturday morning till 4 o'clock yesterday afternoon (Monday), we were visited with repeated and copious showers of rain, accompanied by some loud peals of thunder and lightning, and an extreme dense atmosphere; the wind during the time veered and shifted to almost every point of the compass, when about half past four o'clock yesterday afternoon it came out from almost east, with all the violence and fury of a hurricane, and continued to about half past 8 last evening, throwing down chimnies, unroofing buildings, and prostrating trees in various directions. When the gale was at its height, it presented a most awful spectacle. The falling of slate from the roofs of buildings, and broken glass from windows, made it unsafe, for anyone to venture into the streets.[17]

There were no instruments, of course, to measure the speed of the wind that afternoon and evening as the hurricane moved over New York City. From the evidence of structural damage cited in the press, at least full hurricane force winds were experienced at the surface of the ground with perhaps even greater speeds indicated at higher levels such as church steeples and the topmasts of ships. The famous windmill that served as a landmark at Bergen Point in Jersey City sustained immense damage with all its sail arms blown away, its small fly wheel torn off, and roof and upper structure shattered and splintered in many places.[18] Unfortunately, no complete set of satisfactory meteorological records for 1821 have been located for New York City. One barometer, as reported by Redfield, dropped from 30.13″ at 0600 to a low of 29.38″ at 1930.[19] Rainfall for the day measured only 0.87 inches in the Laight records.[20]

A brief resume of the damage in the New York area condensed from press reports follows:

> All wharves along the North River were damaged and many of their frames pulled from their foundations—the steamboat dock at Marketfield Street destroyed—the Battery inundated and the earth along the embankment washed away back to the first row of trees—Mr. Rabineau's bath houses on White's wharf were torn to pieces—Whitehall Dock had a sloop dismasted and a ferry boat sunk—at Coenties Slip a sloop sank and two other were dismasted and lost their bowsprits—only two vessels out of twelve at Quarantine rode out of the gale, the others went ashore—at the Public Store Dock there were 12 large vessels ashore, at the Fountain Ferry four, and at the Kilns five—in the upper part of town some houses were unroofed and others blown down—ten cows were killed on Broadway when a building collapsed—one wing of the Ball Alley crumbled before the wind pressure—many trees in the Park were prostrated.[21]

To the eastward on Long Island, in the violent eastern semicircle of the storm, the scene was "the most awful and desolating ever experienced," according to the Brooklyn *Long Island Star*. Damage in the City of Churches was on a par with that across the river in New York City, and eastward in Nassau and Suffolk Counties the havoc increased.[22] All along the South Shore, exposed to the southeasterly hurricane, ships were either completely wrecked or driven high and dry on the sandy stretches of Long Beach or Rockaway Beach. A disaster "of the most melancholy nature" occurred when a coastal sloop sank with a loss of 17 passengers and crew. A Huntington editor reported so much devastation in that area, with everyone having some loss to report, that he thought it useless to chronicle individual calamities. Many barns and smaller buildings were laid completely flat by the wind blasts. Crops suffered severely.[23]

The center of the storm moved across Long Island Sound and into Fairfield County of Connecticut, inflicting great damage in the Stamford and Greenwich area where several houses were blown down. At Bridgeport, on Long Island Sound 45 miles east of Manhattan Island, the high winds commenced about 1800 from the southeast, to reach a climax at 2100 "when the tempest raged with a degree of fury the most awful and destructive," according to the local *American Farmer*.[24] Shipping along the Sound near Bridgeport was spared complete destruction by the happy coincidence at 2100 of low tide at the time of the peak wind gusts. A shift to southwest occurred soon after 2300, and then the tidal threat commenced to subside. Damage also extended up the Hudson Valley some distance where the *Westchester Herald* told of strong winds and extremely high tides along the river.[25]

In his original article describing the storm, William C. Redfield believed that the storm center came ashore east of Bridgeport, but he later admitted that he had experienced difficulty in getting reliable wind data from any point between New York City and New Haven. In a letter written to Redfield in 1839, a minister living at Stratford on the Sound described how all of his many poplar trees fell almost due north toward the conclusion of the storm which came there "a little before 10 P.M." [26]

The impact of the hurricane in the Bridgeport area is well described by the editor of the *American Farmer*:

> The storm at this place lasted until near 1100 o'clock at night. Several buildings have been blown down and unroofed—chimneys and windows demolished—the largest trees torn up by the roots—fences destroyed—and the roads rendered impassable in many places by the trunks and branches of prostrate trees. The spire of the steeple of the Presbyterian meeting house, together with the lightning rod, vane, and ball, were blown down. On the water the scene of desolation is yet more appaling. . . . We are sorry to learn that the Lighthouse at Black Rock is destroyed. Many years must elapse before Bridgeport regains its former appearance.
>
> We hear from Trumbull, Newtown, Washington &c. that the gale was equally destructive in the interior. The destruction of fruit and forest trees was immense.[27]

Prof. Benjamin Silliman at New Haven described the southeast gale "as gradually increased from noon to dark, when it raged with tremendous violence, and continued until nearly midnight. It terminated very abruptly, and passed in a very short time from hurricane to a serene and starlight night." The wind shift there also was through south to southwest.[28] Just to the north, at Middletown, extensive damage was reported:

> So far as our own observation has extended, we have never known a wind of equal duration, by any means so violent. The wind commenced about 7 o'clock and continued to nearly 12. It blew down the high spire steeple of the Episcopal Church—it took about one half of one side of the roof of the Methodist Meeting House—it partially unroofed and blew down several barns and outhouses in the city and its vicinity. The gale was principally from the south-east quarter, veering ocasionally to different points between south and east. The Methodist Church at North Haven was partially destroyed—the brick school at Wethersfield down—at Newington the steeple of the Episcopal Church blown down.[29]

Farther west the storm spread havoc through northern Fairfield and Litchfield counties as it rushed northward. The editor of the *Litchfield Miscellany* described the event:

> The gale of wind on Monday night last was probably the severest ever experienced in this vicinity. It blew from the South East, and everywhere left marks of its rage in prostrate fruit and forest trees, fences, sheds, etc.
>
> In the Society of Northfield, in this town, it blew down five or six sheds in a short distance of each other. The steeple to the South Farms Meeting House [now Morris in the southern part of Litchfield Township], was almost blown over. The top of it was bent some six or seven feet, so that the whole appeared considerably jostled from its fastenings.[30]

The central New England towns of Hartford, Springfield, and Amherst were all east of the track of the center and close enough to the path to receive maximum force winds. The editor of the *Connecticut Mirror* at Hartford analyzed the reports of the storm in his vicinity:

> We were visited by a gale of wind on Monday evening which was more severe than any which we recollect to have witnessed before, not excepting the September gale of 1815. It began to blow in this city at 7 o'clock in the evening—the wind to S.E. In New Haven it began an hour earlier, in New York it commenced at 5 o'clock with the wind to N.E. In Philadelphia the storm began as early as 2 o'clock in the afternoon. It appears not to have been very severe farther south than New Jersey, nor farther east than New London. In Worcester and Boston the gale was not violent and did but little damage. In the western part of Connecticut there was little wind, but a very heavy fall of rain which did great damage to bridges and roads. We have heard from most of the towns on the Connecticut River for the distance of 50 to 60 miles and the storm seems to have been about as violent as it was here.[31]

An observer at Springfield described the storm, which lasted there from 2100 to 2400, as "principally wind there being little rain during the night." [32] At Ellington in Connecticut northeast of Hartford, John Hall, a veteran weather observer, had "a most violent south wind from 8 o'clock till 2 at night with some rain." He added: "A great deal of damage by it to fences, orchards, & forests & considerable to many buildings. Almost all fruit blown off fruit trees. Great scarsity of winter apples in consequence." [33] And up in Massachusetts, Noah Webster noted that "a violent hurricane from the SE commenced in evening. It prostrated trees and overset some sheds and houses. It was violent at Amherst and as far north as Brattleborough." [34] Farther north the editor of the *Vermont Republican & American Yoeman* at Windsor commented: "In this vicinity the wind did not blow so violently, as there described; but the rain fell in great abundance, and the roads were much damaged." [35]

The farther course of the storm northward of Massachusetts cannot be traced. The Windsor report would indicate that the main effect of the storm came from rain, and this is confirmed by observers at Bennington, Vermont, and Williamstown, Massachusetts. At Williams College a "violent rain in the evening" fell on the 3rd with wind at both 1400 and 2100 out of the southeast—2.97 inches of rain fell during the storm period which lasted two days.[36] At Bennington, Benjamin Harwood thought the weather "threatening," and "very gloomy," but mentioned no wind.[37] It is likely that the storm had commenced to lose its tropical structure when over coastal New Jersey and its passage over western New England completed its transformation to an extra-tropical type. The dying hurricane could still cause gale force winds as far north as northern Massachusetts with very heavy rains continuing northward.

In eastern Massachusetts at Boston a short but severe blow was noticed in the late evening; damage was limited to fruit on the trees awaiting harvest.[38] Edward Holyoke at Salem listed "a most violent squall" at 0300 on the 4th, but made no mention of storm damage.[39] In southeastern New Hampshire, at Epping, Governor Plumer's records told of a gale from the southeast commencing at 2200 and lasting five hours. It blew down corn, but spared trees as the soil was dry and the roots held.[40]

To return to Connecticut and the northern part of Litchfield County, it was there that William C. Redfield noticed that the fallen trees near North Canaan were lying toward the southeast, while in the central part of the state their tops were facing northwest.[41]

> In reviewing these facts, we are led to inquire how, or in what manner it could happen, that the mass of atmosphere should be found passing over Middletown for some hours, with such exceeding swiftness, toward a point apparently within thirty minutes distance, and yet never reach it; but a portion of the same or a similar mass of air, be found returning from that point with equal velocity? and how were all the most violent portions of these atmospheric movements which occurred at the same point of time, confined within a circuit whose diameter does not appear to have greatly exceeded one hundred miles? To the writer there appears but one satisfactory explication of these phenomena. *This storm was exhibited in the form of a great whirlwind.*[42]

Thus was stated the proposition that started the serious investigation of the dynamics of our American storm systems.

[1] James P. Espy. Facts collected by Mr. Espy, taken from newspapers of the time. *Jour. Franklin Institute* (Phila.), ns., 23, 1839, 156.

[2] William C. Redfield. Remarks on the prevailing storms of the Atlantic Coast of the North American States. *Amer. Jour. Sci.,* 20, July 1831, 17–51.

[3] *Nat. Int.* (Wash., D. C.), 13 Oct 1821.

[4] Washington, D. C., 14 Sept, in *Nat. Int.*, 22 Sept 1821.

[5] *Edenton Gazette,* 10 Sept, in *Amer. Beacon* (Norfolk), 18 Sept 1821.
[6] *Idem; Nat. Int.,* 17 Sept 1821.
[7] *Norfolk Herald* (extra), 4 Sept 1821.
[8] *Amer. Beacon,* 5 Sept 1821.
[9] *Ibid.,* 13 Sept 1821.
[10] *Nat. Int.,* 10 Sept 1821.
[11] *U. S. Gaz.* (Phila.), 8 Sept 1821.
[12] Charles Tomlin. *Cape May Spray.* Phila., 1913. 70–71.
[13] Letter from Tuckerton, 6 Sept, in *N. J. Mirror* (Mt. Holly), 12 Sept 1821.
[14] Redfield, 25.
[15] *Nat. Int.,* 4 Sept 1821; *Niles National Register* (Balto.), 8 Sept 1821.
[16] *N. Y. Post,* 4 Sept 1821.
[17] *Centinel of Freedom* (Newark), 11 Sept 1821.
[18] *Idem.*
[19] W. C. Redfield. Remarks on Mr. Espy's theory of centripetal storms . . . *Jour. Franklin Institute,* ns. 23, 1839, 331n.
[20] Henry Laight, Ms. Met. Reg. (National Archives).
[21] *N. Y. Post,* 4 Sept 1821.
[22] *Long Island Star* (Brooklyn), 6 Sept. 1821; also 13 Sept.
[23] Huntington. 6 Sept, in *N. Y. Post,* 7 Sept 1821.
[24] *American Farmer* (Bridgeport), 5 Sept 1821.
[25] *Westchester Herald* (Mt. Pleasant, N. Y.), 11 Sept 1821.
[26] W. C. Redfield. *Jour. Franklin Inst.,* ns., 23, 331n; W. C. Redfield. Ms. Letter Book, 1, 157. (Yale).
[27] *Amer. Farmer,* 5 Sept 1821.
[28] Benjamin Silliman. The Tempest of Sept. 3d, 1821. *Amer. Jour. Sci.,* 4, 1822, 171–72.
[29] Middletown, Conn., 6 Sept, in *Haverhill* (Mass.) *Gazette,* 15 Sept 1821.
[30] *The Miscellany* (Litchfield), 8 Sept 1821.
[31] *Conn. Mirror* (Hartford), 10 Sept 1821.
[32] *N. Y. Spectator,* 8 Sept 1821.
[33] John Hall. Ms. Met. Reg. (Yale).
[34] Noah Webster. *Diary.* 2, 182.
[35] *Vermont Republican & American Yoeman* (Windsor), 17 Sept 1821.
[36] Ms. Met. Reg. (Williams College).
[37] Benjamin Harwood. Ms. Diary. (Bennington Museum).
[38] *Columbian Centinel* (Boston), 5 Sept 1821.
[39] Edward Holyoke. Ms. Met. Reg. (Harvard).
[40] William Plumer. Ms. Met. Reg. (Harvard).
[41] J. H. Redfield. *Recollections of John Howard Redfield.* Phila., 1900. 45–46.
[42] W. C. Redfield. *Amer. Jour. Sci.* 20, 21.

THE EARLY JUNE HURRICANE OF 1825—II

A severe storm of tropical origin swept up the Atlantic Coast during the first week of June 1825 with reports of major damage spread all the way from Florida to New York City. Shipping logs told of a disturbance at Santo Domingo on 28 May and at Cuba on 1 June.[1] No doubt, these, especially the second, were directly related to the disturbance that moved northeastward along the United States seaboard. All indications point to the storm's having achieved full hurricane force winds at many points along the coast.

Gales were reported at St. Augustine on the 2d, but they did not commence at Charleston until about 1000, Friday morning, June 3d.[2] Trees and fences were leveled at the South Carolina city, but, unfortunately, we do not have a full meteorological report. At Fort Johnston at the mouth of the Cape Fear River in North Carolina, however, the weather log mentioned a south wind lasting 30 hours, a direction indicative that the center of the disturbance passed to the west of that point. Very heavy damage reports appeared in the North Carolina press. The tide rose 14 feet at some places—only winds with an easterly component could have caused such a storm tide on the Carolina coast.[3]

Farther north at Elizabeth City, the *Star and Intelligencer* noticed that the storm commenced on Friday afternoon from the northeast, increased until midnight when it was blowing a "heavy gale," and continued until about 1000 on Saturday morning, the 4th. Trees in the vicinity of the northeastern North Carolina commercial center suffered severe damage, some being pulled right out of the ground by their roots. The prevalence of northeast winds at Elizabeth City would place the center of the disturbance as curving to the northeast and heading seaward.[4]

As in the disastrous hurricane of 1821 which followed a track very similar to the present storm, the port of Norfolk lay close to the path of the oncoming center. We have a very informative account of the day from the columns of the *Norfolk and Portsmouth Herald*:

REMARKABLE STORM

It is uncommon to hear of violent storms and hurricanes on any part of our extensive coast, in the month of June; but we have to notice a visitation of stormy weather, which commenced about 9 o'clock on Friday night, rarely, if ever equalled within the memory of the oldest inhabitant. The storm of the 3d of September, 1821, was perhaps rather more violent, but it lasted only three or four hours, while this continued with undiminished violence, from the hour we have stated until 12 o'clock on Saturday night, or about 27 hours. It commenced raining early on Friday morning, and continued without any intermission the whole of the day, and during the prevalence of the storm. The wind at the commencement of the storm was at N.E., and so continued until about 12 o'clock on Saturday, when it began to haul round gradually to the Northwest and Westward, and held up at S.W. Sheltered as we are from the Northerly and Easterly winds, neither the town, nor the shipping in the harbor suffered much from the fury of the tempest; a few fences blown down, trees uprooted, and gardens spoiled, comprise the whole of its ravages, but considerable damage was done by the high tide which rose at least eighteen inches higher than it was ever known to be within the last 40 years. The highest pitch of the tide was at 12 o'clock on Saturday, at which time the stores on the wharves generally were inundated from the depth of 3 to 5 feet, and the water extended up to the doors on the N. side of Wide Water street. The whole Town Point to within a few feet of Main street was overflown, as also was that part of the town extending Eastward from Market Square to the Draw Bridge, the water rising considerably above the line of Union street. In most of the stores on the wharves, all articles liable to be damaged by the tide were lifted, as it was thought, above its reach, but in every instance it was found (too late for remedy) that the precaution was unavailing in consequence of the unusual rise of the tide, and the articles were of course damaged. We understand, however, that the loss in no individual instance is very considerable amount.[5]

The effect of the storm reached well inland. Washington had a cold, heavy rain all day Saturday with high winds laying the crops in the vicinity prostrate. The wind rush at Philadelphia tore up trees by the roots in front of the State House.[6]

The northward drive of the hurricane carried along the New Jersey Coast and into the New York and Long Island area. In the metropolitan suburbs the press reported the storm as commencing at 0700 on Saturday morning, reaching its most violent phase between 1900 and 2300, and not subsiding until 0400 on Sunday morning. A heavy rain fell constantly throughout. The height of the gale just preceded the time of high water, though the indicated rise of two feet above normal was not exceptional. Nevertheless, considerable damage to shipping in the harbor came from the combination of wind and tide. A Columbian frigate *Venezuela* was driven ashore as were many smaller boats. The largest loss of life occurred along the Long Island shore when the schooner *Hornet* out of Truro on Cape Cod foundered with the loss of her entire crew of seven.[7]

Throughout the New York area countless trees were prostrated by the hurricane force winds. One editor commented about the unusual phenomenon: "We never recollect a storm in June of equal severity or duration. It was like a regular and furious equinoctial, than any thing else; and being unlooked for, has, no doubt, been more extensively calamitous."[8]

The Nantucket weather report had rain all day Saturday with falling barometer and northeasterly winds. At sunrise on Sunday, however, the barometer had reached its lowest point at a stated observation, 29.57", and the wind had shifted into the southeast and by noon was at south-southeast. Temperatures had risen from the mid-50's on Saturday to 61° at sunrise. All would indicate that the center of the June hurricane passed well to the west of that island outpost.[9]

[1] *N. Y. Ship List,* 25 June & 9 July 1825.
[2] SG; *Nat. Jour.* (Wash.), 14 June 1825.
[3] SG.
[4] *Star and Intelligencer* (Elizabeth City) in *Nat. Jour.,* 9 June 1825.
[5] *The Norfolk and Portsmouth Herald,* 6 June 1825.
[6] *Nat. Jour.,* 9 June 1825.
[7] New York, 5 June, in *Nat. Jour., 7* June 1825.
[8] *Idem.*
[9] *Nantucket Inquirer,* 13 June 1825.

TWIN HURRICANES FEATURE THE SEASON OF 1830—II

The hurricane season of 1830 produced a goodly number of storms. Most of the activity, however, centered off the coast where shipping suffered severely. Every issue of the *New York Shipping List* carried a large number of disaster accounts.

The first major storm of the season cut across eastern North Carolina on Tuesday, August 17th, raising havoc with shipping on the bays and rivers of the Albemarle Sound region and also spreading destruction on land to buildings and crops.[1] The center appeared to veer eastward and passed out to sea south of Norfolk. Off the Delaware Capes we have reports that three ships were found capsized at the end of a "tremendous gale" which commenced at 1300/17th and continued for six hours before shifting to northwest. The lightship at Five Fathoms Bank off the Delaware Capes broke its moorings early on the 17th.[2]

William Redfield observed the northeast storm effects at New York City. He noticed at the height of the gale there in the late afternoon that the western horizon was clear and that the setting sun shone on the canopy of clouds to the eastward, making a fine spectacle. This observation furnished his first intimations as to the inland extent of a typical coastal northeaster. On Long Island Sound, at Stratford, Connecticut, although promising cloud banks appeared, no rain fell at this time to alleviate a growing drought.[4]

The gale force winds next swept the islands and capes of southeastern Massachusetts. The *Nantucket Inquirer* thought the storm on the 17–18th "one of the most severe gales experienced for a long time." The peak winds occurred from midnight to 0100 and "almost amounted to a hurricane." Some small buildings were blown down and trees uprooted. Winds backed from NE to NNE, and to N. The lowest barometer at an observation time occurred at 0700/18th with a reading of 29.66".[5] Damage to trees, fruit, and buildings also was reported from Barnstable on Cape Cod.[6] The *New Bedford Courier* carried a dispatch from Holmes Hole: "We have seldom been visited with a more severe gale than was experienced here last night. It commenced blowing fresh from about 9 o'clock and continued to increase to about 12, when it blew a heavy gale. Roof blown off house."[7]

Redfield listed further accounts of the hurricane on the night of the 18th when it smashed into Nova Scotia.

The second coastal hurricane followed a course somewhat to the east of its predecessor, with its arc of curvature also to the north and east. A gale was noticed in the press at St. Barts in the northern Windward Islands on 20 August.[8] Redfield placed the storm near Turk's Island on the 22nd and next day in the northern Bahamas.[9] From the various ship logs he collected, it would appear that the center of the hurricane passed northward close to the 70° W meridian when off Cape Hatteras on the 24–25th. A ship near there reported a severe gale at east-northeast of 42 hours duration, an indication that the storm was now rather slow-moving and of considerable size.[10] Another ship off Cape May gave the continuance of the gale at 40 hours on the 25–26th. No damage reports have been noticed from any mainland location prior to its arrival in southeastern New England.[11] Thomas Robbins at Stratford on western Long Island Sound entered in his diary: "We had a hard storm of wind and a little rain."[12]

The storm's curving path to the northeastward carried quite close to the extremities of southeastern New England. The *Nantucket Inquirer,* without giving specific details, judged the storm "very severe."[13] Mr. Folger's barometer on the Island dropped to 29.26" at the 2200 observation on the 26th and had risen to only 29.33" at 0700 the next morning. Wind had gone from east-northeast to northwest at this time, showing that the center had accelerated and was now well past Nantucket.[14] Samuel Rodman's barometer at New Bedford also reflected the same relative curvature. Rainfall at New Bedford amounted to 2.76".

The *Barnstable Patriot* commented that the second severe storm within a week had completed the job of "the fury of the gale of the 17th by knocking the remaining fruit from the trees."[16] Cape Ann and the Salem area also received the sweep of winds; at the latter there were "very high winds yesterday and last night. It blew off most of the fruit."[17]

These August storms served as an introduction to a very active tropical storm season on the Atlantic. The southeastern area of the Grand Banks and Newfoundland had stormy weather on 20 September and again on the 24th.[18] Whether these were true hurricanes is not known. Another disturbance took form in the Windward Islands late in the month, passed northwestward near Bermuda, and struck the Grand Banks another blow on 2 October.

In October a third coastal hurricane swept the Atlantic seaboard. It was noticed at Fort Johnston in North Carolina on the 6–7th, as a east-northeast gale in Chesapeake Bay on the 9th, and again at Barnstable on Cape Cod on the 11th.[19]

[1] *N. Y. Mercury,* 1 Sept 1830.
[2] *N. J. Mirror* (Mt. Holly), 1 Sept 1830.
[3] W. C. Redfield. *Amer. Jour. Science,* 20, July 1831, 35.
[4] Thomas Robbins. *Diary,* 2, 189.
[5] *Nantucket Inquirer,* 21 Aug 1830.
[6] *Barnstable Patriot,* 21 Aug 1830.
[7] *New Bedford Courier* in *Patriot,* 28 Aug 1830.
[8] *N. Y. Ship List,* 29 Sept 1830.
[9] W. C. Redfield. *Am. Jour. Science,* 31, Jan 1837, 126.
[10] W. C. Redfield. Ms. Record Book. (Yale).
[11] *Idem.*
[12] Thomas Robbins. *Diary,* 2, 191.
[13] *Nantucket Inquirer,* 28 Aug 1830.
[14] *Idem.*
[15] Samuel Rodman. Ms. Met. Reg. (Harvard).
[16] *Barnstable Patriot,* 28 Aug 1830.
[17] J. B. Felt. *Annals of Salem,* 2, 121.
[18] W. C. Redfield. *Amer. Jour. Science,* 20, July 1831, 41.
[19] SG; *N. Y. Ship List,* 16 Oct 1830; *Barnstable Patriot,* 20 Oct. 1830.

THE ATLANTIC COAST HURRICANE OF LATE AUGUST 1839

A vigorous hurricane moved up the Atlantic Coast on the 28th, 29th, and 30th of August 1839. Though it did not cross the coastline at any point, its path was not far offshore and the sweep of winds in its western semicircle were of such force to cause considerable crop damage in North Carolina and eastern New England, in addition to buffeting coastal shipping from Charleston to Nova Scotia. On Cape Cod the storm was "the most severe for many years," and in eastern Maine it was thought the greatest blow since the Great September Gale of 1815.

Willis Gaylord, a scientific agriculturalist and associate editor of the *Genesee Farmer* of Rochester, became the self-appointed historian of this hurricane. He had contributed many interesting articles on meteorology to his magazine as an avid follower of Redfield. Drawing his material from the contemporary press, he published a timetable of the storm's onset and conclusion in his rural journal.[1]

His first check-point for the gale came at Charleston, South Carolina, where the northeast winds commenced at 1900/28th, shifted to northwest at 0600/29th, and lulled finally at 1330/29th. No damage to shipping resulted from the 18-hour storm.

The center apparently passed very close to Cape Hatteras about noon of the 29th. Gaylord's timetable put Wilmington, North Carolina, about four hours after Charleston in time of the storm's commencement and in the wind shift from northeast to northwest.

The Washington (North Carolina) *Republican* had east-northeast winds on the 28th which gradually increased until the morning of the 29th when they "blew with Tremendous violence."[2] The wind, then shifting to north-northwest, drove the piled-up waters of Albemarle Sound back completely over the low-lying islands that form the Outer Banks of North Carolina. At Ocracoke Bar the gale was "more severe than for many years past." Of the 15 vessels there at the onset of the gale, only three escaped unharmed.[3] Just to the north the Elizabeth City *Phoenix* reported the height of the gale on Friday morning, the 29th: trees were down and bridges out in that area with no stages operating to Norfolk.[4]

At Norfolk the storm raged from just after midnight of the 29th until 2100/30th, indicating an expansion in the size of the disturbance. The newly-built breakwater at Delaware Bay sheltered 66 ships during the period of high winds. There were no accounts of damage at either Philadelphia or New York, but exposed places on Delaware Bay such as Lewes and Cape May reported ships ashore. At Sandy Hook the floating light or lightship was set adrift on the 30th when its anchoring cables snapped under the stress of wind and wave.[5] Gaylord noted that the high winds were confined to the coastal plain, that "eighty miles inland it was scarcely felt."[6]

As the storm curved eastward with the coastline, southeastern New England received a vigorous punch from the northeasterly gales when the center sideswiped Cape Cod and the Islands. At sunrise of the 30th a cold mist was falling at Providence with a chill east wind. Rain began at 1000 as the wind freshened. With the rain continuing throughout the afternoon, the wind became "very heavy" and at night reached "almost a gale." Alexis Caswell's barometer on the Brown College campus dropped as low as 29.64" by sunrise of the 31st. His rainfall catch totaled 3.00".[7] At nearby Mattapoisett on Buzzard's Bay, Rev. Thomas Robbins noted "the wind last night was very severe . . . cold and had fire in my room. The late storm has been severe and disastrous."[8]

Out at Nantucket, much closer to the path of the tropical whirl, the wind came out of the northeast all day on the 30th with heavy rain falling and the temperature hanging in the cool fifties. William Mitchell thought: "we have not had so severe a gale for many years."[9] The *Inquirer* reported a long list of marine disasters with many ships driven ashore both on the Islands and on Cape Cod. At his 0700 observation on the 31st, Mitchell's barometer was down to 29.48", the temperature stood at 54°, and the wind had backed to north-northwest with a drizzle continuing to fall. "On land the trees, shrubbery, corn, and other vegetables have suffered greatly," commented the *Inquirer*.[10]

The Salem historian, J. B. Felt, witnessing the scene while compiling his monumental account of local events, thought it "a more violent northeast storm than has been experienced for a considerable period. It was destructive to trees, fences, and shipping."[11] Boston Harbor, fortunately, escaped any serious damage to shipping. Down East it was another matter. The course of the storm and the curving coastline tended to converge. Whole trees were blown down and the fine Maine apple crop beaten from the trees. Corn and other crops were laid low. The *Maine Farmer* was of the opinion that the storm caused more damage than any since 1815.[12] Farther along at Yarmouth in Nova Scotia strong easterly gales did their work of destruction and drove several ships ashore.[13]

Gaylord concluded his study with a significant observation on the chilly condition of the atmosphere during this intense coastal hurricane: "One of the singular facts connected with this gale was the cold it produced as was evidenced by the fall of snow in many places along the coast and some distance in the interior. The Catskills were whitened; and considerable fell at Salem, and other places."[14]

[1] *Genesee Farmer* (Rochester), 9–41, 12 Oct 1839, 323.
[2] Washington (N. C.) *Republican* in *N. Y. Ship List,* 7 Sept 1839.
[3] *N. Y. Ship List,* 11 Sept 1839.
[4] Elizabeth City *Phoenix* in *National Gazette* (Phila.), 7 Sept 1839.
[5] *N. Y. Ship List,* 4 Sept 1839.
[6] *Genesee Farmer,* 12 Oct 1839.
[7] Alexis Caswell. Meteorological Observations made at Providence, R. I. *Smithsonian Contributions to Knowledge*. Wash., D. C., 1860. 38.
[8] Thomas Robbins. *Diary*. 2, 542.
[9] William Mitchell. Ms. Met. Obs. (Maria Mitchell Observatory, Nantucket).
[10] *Nantucket Inquirer,* 4–7–11 Sept 1839.
[11] J. B. Felt. *Annals of Salem.* 2d ed. Salem, 1845–49. 2, 122.
[12] *Maine Farmer* (Augusta), 7–14, Sept 1839.
[13] *N. Y. Ship List,* 25 Sept 1839.
[14] *Genesee Farmer,* 12 Oct 1839.

THE MEMORABLE OCTOBER GALE OF 1841

One of the most intense hurricanes of the century raced northward through the shipping lanes off the Middle Atlantic seaboard on the morning of 3 October 1841, and arrived almost unheralded soon after noon on that fateful Sunday to catch a large portion of the Cape Cod fishing fleet on their favorite grounds southeast of Nantucket. It is still known as "the October Gale" along coastal Massachusetts. Its terrible reputation has lived long in the memories of Cape Codders since so many seafaring youths of the small ports along Massachusetts Bay and Nantucket Sound went to watery graves on that wild night on the Georges Bank. The easterly gales, quickly rising to full hurricane strength, prevented the small vessels from turning the Cape and making safe port behind Provincetown. They were either overwhelmed and submerged by the gigantic seas or were dashed to pieces in the cruel breakers along the beaches.

When one visits the down-Cape town of Truro today, he will immediately see an impressive, plain marble shaft rising from a solid brownstone base. On its face is inscribed the simple statement:

Sacred
To the memory of
FIFTY-SEVEN CITIZENS OF TRURO
who were lost in seven
vessels which
Foundered at sea in
the memorable gale
of October 3, 1841

Nantucket, the closest land mass to the path of the storm, bore the full brunt of the hurricane winds on the 3d and 4th. Fortunately, we have the excellent record of the local weatherman, Hon. William Mitchell, who with his daughter Maria maintained an astronomy observatory on the island and took a keen interest in all current scientific matters. Mitchell wrote on the day following the event to his friend and frequent correspondent, William C. Redfield of New York City:

> Our island having been visited with far the most disastrous storm known in its whole history, I send herewith, before learning anything from the continent, my meteorological observations during the whole period. It commenced at 8 A.M. of the 3rd, and its termination may be stated at 8 P.M. of the fourth. The height of the wind, the duration of its violence, & its disastrous effects are without parallel in this region. Sixteen wrecks have already been described from the observatories, though the weather is still very thick.[1]

The center of the hurricane remained offshore during its entire advance along the Atlantic coast. It passed close enough to the Outer Banks of North Carolina to drive several ships ashore on the 2d and to cause similar beachings at Cape Henry in Virginia on the same day.[2] The western portion of the circulating winds raked the coastal shipping lanes with destructive gales. The storm track ran approximately parallel, but many miles to the east of the course of the Great September Gale of 1815. Thus, no major New England city received the full force of the intense disturbance. But Cape Cod, the Islands, and Cape Ann, all suffered severely, much more than they had in the celebrated gale 26 years before.

Writing in 1924 in his classic *The History of Nantucket,* Alexander Starbuck stated that "on October 3, 1841 occurred what was doubtless the severest gale recorded in the history of the island." [3] This judgment had been documented by Arthur H. Gardner in his *Wrecks Around Nantucket* (1915): "On the morning after the gale, nineteen vessels lay stranded on or near the island, while within sight of shore, the masts of two others protruded from the water—such a sight was never witnessed before or since upon the island." [4]

The editor of the *Nantucket Inquirer* also took notice of the effect of the storm on the island and islanders:

> During the night of Sunday, especially, when every building trembled under the pressure of the furious elements, there were but few families free from alarm and consternation. On that night, literally, not many slept without rocking.
>
> * * *
>
> A great number of chimneys, some of them from buildings nearly new, were thrown down by force of the wind. The walks upon the roofs of some thirty dwelling houses in various quarters of the town were blown off.—Trees of large dimensions, flag-staffs, fences, and other exposed objects, were prostrated. The tide rose to a height almost unprecedented—reaching from two to three feet above the surface of the wharves, and extending into most of the lower streets, strewing in various directions quantities of lumber, cord wood, and other buoyant objects.[5]

Mitchell also transmitted his complete meteorological log to Redfield. It showed the storm at its height from soon after midnight to sunrise (the town clock stopped at 0129 on the 4th), as the wind slowly backed from northeast to north and the barometer reached its lowest point:

Oct. 3	0800	30.00	51	ENE	Very squally
	0900	29.91	50	ENE	Gale
	1000	29.80	51	ENE	Gale increasing
	1200	29.74	52	ENE	Gale very high
	1300	29.72	52	ENE	Increasing
	1400	29.70	51	ENE	Unabated
	1600	29.68	50	ENE	Unabated
	1700	29.63	49	ENE	Unabated, heavy rain
	1800	29.63	49	NE	Unabated, heavy rain
	1900	29.64	49	NE	Gale increasing, heavy rain
	2000	29.59	48	NE	Unabated heavy rain
	2100	29.58	48	NE	Increasing heavy rain

Oct. 4	0200	—	—	NNE	Gale terrible
	0600	29.18	49	N	Gale terrible
	0800	29.32	50	N½W	Abates
	1000	29.42	51	N½W	Abates
	1200	29.50	52	N½W	Ordinary heavy
	1400	29.55	50	N½W	—
	1600	29.68	49	NbyW	NbyW now moderate
	1800	29.73	49	NbyW	Abating
	2000	29.80	48	NbyW	Terminates[6]

Both Martha's Vineyard and the Cape experienced the same type of destruction as was meted out to Nantucket. At Edgartown the causeway joining the lighthouse with the main island was carried away, and shipping suffered extensively though to a lesser degree than on the exposed windward shores of Nantucket and outer Cape Cod.[7] The beach from Chatham to near Provincetown was literally strewn with parts of wrecks. Between forty and fifty vessels went ashore on the sands there, and fifty dead bodies were picked up. At Hyannisport several vessels were cast on the shore; in one of the boats the entire crew, eight in number, were found in the cabin. From Dennis twenty-six young men were lost in various wrecks during the two-day storm.[8] The Barnstable salt works were severely injured by wind and tide, and the wall of an oil works back at New Bedford collapsed during the height of the wind pressure.

An interesting account of the impact of the storm on his home town was given by "B" of Harwich, a middle-Cape town facing southward on Nantucket Sound which was a lee shore during the northeast gale. This was printed in the *Yarmouth Register* in 1869 when the next major gale in the area's history came along:

> Harwich, Sunday, Oct. 3, 1841.—A very severe storm of wind and rain from the north-east which has continued through the day, steadily increasing in violence. No meeting.
>
> Monday, Oct. 4, 1841.—Storm raging with dreadful fury. Wind N.N.E., constant rain. Clouds driving over have a remarkable appearance—low, black, rolled together like wisps of hay, or fleeces of wool, or more like the waves of the sea in a gale. Trees, fences, and small buildings prostrated. Near night put up the fences to protect garden and cornfield; but cattle keep so still there is no need of fences. Storm reminds one of a dismal winter evening, rather than a season of the year generally mild and pleasant.
>
> Tuesday, 5th.—Wind entirely subsided, but cloudy and wet. We survey the effect of the storm. All vegetables of any height laid flat to the ground; even the large leaves of cabbages torn off and strewed about fences. Everywhere the ground is covered with the green leaves of fruit and forest trees. Roads leading through woods are thickly carpeted with leaves and small branches. Everywhere are seen trees uprooted, or large limbs hanging by a few splinters. But the great disaster is among the shipping. The number of vessels gone clear from our shore—Deep Hole, Marsh Bank, Herring River, and Bass River—dragging anchors or parting cables, is said to be thirty-six.
>
> This afternoon several vessels have put out to go in search of the lost. The prospect of the sea is mornful, a solitary vessel here and there, heading outward on this errand, or coming in slowly disabled.

Farther north where Cape Ann juts out from the mainland the storm sideswipped with strong northeast gales the many small harbors that dot the promontory. The breakwater at Rockport, erected in 1832 at a cost of $17,000, gave way before the onslaught of the waves. Many ships in the vicinity were thrown on the beaches for total losses. According to the *Gloucester Telegraph,* the damage on Cape Ann to ship and shore installations ran to $40,000.[10] At Salem the storm raged for three days without let-up, according to its local historian.[11] The waves at the Isle of Shoals off the New Hampshire coast were thought to have been the highest in the memory of the oldest residents of that exposed spot.[12] At the more protected port of Boston a strong northeast gale prevailed outside on Sunday night but "no damage of consequence was done in the harbor."

Since the October Gale wrought its greatest destruction at sea, it will be fitting to recount the adventures of a small vessel from Truro, the *Garnet,* which met the raging hurricane full on, yet fortunately survived to tell the tale. Sidney Perley has vividly described the nautical nightmare of those caught at sea that fearful day and night. On Sunday morning the *Garnet* was heading for Georges Bank. It soon breezed up and the light sails were taken in. At noon it had increased to a gale and the mainsail was furled. At four the crew took in the jib, and at six double-reefed the foresail though it soon tore away. By sounding they knew they were being carried close to the shoals where the large breakers would be very dangerous. Perley described their predicament:

> . . . The gale increased every moment, and at ten o'clock a heavy sea tore away the boat and davits . . . The foresail again gave out, was repaired and again set, but as soon as it was up the wind was so terrific that it was blown to ribbons. The mainsail soon shared the same fate, and the jib only was left . . . The sea was breaking over the vessel fore and aft, and the captain advised the crew to go below. All but the captain and his brother did so . . . The helm was put up, and just as she began to fall off, a tremendous sea, or breaker completely buried the vessel, leaving her on her broadside or beam-ends. Zach, the captain's brother, was washed overboard, but he caught hold of the main sheet and hauled himself aboard. The foremast was broken about fifteen feet above deck, the strain on the springstay hauled the mainmast out of step, and tore up the deck, sweeping away the galley, bulwarks and everything else, and shifted the ballast into the wing. A sharp hatchet had been kept under the captain's berth, to be used in case of an emergency. This he soon found, and to it fastened a lanyard, which was tied to a rope

that had already been fastened to Zach's waist, the other end being secured to the vessel. Zach went to the leeward, and when the vessel rolled out of the water, he watched his chance, and cut away the rigging. The captain did the same forward, cutting away the jib-stay and other ropes, and by that means relieved the vessel of spars, sails, rigging, sheet anchor and chains. The crew got into the hold through the lazaret, and threw the ballast to windward, so that she partially righted. They were now a hopeless wreck.[13]

The gale abated somewhat by dawn and the crew of the *Garnet* made whatever repairs possible so they could drift with the wind. Even though the hull was almost completely submerged, they made some way toward the southeast and were soon in deep water. All day Monday they somehow managed to keep afloat as the following northwest wind of the dying storm drove them on. On Tuesday morning a vessel was sighted passing nearby, but so low in the water did the *Garnet* ride that they went unnoticed. By mid-morning the seas moderated somewhat and the weather became more pleasant. They were able to open a hatch and found some potatoes floating in the hold—these provided their first meal since Sunday morning. Just before sunset a sail was seen approaching from the east, so a makeshift flag was hoisted on a long pole as a signal. This was soon observed by the lookouts in the mast of the large ship. The storm-weary sailors were then taken aboard and found themselves on the *Roscius,* plying between New York and Liverpool, one of the first merchant ships of her time with 400 passengers aboard. The captain of the *Garnet* learned that his water-filled craft had drifted to a point some 200 miles off Sandy Hook, New Jersey.

Another incident of the storm involved a small coaster which was blown out of Chatham harbor at the onset of the gale from the northeast. Only the cook was aboard at the time. He spent almost 20 days at sea entirely alone before his craft made a safe harbor at Port Mouton, Nova Scotia.[14]

Not only were small ships buffeted by the great gale. The proud Cunarder *Caledonia* encountered the hurricane when off the New England coast on Sunday and put into Halifax to effect repairs. A passenger described the scene:

On Sunday we shipped a heavy sea, which disabled nine of the men, among them the third mate and carpenter, breaking one of the legs of the former in two places, and the thigh of the latter; burst open the hatchway of the fore-cabin, deluging the saloon, and compelling the occupants to evacuate; the other cabins were also wet, although not so badly—but it will be impossible to have them dry and comfortable for the rest of the voyage. One of the boats was carried away, and the life-boat nearly so; part of the bulwarks and covering of the paddle houses were also carried away, and on Monday night the tiller ropes were in the same condition, but their want was supplied in the course of an hour or two.[15]

The duration of the northeast hurricane may be judged by the necessity for the *Caledonia* to stand by for fifty-two hours before she could make any headway toward Halifax.

Though the October gale was primarily a New England storm, the disturbance made itself felt farther south in its sweep up the Atlantic seaboard, remaining far enough offshore so that winds were much less destructive than on the exposed portions of the New England coast. At Washington the 3d was "a remarkably stormy day; the wind blew furiously until midnight." Steamers plying Chesapeake Bay were obliged to heave to and anchor behind a headland near Annapolis all day.[16] At New Brunswick in central New Jersey the storm was "very furious" with rain mingled with hail and snow that night.[17] The wind at New York "blew a complete gale all day and night, carrying away awnings and branches of trees like feathers, and doing much damage to shipping at the wharves and in the harbor." Losses were estimated in the press at $2 million, the greatest storm damage since the Long Island Hurricane twenty years before.[18] At Erasmus Hall in Flatbush the two-day storm brought 3.05″ of rain with cold northeasterly winds. The mercury stood at 38° on Sunday night.[19]

The cold currents flowing southward on the western semicircle of the hurricane probably contributed greatly to the extreme intensity maintained by the tropical cyclone when off the New England coast. At New Bedford, close to the base of Cape Cod, Samuel Rodman's barometer had sunk to a sea level reading of 29.53″ at sunrise of the 4th with a cold north wind blowing. Rainfall there, too, had been heavy—2.87″.[20] Back at Providence, Alexis Caswell recorded the same story. At his first observation on the 4th, at sunrise, with his thermometer reading 34°, he entered the following significant remarks: "Rain and snow. Wind a little west of north and very high. Snow in the air, and also on fences, and the roofs of houses." [21]

The occurrence of snow as early as October 3–4th comprised an outstanding meteorological event. Once before in the century—9 October 1804—a small, but intense hurricane had brought down a blast of polar air from Canada, cold enough to trigger a major snowstorm when supplied with copious amounts of moisture furnished by the tropical storm. Again in October 1841 snow fell to the coast of southern New England under similar circumstances.

At Boston the snow stuck to the ground for only a few minutes, but inland at suburban Waltham one inch was measured and at Concord three inches accumulated.[22] Eastern Connecticut, favorably situated in the cold sector and close to the source of moisture, received the greatest covering. People in the vicinity of Thompson made the event memorable by attending Monday's town meeting in sleighs. Farther south the community of Middletown, located in a singular snowbelt only a dozen miles inland from Long Island Sound, had a fall of 12 inches with depths to 18 inches reported in the surrounding highlands.[23]

Snow fell even at tidewater. In New Haven the rain turned to a "violent storm of snow and sleet" on Sunday evening when the wind hauled to northwest. On Monday morning, although rain had again fallen for several hours preceding, the snow and slush lay three inches deep on the Yale campus and did not entirely disappear until Tuesday morning.[24] Kent in western Connecticut received a 12-inch covering as did most of the hill towns in that area. Down at New York City, William Redfield reported snow mixed with the rain on Sunday night, 3 October, for the earliest snow in the New York records.

[1] Wm. Mitchell to W. C. Redfield. Nantucket, 10th mo. 5th 1841 at 8 A.M. Redfield Mss., Letters Received. 2, 487. (Yale).
[2] *Amer. Beacon* (Norfolk) in *N. Y. Ship List,* 13 Oct 1841.
[3] Alexander Starbuck. *The History of Nantucket.* Boston, 1924. 335–36.
[4] Arthur H. Gardner. *Wrecks around Nantucket.* Nantucket, 1915. 46.
[5] *Nantucket Inquirer,* 9 Oct 1841.
[6] Redfield, 487.
[7] S. Hazard. *U.S. Commercial and Statistical Register* (Phila.), 5–17, 27 Oct 1841, 261.
[8] Sidney Perley. *Historic Storms of New England.* Salem, 1891. 28.
[9] "B," 13 Sept 1869, in *Yarmouth Register,* 17 Sept 1869.
[10] *Daily Advertiser* (Boston), 4 Oct 1841.
[11] J. B. Felt. *Annals of Salem.* 2, 122.
[12] *Maine Farmer* (Augusta), 1841, 334.
[13] Perley, 283–84.
[14] *N. Y. Ship List,* 10 Nov 1841.
[15] *Boston Atlas* in Hazard, 262.
[16] *Nat. Int.* in *Daily Advertiser* (Newark), 12 Oct 1841.
[17] Charles Peirce. *A meteorological account of the weather in Philadelphia.* Phila., 1847. 208.
[18] *N. Y. Tribune,* 5 Oct 1841.
[19] Erasmus Hall. Ms. Met. Reg. (National Archives).
[20] Samuel Rodman. Ms. Met. Reg. (Harvard).
[21] Alexis Caswell. Meteorological Observations made at Providence, R. I. *Smithsonian Contributions to Knowledge.* Wash., 1860. 51.
[22] Boston dispatch in *Daily Adv.* (Newark), 6 Oct 1841.
[23] Dept. of Agriculture. *Monthly Report.* Dec 1867. 395.
[24] *Columbian Register* (New Haven), 9 Oct 1841.

THE GREAT HURRICANE OF 1846—II

The Great Hurricane of 1846 possessed extreme energy in the Key West area, and much of this physical power was sustained and carried northward in its course over Florida, Georgia, and the Carolinas. After passing west of all the principal ports of the South Atlantic States, the storm center traversed east-central North Carolina before bursting into the tidewater areas of Chesapeake Bay. From here northward the gale of 1846 resembled later severe hurricanes such as October 1878, late September 1896, and October 1954.

Wind reports at the peak of the storm at Norfolk, Baltimore, and Philadelphia were from the southeast, indicating that the storm track passed to the west of those points. A southeasterly gale in this region is particularly dangerous since it can raise the waters of Chesapeake and Delaware Bays to their maximum heights. At Baltimore all the commercial wharves were flooded at the height of the storm and much of Pratt Street along the harbor put under water. The Potomac at Alexandria and Washington reached its highest tidal flood mark in twenty years. The lowlands along the Delaware near New Castle were overflowed in the greatest storm surge in 70 years, probably a reference back to the September hurricane of 1775. The waters rose high enough to put out the fire in a locomotive which had been stalled by the rising waters. The wind force at New Castle was sufficient to topple the steeple of the historic Episcopal Church there.[1]

The *New York Sun* gave a timetable for the commencement of the gale at northeastern points: Washington at 0200, New York City at 0800, and Boston at 1400 on the 13th.[2]

At Philadelphia the gale proved "the most destructive storm in 30 years" as the peak winds from the southeast backed up the waters of the Delaware River to flood all the wharves. "The Delaware was lashed into a perfect fury and its roar would have drowned the thunder of Niagara itself." [3] Some structural damage resulted throughout the city though the losses were less than expected from the strength of the gale. The weatherman at the Pennsylvania Hospital noted "a tremendous gale" out of the southeast from 1300 to 1600. His barometer at 1500 read 29.25". The skies cleared by 2000. The rain gage caught 1.25 inches during the storm period.[4] In southern New Jersey many buildings were blown down, trees felled, salt marshes flooded, and cattle and sheep drowned.[5]

Though New York City lay well to the east of the storm track, the wind there was judged the greatest in seven years, or since the famous January storm of 1839. Wind conditions reported in the Metropolitan area indicated that the storm center, after its long traverse over land, might be in the process of becoming extra-tropical in structure. It seemed to be spreading out latitudinally in much the same manner exhibited by Hurricane Donna over Long Island in September 1960. The observer at Erasmus Hall in Brooklyn on the 13th noted in his weather log: "violent gale in the afternoon from the Southeast. Lulled somewhat at sunset. Blew very hard again at 1900 and continued for an hour." [6] At Newark, where the wind was "more violent than any experienced in years," the gale commenced at the southeast. William Whitehead, the local weather observer, called it "a disastrous gale." [7] The waters of Upper New York Bay, spurred by the southeasterly flow, lashed at the wall of The Battery and washed away 100 yards of its bulwark. Trees were uprooted at the height of the gale throughout the city.[8]

Barometers in the Metropolitan area dropped to relatively low readings during the storm's passage, even though at a distance of 100 to 125 miles east of the supposed path of the center. At Erasmus Hall the lowest pressure of 29.15″ was observed "on the evening" of the 13th, while at Newark the low mark of 29.15″ occurred at 1700. The accompanying rainfall of 1.25 inches measured at Newark fell in ten hours following 0800/13th, and ceased shortly after the time of lowest barometer. Farther east at New Haven the mercury column dipped to 29.30″ at 1800 as the wind veered from northeast to southeast and south and finally around to west later in the night. The highest winds occurred around 1800 from the southeast.[9]

In New England it was mainly a severe wind storm. The wide extent of the eastern sector is shown by the comment of the *Yarmouth Register* out on Cape Cod: "The South East gale on Tuesday night was very severe, but we have heard of little loss sustained from shipping." [10] The *Barnstable Patriot* thought it the "most severe and extensive in five years since the great gale of October 1841." Mention was made of the light rainfall accompanying the high gales. At Barnstable the mounting wind reached gale force by 1900/13th and continued high until midnight when it commenced to lull away and wear to the south.[11] At New Bedford the rainfall on the 13–14th amounted to only 0.33 inches. The low point of the barometer there was reached at 2200 when the wind was raging out of the south-southeast at force 9. By sunrise it had dropped away to a mere southwest force 1.[12]

An interesting phenomenon accompanied the gale in west-central Massachusetts. On the road from Springfield to Amherst, just east of the Connecticut River, there is a notch north of the town of South Hadley through which present Route 116 passes northward. Trees on each side of this natural defile were either felled, pulled up by the roots, or their trunks broken off by the funneling effect of the strong southerly gale.[13]

At Norwich in eastern Connecticut the railroad was blocked by numerous downed trees, and at Worcester several factories were blown down. In the Boston area the principal destruction occurred to trees, many being prostrated or uprooted by the dry gale. At Hartford the force of the wind was sufficient to pick up the railroad bridge crossing the Connecticut River, set it down on the surface of the water, and then continue to blow the structure upstream until it grounded.[14]

Throughout western Pennsylvania and western New York very heavy rains accompanied the hurricane's passage as is usual to the west of the center's path. At Buffalo the floods caused a breach in the bank of the Erie Canal and traffic was interrupted. The downpours were especially heavy in the Danville, New York, sector.[15]

[1] An excellent summary of press reports in *U. S. Gaz.* (Phila.), 2 Nov 1846.
[2] *N. Y. Sun* in *Daily Adv.* (Newark), 16 Oct 1846.
[3] *North American* (Phila.) in *Daily Adv.*, 14 Oct 1846.
[4] John Conrad. Ms. Met. Reg. (Penna. Hospital, Phila.).
[5] *Bridgeton Chronicle* in *Daily Adv.*, 19 Oct 1846.
[6] Erasmus Hall. Ms. Met. Reg. (National Archives).
[7] *Daily Adv.*, 2 Nov 1846.
[8] *N. Y. Tribune*, 14 Oct 1846.
[9] *New Haven Palladium*, undated clipping in National Archives.
[10] *Yarmouth Register*, 15 Oct 1846.
[11] *Barnstable Patriot*, 21 Oct 1846.
[12] Samuel Rodman. Ms. Met. Reg. (Harvard).
[13] Maine Farmer (Augusta), 29 Oct 1846.
[14] *U. S. Gaz.*, 2 Nov 1846.
[15] *Rochester Telegraph* in *Daily Adv.*, 16 Oct 1846.

THE OCTOBER HURRICANE OF 1849

A coastal hurricane moved close to Long Island and the New England shore at the conclusion of the first week of October 1849. New York City, New Haven, Providence, and Boston reported a severe northeasterly windstorm on Saturday night, October 6th, with very heavy rain falling, especially on the New England coast. In the shipping channels the usual losses were chronicled in the press. The principal marine disaster occurred near Boston when the brig *St. John*, loaded with Irish immigrants from Galway, struck the rocks off Cohasset and foundered during the stormy night with the loss of 27 passengers. Henry Thoreau chanced to be in Boston at that time on his way to spend a sojourn at the seashore. He visited the place of the wreck on Tuesday morning. His description of the pathetic beach scene formed the first chapter in his celebrated travelogue, *Cape Cod.*[1]

In New York City the rainstorm set in about 1000/6th and increased in severity during the day. At Erasmus Hall, with a wind out of the southeast increasing from fresh to strong, the barometer declined steadily to 29.60″ at 2100.[2] The height of the gale lasted from 2000/6th to 0200/7th, according to *The New York Tribune,* and it was during this time that considerable structural damage resulted throughout the city. By Sunday sunrise the Erasmus barometer read 29.30″ and wind overnight had backed into the northeast. Rainfall there for the two days amounted to 2.43 inches.

The *Tribune* on Monday morning ran a summary of the reported damage and followed up the next day with the statement that the original estimates had not been exaggerated. Probably the most important structure to feel the storm's fury was the new Presbyterian Church on 29th Street which collapsed under the wind pressure. Another victim was the exhibition pavillion at 8th Street and 4th Avenue where two large dioramas by Daguerre were on display. The wind blew in the east side of the flimsy structure, then, rushing through, tore down the large canvas representations. They were said to be 1,000 feet long and worth $5,000.[3]

Out on eastern Long Island the Sag Harbor weatherman for the Smithsonian made four interesting observations on the 7th: his winds backed from east-by-north at sunrise, to east-northeast, north-northeast, and to north at the stated observation times. His successive barometer readings were 29.14″, 29.13″, 29.11″, and 29.35″—the lowest being at 1500. His rainfall for the 33-hour period totaled 6.50 inches.[4] Thus, the storm center must have passed near the eastern tip of Long Island. The same general conditions were repeated across the Sound at Fort Trumbull near New London where the northeast winds mounted to force 6 during the day and a total of 6.15″ of rain fell.[5] The storm also extended inland—Amherst and Framingham in Massachusetts noted it.[6]

To the east at Providence northeast gales were experienced on the night of 6–7th with the heaviest part of the storm coming during the early morning hours of the 7th, though the barometer did not reach its lowest point of 29.09″ until 1500 in the afternoon—the same condition as reported at Sag Harbor.[7] Boston had a severe gale on Sunday. Trees were uprooted and chimneys blown over. All telegraph communication to the southward went out as wires were downed by the winds.[8]

Out at Nantucket the *Inquirer* reported the storm as "very severe" for an hour or two after midnight of the 6–7th. Heavy rain fell. It continued windy all day Sunday as the barometer continued to slump, though the main part of the blow had already passed. No material damage occurred to structures on the island.[9] At Provincetown a severe easterly storm on the 7th took a toll of small fishing vessels which broke loose from their moorings.[10]

All reports mentioned the cold temperatures accompanying the storm. Apparently a very cold air mass overlay Canada as two inches of snow fell at Temple, New Hampshire, on the 7th—Temple lies only 10 miles north of the Massachusetts border.[11]

[1] Henry D. Thoreau. *Cape Cod.* New York, 1908. 3–20.
[2] PR.
[3] *N. Y. Tribune,* 8 Oct 1849.
[4] SI.
[5] SG.
[6] SI.
[7] Caswell, opt. cit., 99.
[8] *N. Y. Ship List,* 10 Oct 1849.
[9] *Nantucket Inquirer,* 10 Oct 1849.
[10] *Barnstable Patriot,* 17 Oct 1849.
[11] *Idem.*

THE TRIPLE TROPICAL STORMS OF 1850

The season of 1850 brought three storms of likely tropical origin into the Northeastern States—each had its own physical characteristics and each created quite different wind and rain patterns. The occurrence of three storms in one season so far north is almost unique—the only other such year that comes to mind occurred in 1954 when Carol, Edna, and Hazel each spread its own variety of disaster from the Carolinas to Canada. This northward shift of the hurricane belt in 1850, as in 1954, spared Florida from its normal summer and autumn tempestuousness or threats of such. The Medical Corps observer at Fort Dallas (Miami) mentioned in his 1850 annual summary that no great windstorms had been experienced on his coast during the usual season.[1] It is perhaps significant to note that the summer of 1850 at Newark, New Jersey, was the warmest in the long span of early records there. Tropical air and tropical disturbances dominated the entire season.

In pursuing our hurricane studies for the year 1850 we are somewhat handicapped by the lack of Smithsonian reports for this year. The admirable system of meteorological reports had been instituted in 1849 under the leadership of Joseph Henry and was greatly expanded the following year. Unfortunately, most of the manuscript records for 1850 were destroyed in the disastrous fire at the Institution in 1865.[2] Thus, we shall have to revert to our former practice of dependence on press reports and what private meterological registers have survived.

THE JULY STORM

A tropical depression moved into the Carolina-Virginia area on 18 July. The weather-watcher for the Philadelphia *North American & United States Gazette,* who compiled many accounts of the storm, stated that the disturbance came inland south of Cape Hatteras, but in view of the absence of any meterological reports for 1850 in the area, we cannot pinpoint the landfall.[3] Ships off Cape Fear and Cape Lookout had experienced a three-day hurricane from all directions of the compass from the 15th to the 18th.[4] Dispatches from Wilmington and Elizabeth City in North Carolina spoke of a "tremendous storm" and "great damage" on the 18th, though meteorological details were lacking.[5]

Maintaining its vigor as it moved northward, the storm gave Chesapeake Bay and Delaware Bay a good lashing with veering gales kicking up high waves and pushing tides to flood stage. At Burlington, on the Delaware above Philadelphia, the press reported strong winds with heavy rain that caused much damage to crops and roads on the night of the 18–19th. The Delaware broke through its embankments at Burlington to flood the lowlands along the river.[6] The heavy rains sent the Lehigh River into its worst flood since 1841, and on the Schuylkill River twenty lives were lost when river boats capsized and sank.[7]

Our Philadelphia weatherman kept close local check on the progress of the storm. On the morning of the 18th the atmosphere was almost calm with a light air from the southeast. By 1400 a steady rain was falling with strong wind from the northeast; by 2200 the blow had increased to "very strong." Sunrise of the 19th saw the wind already shifted to the east-northeast, from which point it continued to veer all morning, reaching south by noon when the storm was probably centered in central Pennsylvania. Over four inches of rain fell at the Quaker City.[8] The observer at the Pennsylvania Hospital noted "a tremendous gale last night, and before morning abated." At 0800 his barometer still read only 29.33" and it must have been much lower during the height of the storm in the early morning hours.[9]

At Newark we have the meteorological account of William Whitehead who commenced his excellent records in 1843 and published a monthly summary in the local press over a 44-year span without missing a single month—528 summaries in good literary style can now be found in the columns of the *Newark Daily Advertiser* from 1843 until 1887. We have met Whitehead before. In 1835, when he was collector of customs at Key West, he maintained meteorological records and witnessed the celebrated hurricane of that year. His 1850 notes follow: "July 18—sprinkling with wind at East—about 2000 a steady rain set in with fresh wind which gradually increased to the fitful gusts of a hurricane, more violent than experienced in July for thirty years or more." His rain gage overflowed its capacity of 4.75 inches, but Whitehead believed that the additional quantity was small from comparison with other gages in the vicinity.[10]

The New York Tribune under the headline, "The Hurricane," described the storm's effect there. The period of highest winds had lasted from 2200/18th to 0600/19th and heavy rain had accompanied the southeasterly blasts. Houses were unroofed and trees stripped of branches. In the harbor all ships dragged anchors, some were driven ashore, and along Long Island many small coasters were wrecked. At Coney Island the Big Tent and all the bath houses were demolished by the driving force of wind and tide together.[11]

The track of the storm carried westward and northward over Pennsylvania and central New York State. All stations in New England reported southeasterly winds and gales and a great rain. There were property and crop damage reports from Connecticut, but not in eastern Massachusetts.[12] Fort Trumbull near New London on Long Island Sound provided one of the few meteorological reports we have. The Medical Corps observer there noted a "great storm & heavy rain" on the 19th. Rain had commenced at 0200—by sunrise the wind was coming out of the southeast at force 5, rising at 0900 to force 8. It remained high from the southeast the remainder of the day and evening with temperatures in the middle and low 70's. When the rain had ceased by 1700, a total of 3.24 inches had collected in the rain gage.[13] A New Haven report stated that the very high winds had commenced to abate soon after 0900/19th with moderately high winds the remainder of the day.[14]

To the northeast at Providence the rain commenced between 0500 and 0600 at a moderate rate, but became heavy accompanied by high winds between 1000 and 1400. By sunset the clouds had broken with patches of clear sky visible. Winds remained in the southeast at all three observations on the 19th. Total rain was given as 1.00 inch.[15]

Farther inland in western New England the rainfall proved excessive, with Vermont being especially hard hit: Burlington had 3.23 inches and Montpelier 5.00 inches. The Windsor *Vermont Chronicle* summarized the flood situation: "In this section of the country, it appears that North and East the flood was somewhat less than in 1830. South and Southwest it was much heavier, and thus was therefore, taken together, the greatest storm and flood that ever occurred here within the memory of man." At the mouth of the Dog River on the Connecticut the waters rose two to three feet higher than in the historic flood in the Vermont hill country of July 1830.[16] When the flood waters reached Hartford, Connecticut, they crested at 16′ 7″, causing $100,000 damage in the city and vicinity.[17]

[1] SG.
[2] Lewis J. Darter, Jr. *List of Climatological Records in the National Archives.* Special List No. 1. The National Archives, Washington, 1942. xlvi.
[3] *North American & U. S. Gazette* (Phila.), 2 Aug 1850.
[4] *N. Y. Ship List,* 31 July 1850.
[5] *N. Y. Tribune,* 22 July 1850.
[6] *Cultivator* (Albany), Aug 1850, 281.
[7] *N. Y. Tribune,* 22 July 1850.
[8] *North American,* 2 Aug 1850.
[9] John Conrad. Ms. Met. Reg. (Penna. Hospital, Phila.).
[10] *Daily Advertiser* (Newark), 23 July 1850.
[11] *N. Y. Tribune,* 20 July 1850.
[12] *N. Y. Ship List,* 31 July 1850.
[13] SG.
[14] *N. Y. Tribune,* 20 July 1850.
[15] Alexis Caswell. *op. cit.* 103.
[16] *Vermont Watchman* (Montpelier) in *Vermont Chronicle* (Windsor), 30 July 1850.
[17] *Daily Adv.,* 23 July 1850.

THE AUGUST STORM

A hurricane, after passing near the western end of Cuba on 22 August, moved northward through the eastern Gulf of Mexico, and struck the Apalachicola region with exceptionally high storm tides on the 23rd. Its course then carried overland through Georgia and the Carolinas. Both Charleston and Wilmington felt the lash of high winds on the 24th, but sustained no material injury. Several ships were in distress off Cape Hatteras where the gale out of the east-southeast made rounding the Cape a difficult adventure for sailing ships. All reports from coastal shipping speak only of southeasterly gales. The center must have moved well inland over eastern North Carolina.[1]

The Norfolk and entire lower Chesapeake Bay region took a hard beating on Saturday night, the 24th, with southeasterly gales prevailing. It was judged the worst storm in 30 years on the bay, probably a reference back to the great September hurricane of 1821.[2] The wheelhouse of the ship *Osceola,* plying the bay waters, was stripped from the deck and blown overboard by a wind gust. Norfolk's winds commenced in the southeast, but later veered around as the storm moved to the westward, ending up in the west.[3] Farther north along the shores of the bay the gales laid low crops and small buildings. At Baltimore trees were blown down. At the entrance to Delaware Bay a southeast gale commenced at 0400/24th and continued all day and into the night until a shift to northwest took place.[4]

At Philadelphia the fringe of the hurricane was not felt until soon after midnight of the 24–25th. Gale force winds howled out of the northeast until 1100/25th when they backed to north.[5] At Newark in northern New Jersey the winds also came out of the northeast and north, so the center of the disturbance must have curved northeastward and seaward after crossing inside the Delaware Capes and moved off the Jersey Coast toward New England. Newark had a very wet day on the 25th with over three inches falling from soon after midnight until 1400.[6]

We have no meteorological reports from Long Island that would indicate the path of the center of this disturbance. Shipping in Long Island Sound was reported in the press to have suffered, but no detail of the winds was given.[7] Both New London and Providence in New England had southeasterly gales at the mid-afternoon observations on the 25th, with excessive rains falling during the day: Ft. Trumbull 4.43 and Providence 2.50 inches.[8] Later on the winds at both places veered through south to west, so the center certainly passed to the westward over Connecticut and brought very heavy rains to all the western New England region. The observer at Litchfield, in the hills of Connecticut, described August "as distinguished for great rains." [9]

[1] See "Severe Storm at Apalachicola."
[2] *Daily Advertiser* (Newark), 27 Aug 1850.
[3] *N. Y. Ship List,* 4 Sept 1850.
[4] *Ibid.,* 31 Aug 1850.
[5] *North American & U. S. Gazette* (Phila.), 2 Sept 1850.
[6] *Daily Adv.,* 2 Sept 1850.
[7] *Ibid.,* 27 Aug 1850.
[8] SG; Caswell, *op. cit.,* 104.
[9] J. L. Hendrick. *Report of the Commissioner of the U. S. Patent Office, 1850.* 378.

THE SEPTEMBER STORM

The third major disturbance of the season passed up the coast, apparently well offshore, on 7–8 September 1850. Many ships were in distress off the Delaware Capes as a result of the gale there on Sunday, the 8th, "more violent than for several years." [1] In New York City the winds were of sufficient strength to strip trees of branches and limbs.[2] Newark had high winds and a total of 2.60 inches of rain, to raise the season's total toward record levels.[3]

Offshore the *U. S. Relief* apparently passed through the eye of the system on the night of the 7–8th when at 39 02N, 72 57W, or approximately 140 miles east of Atlantic City, New Jersey. The shipboard barometer read 29.18″ as the wind hauled from east-southeast to west-northwest.[4]

The center must have passed just south of Nantucket near noon of the 8th. A strong blow from the southeast in the morning suddenly shifted to north at 1000 and increased to "a perfect gale" at noon. The storm abated toward evening and was almost calm by 1830. Minor damage occurred, but no serious shipwrecks.[5] Over on the Cape at Barnstable the wind rose early Saturday evening and "the line storm" began. "For a few hours Sunday it blew a gale and rained in torrents," reported the *Patriot,* but damage was confined to crops.[6]

Inland at Providence the wind during the period came from the northeast, the barometer dipped to 29.43″, and rainfall totaled 2.00 inches.[7]

The offshore hurricane, bypassing all United States land areas, apparently moved inland over Nova Scotia on Sunday night. Much damage was reported in the press there after the telegraph wires, downed in the storm, had been repaired.[8]

1 *Yarmouth Register* (Mass.), 26 Sept 1850.
2 *N. Y. Tribune,* 9 Sept 1850.
3 *Daily Advertiser* (Newark), 9 Sept. 1850.
4 *N. Y. Ship List,* 14 Sept 1850.
5 *Nantucket Inquirer,* 9 Sept 1850.
6 *Barnstable Patriot,* 17 Sept 1850.
7 Caswell, *op. cit.,* 104.
8 *Yarmouth Register,* 12 Sept 1850.

THE COASTAL HURRICANE OF SEPTEMBER 1854

A strong northeasterly hurricane sideswiped the Middle Atlantic and New England coasts on 10–11 September 1854. This was the same disturbance that had lashed the Savannah and Charleston areas with their largest storm since 1804. The center moved out to sea in the Hatteras region in pursuing a northeasterly course.[1]

Winds at all land points in the Northeastern area were easterly. The gale proved of sufficient force to uproot trees at Philadelphia and New York and to litter the streets with branches and limbs. In the Williamsburg sector of New York City a building was flattened at the height of the blow.[2] At Newark, New Jersey, William Whitehead recorded that the gale commenced at 2130/9th and blew fresh all day the 10th accompanied by a heavy rain until 1300. A total of 2.60 inches were measured.[3]

Along Cape Cod and Long Island winds were east-northeast to northeast, at gale force. Sag Harbor near the easterly tip of Long Island had force 6 at the 1500 and 2100 observations on the 10th. The barometer there read 29.60″ at 2100, the lowest at any of the three daily observations.[4]

At Nantucket winds mounted on the 10th to force 8 from east-northeast at 2100. The barometer stood 29.78″ at that time, also the lowest at any observation time. By next morning it was still blowing force 6 out of the northeast though the rising barometer now stood at 29.95″. The 10th on Nantucket was described as "a stormy day." Thunder occurred around 1400.[5] The Smithsonian weather observer measured 6.06 inches during the storm.[6] A ship on Nantucket Shoals about 40 miles east of the island was thrown on its beam-ends at 2300/10th by the northeast hurricane then raging.[7]

The wind effects of the offshore storm did not reach far inland. Providence had a heavy rain all day the 10th, but force 3 from the northeast (on a scale of 1 to 4) was the highest wind rating achieved in Alexis Caswell's records. His rainfall, amounting to 3.45 inches, had ended by 2100/10th.[8]

1 See "The Great Carolina Hurricane of 1854."
2 *N. Y. Tribune,* 11 Sept 1854.
3 *Daily Advertiser* (Newark), 2 Oct 1854.
4 SI.
5 *Idem.*
6 SI; also *Nantucket Inquirer,* 11 Sept 1854.
7 *N. Y. Ship List,* 16 Sept 1854.
8 Caswell, *op. cit.,* 128.

THE CHARTER OAK STORM OF AUGUST 1856

This tropical storm bears some resemblance to Hurricane Diane of 1955. In fact, both occurred on identical days of the same month, just ninety-nine years apart. The disastrous floods of 1856 were repeated in the equally disastrous inundations of 1955, though not in exactly the same pattern.

The disturbance of 1856 was first noticed in the press over Virginia. Washington, D. C., had east and southeast winds all day Tuesday, the 19th, with a heavy downpour of rain. Wind speeds, however, were not worthy of special mention. New York City also had heavy showers of a tropical type on the 19–20th; a catch of 3.32 inches was measured.[1]

At Sag Harbor on eastern Long Island winds were light to moderate at all three observations on the 19th (east force 3 was the highest), and the same held for the 0700 and 1400 readings on the 20th. During this period of light air flow from the southeast, the barometer dipped steadily to the low point of 29.06".[2] Over on the South Shore of Long Island, however, press reports spoke of ships being driven ashore in the Quogue area by southeast gales.[3] Nantucket, too, had "a heavy gale on the night of the 19–20th." At the 0700/20th observation the wind was raging out of the southeast at force 7 and remained in that quarter until after 2100. The barometer at reading time dropped to 29.50" and then perhaps lower.[4] A southeast gale swept Horse Shoe Shoals off Cape Cod at 0800/20th.[5]

Inland Providence had southeasterly winds on the night of the 19–20th with a steadily falling barometer. A continuous rain setting in late on the 19th dropped a total of 2.60 inches before ending on the 21st. In the evening of the 20th the wind at the Rhole Island capital backed around to northeast, with the barometer continuing to descend. By 0700/21st it had reached a low point of 29.13".[6] At Boston, too, the barometer dropped to a low figure for August: 29.36". Northeast winds prevailed there at the time of lowest pressure.[7] Southeasterly winds on the 19th at both Pomfret, Conn., and Mendon, Mass., shifted into the northeast on the 20th. Heavy rains fell: Pomfret 3.23" and Mendon 2.90".[8]

Analyzing the observations from Long Island and Nantucket and comparing the wind behavior there with that on the mainland would indicate that a tropical low pressure area, without a well-organized center or a very steep pressure gradient, moved northward from Virginia to New Jersey and then eastward close to the shore of Long Island, probably continuing east-northeastward to cross the neck of Cape Cod into Massachusetts Bay and beyond. Nantucket definitely remained in the southern sector of the system, while the backing winds at Providence and Boston would place those points in the northern sector. Winds were unusually light, considering the low state of the barometer, on the northern side of the storm; only on southern Long Island, Nantucket, and Cape Cod did gale force winds prevail. Cape Ann northeast of Boston also reported squally weather with several shipwrecks taking place offshore. The main effect of the storm over the mainland came with very heavy rains that contributed to a growing flood situation.

The major casualty of the storm involved the famous Charter Oak at Hartford, a witness to the founding of the Connecticut Colony in 1636 and the secret repository of the Connecticut Charter during the troublesome Andros regime in the 1680's. This sturdy oak was thought to have dated from pre-Columbian days and to have withstood many a New England hurricane and storm. It is ironic that it fell victim to a relatively mild hurricane situation. The tree was reported to have snapped off about six feet from the ground at 0045/21st—perhaps the winds were higher at this midnight hour than our available weather reports at 2100/20th and 0700/21st indicated.[9]

[1] Lorin Blodget. *Climatology of the U. S.* 395.
[2] SI.
[3] *N. Y. Ship List,* 23 Aug 1856.
[4] SI.
[5] *N. Y. Ship List,* 23 Aug 1856.
[6] A. Caswell, 140.
[7] *Evening Traveller* (Boston), 21 Aug 1856.
[8] *Conn. Courant* in *Daily Adv.* (Newark), 21 Aug 1856.
[9] SI.

THE NEW ENGLAND TROPICAL STORM OF 1858

Ships off the Carolina Capes experienced heavy gales from the northeast and east-northeast on 14–15 September 1858.[1] No other traces of this disturbance have been uncovered except from points on Long Island and New England.

The storm center made a landfall somewhere in central Long Island. The Smithsonian weather observer at Sag Harbor near the eastern tip noted his wind out of the northeast all daylight on the 15th, but during the evening it commenced to veer to east and by sunrise next morning, the 16th, was out of the southeast and blowing strong. By 1500/16th his barometer had dropped to the very low point of 28.87" with the wind still out of the southeast. A 22-hour rain ended soon after that time. By 2100 the wind was around to the west and the sky clearing.[2] The storm center had passed to the west of Sag Harbor and was now in New England.

At Providence, Rhode Island, much the same conditions were experienced on the 16th. Alexis Caswell's notes follow: "At sunrise the wind was very heavy at about ESE and raining. The wind was fitful, very heavy at intervals, hauling to SE. At 1700 barometer had fallen to 28.90" and the wind had lulled. Before 1800 wind came to NW with blustering and heavy gusts, and the barometer rose very rapidly." The center of the disturbance must have passed very close to Providence.[3]

Boston's highest winds were experienced early in the morning of the 16th when the direction was east-northeast. By 1000 a shift to southeast had taken place, but thereafter the intensity of the gusts decreased although the barometer continued to drop to a low of

29.03″ reached early in the evening.[4] A comparison of the Boston and Providence reports illustrates the complexity of organization that a tropical storm can have when it passes inland over southern New England. In the Boston area minor damage resulted to shipping in the harbor, and much fruit was blown off trees. The only structural damage noticed in the press was the leveling of an unfinished house in Chelsea.[5]

Nantucket and Cape Cod, along with other exposed islands and capes, felt the full sweep of the gales in the eastern sector of the storm system. At Nantucket: "A terrible gale on the 16th prostrated chimnies and fences throughout the town. The greatest force was from 3:30 P.M. to 4:30 P.M. Wind S-by-E to SSW." The Nantucket barometer at the 1500 observation was down only to 29.42″, indicating the center lay much farther to the west.[6] At West Harwich on Cape Cod the gale was "very severely felt." An unusually high tide accompanied the peak winds driving many small vessels ashore that afternoon.[7]

Northward in Maine the storm at Bangor proved "one of the heaviest in years." As it continued until midnight, trees were uprooted, chimneys blown down, and awnings rent at the Maine city.[8] On the seacoast at Belfast the gale was reported as "heavy" with much minor damage to shipping. Apparently central and eastern Maine felt the strong sweep of the gales in the eastern sector of the storm that had passed over Nantucket earlier.

[1] *N. Y. Ship List,* 22 Sept 1858.
[2] SI.
[3] Caswell, 152.
[4] *Evening Traveller* (Boston), 17–20 Sept 1858.
[5] *Idem.*
[6] SI.
[7] *N. Y. Ship List,* 22 Sept 1858.
[8] *Bangor Times* in *Eve. Traveller,* 18 Sept 1858.
[9] *N. Y. Ship List,* 29 Sept 1858.

THE "EXPEDITION" HURRICANE OF NOVEMBER 1861

One of the main objectives of President Lincoln's military plan for the first year of the war centered in the seizure of strategic harbors, forts, and promontories along the Atlantic and Gulf coasts, positions which would help to apply the growing blockade of Confederate seagoing commerce. After the frustrations of the first summer that saw the Federal retreat after Bull Run in July, much enthusiasm was aroused in Northern quarters by the so-called "Expedition." During October the "largest fleet of war ships and transports ever assembled" gathered in the vicinity of Fortress Monroe inside the entrance of Chesapeake Bay. Though its destination was supposedly a military secret, it was obviously preparing to attack Confederate positions in the Carolinas or Georgia. During the assembly period high winds and gales hindered the orderly assembly of the fleet.[1] Finally, with much fanfare in the Northern press, the "Expedition" set sail on 29 October. During the next few days many wild rumors reached the press as to the progress and ultimate destination of the ships.

Great concern soon was expressed for the fate of the fleet since it was known that one or more tropical disturbances had moved northeastward along the coast during the week following 28 October, but there existed no official weather service or means of communication to inform as to the extent and severity of the storms.

We are now able to assemble the scattered meteorological data and learn just what happened. The weather report made by the Board of Health at Charleston, South Carolina, indicated that November 1st was "rainy and windy—disagreeable day" with east winds and a falling barometer. Rainfall totaled 3.41 inches. Next day a strong northwest wind swept the Carolina city with the glass rising rapidly from a 0700 low of 29.74″.[2]

The full force of the storm lay farther east. It hit Cape Hatteras early in the morning of the 2d. The fort guarding Hatteras Inlet had been captured earlier and was now garrisoned by Federal troops. *The New York Tribune's* special correspondent noted the gale as increasing on the evening of the 1st to reach a "perfect whirlwind" by early morning. At 0300 the storm-driven waves commenced to wash over the sandy strip making up the island, completely covering all dry land except the position of the fort itself. For four hours, probably coinciding with high tide, the island was submerged, but by 0700 the waters commenced to subside and had entirely withdrawn by noon.[3] Four sentries of the Twentieth Indiana regiment were drowned.[4] Out at sea the Federal fleet was just rounding the dangerous Carolina Capes when the storm struck. Two vessels were sunk. Though greatly disorganized, the fleet continued on to its destination and seized Port Royal between Charleston and Savannah on 7 November.[5]

The Expedition Hurricane proved noteworthy for the excessive tides produced from Hatteras northward to Maine. In New York City the harbor waters rose to the highest point since 1833; at Newport they were last exceeded by the Great September Gale of 1815; at Boston they approached the record heights of the Minot Light Storm of 1851; and at Portland only the tremendous tide accompanying the March 1830 storm proved superior.[6]

In New York City the storm commencing on Saturday morning, the 2d, continued for about 20 hours. The wind was mainly easterly and northeasterly and raised a tremendous tide. When normal high water came at 0837, the waters overflowed the wharves along both the East and North rivers. On Broad Street the water extended five blocks inland as far as Beaver Street. A popular bar in the Northern Hotel on Cortlandt Street, being isolated by the rising waters, relied for communication and customers on an enterprising gentleman who procured a boat and ferried customers to and from the bar at 2¢ per ride. Next morning a great rat chase ensued as "sports" sicked their favorite hunting dogs on the rodents who had been flooded out of their usual haunts. In Brooklyn the force of the wind on Saturday knocked out the telegraph wires connecting the City Hall with police and fire systems. A number of trees were uprooted there.[7]

The Jersey meadows between Jersey City and Newark were inundated, the tracks of the New Jersey Railroad undermined, and the Newark Turnpike and Plank Road became impassable due to high water. All steamers on Long Island Sound either sheltered behind headlands or were held in port all day Saturday as the gales swept that body of water. The Shore Line R.R. at Bridgeport went under water, cutting rail transportation between New York and Boston. Flooding took place all along the shores of Long Island. The Hempstead Plank Road had so much water that vessels could float along the erstwhile roadway. Many local mills were flooded and put out of action. The northeasterly gale also drove several ships on the beaches along the North Shore of Long Island.[8]

The eastern side of the storm gave the southeastern New England shore a good lashing on the 2–3d. Out at Provincetown over 150 fishing boats sustained injury from the southeasterly gale and 20 were driven ashore. At Wareham at the neck of Cape Cod the gales shifting from northeast to southeast drove the waters of Massachusetts Bay into the village and cut the railroad.[9] Samuel Rodman at New Bedford had "a severe gale from the southeast causing a very high tide morning of the 3d submerging all the wharves." His barometer at 0700/3d read 29.52" with wind ESE force 5.[10]

The storm in the Boston area did not commence until about 2100/2d when an east-northest wind reached force 4. The blow increased to "very strong" in the early morning and remained so at sunrise. Though the night had been stormy, it was mild. Heavy rain fell until 1000/3d, with a very heavy downpour coming between 0800 and 0900. The meteorological conditions during the remainder of the day were as follows: at 0900 the wind suddenly changed to southeast and soon subsided to a good breeze (force 2), and after noon to light southwest to south. The rain ended at 1000, and it was nearly clear by 1130. Nevertheless, the barometer continued to fall, not reaching its lowest point until 1400—an even 29.50". The damaging tides in the Boston area came at noon on the 3d, well after the high winds had subsided. The gale during the night of the 2–3d was judged the worst northeaster since 1858.[11] The ship *Maritania,* bound Liverpool to Boston, struck a rock during the height of the storm about one mile east of Boston Light. Twenty-two were drowned in the sinking.[12]

[1] Press clipping in Redfield Log Book. (Yale.)
[2] PR.
[3] *N. Y. Tribune,* 7 Nov 1861.
[4] *Harper's Illustrated Weekly* (New York), 23 Nov 1861.
[5] David Stick. *Graveyard of the Atlantic.* 51.
[6] *Evening Traveller* (Boston), 4 Nov 1861.
[7] *N. Y. Trib.,* 4 Nov. 1861.
[8] *Ibid.,* 5 Nov 1861.
[9] *Eve. Traveller,* 5 Nov 1861.
[10] Samuel Rodman. Ms. Met. Reg. (Harvard.)
[11] *Eve. Traveller,* 4 Nov 1861.
[12] *Idem.*

THE EARLY AUGUST OFFSHORE HURRICANE OF 1867

A hurricane of moderate intensity curved up the Atlantic seaboard on the first two days of August 1867. Though remaining offshore at all times, it ravaged the shipping lanes and gave Long Island, the Massachusetts Islands, and Cape Cod a threatening sideswipe.

Our first report came from a point about 140 miles east of Norfolk where a ship took the blow from the southeast at 0200/2d. The barometer aboard sank to 28.60" two hours later, indicative of a well-developed storm area. By 1000 the worst of the tempest was over and the wind had hauled to northwest.[1] The brig *Chanticleer* off the lower Delaware coast first felt the lash of a heavy southeast gale at 0400 which then increased to a full hurricane by 0800. Soon the vessel was on its beam-ends. The highest winds continued for about four hours. After noon the wind gradually worked around to south and then to west, moderating by 1400. This wind pattern would place the center of the disturbance between the *Chanticleer* and the coast, or within the 150 miles of the shore.[2]

Another small ship, the bark *Oak Ridge,* passed directly through the center of the severe storm about 1430.

She had been running for some time under a hurricane from east-southeast when the atmosphere became almost calm. Suddenly the wind rose from the west with renewed hurricane blasts. Caught in the cross seas created by the new wind direction, the vessel was finally overwhelmed by gigantic waves and went down about 1600. All perished with the exception of the captain who floated on a small raft for five days until rescued by a passing transatlantic passenger ship. His account of near-rescues and approaching starvation during the five days and nights made a notable contribution to raft literature. The exact position of his ship when in the eye was not given, but she was enroute from Delaware to Boston on a coastal journey when the storm struck.[3]

Another ship just south of Montauk Point at the eastern tip of Long Island took the gale on the northern side. For three hours the wind came out of the east at estimated hurricane force. The barometer fell 0.40″ in 90 minutes.[4] This wind action close to Long Island would indicate that the center of the hurricane was now curving to the northeast or east-northeast. Our nearest weather station on land was maintained at Moriches, 50 miles to the west of Montauk Point: at 1500 the wind veered from northeast to northwest, and blew at force 4 & 5. The rain continued until 1630, a total of 1.21 inches falling since 0945. The wind died out before sunset when the sky had become nearly clear.[5] At New York City the wind had backed around from southeast to northwest. The lowest barometer was only 29.99″; there were showers in the morning, but no rain in the afternoon when the storm center was perhaps 175 miles to the southeast.[6]

Both New London and Newport experienced gales from the northeast during the latter half of the afternoon with very heavy rain: Ft. Trumbull 3.85 inches and Newport 3.60 inches. The wind at the latter attained force 6 at times.[7] The effect of the disturbance reached inland principally in the form of a hard northeast rainstorm: Worcester 3.10; Mendon 3.40; Kingston, Mass., 4.25, and Boston 3.24 inches. At the latter, 2.58 inches fell between the hours of 1300 to 2100. The wind at the Hub City backed from east at 1500 to northeast at 1600 and reached north-northwest by 1800, when it was blowing at force 5 and was "quite stormy," according to the *Traveller's* weatherman.

Down on the Cape, closer to the path of the offshore storm center, winds rose higher although no material damage was reported from any land station. New Bedford was buffeted by a southeast gale until mid-afternoon when a shift to northwest took place as the center sped to the east. The barometer then stood at 29.70″.[9]

Nantucket Island had a hard blow on the afternoon and evening of the 2d. Many disabled vessels came into port during the next few days. The south shoals lightship broke loose from her moorings during the height of the gale.[10] From Nova Scotia came reports that the hurricane had been severe there early on the 3d.[11]

[1] *N. Y. Tribune,* 6 Aug 1867.
[2] *Ibid.,* 8 Aug 1867.
[3] *N. Y. Ship List,* 10 Aug 1867.
[4] *N. Y. Tribune,* 5 Aug 1867.
[5] SI.
[6] *N. Y. Tribune,* 3 Aug 1867.
[7] SG & SI.
[8] *Eve. Traveller* (Boston), 3 Aug 1867.
[9] SI.
[10] *Nantucket Inquirer,* 17 Aug. 1867.
[11] *Idem.*

THE SEPTEMBER GALE OF 1869 IN EASTERN NEW ENGLAND

On the late afternoon of 8 September 1869, a hurricane of unique size and structure raised full hurricane blasts over parts of southeastern New England. It was the first time since September 1815 that inland points had experienced the full fury of a tropical storm.

The central path of the northward-rushing storm probably cut across the extreme eastern tip of Long Island before striking the mainland close to the Connecticut-Rhode Island border near Stonington and Watch Hill. Continuing north-northeastward, the path traversed the length of western Rhode Island, to the west of both Newport and Providence, moved through eastern Massachusetts between Worcester and Boston (probably on a Milford-Framingham-Concord-Lawrence axis), rapidly crossed southeastern New Hampshire, and rushed into western Maine.

The meteorological evidence now available for analysis points to a rather unusual wind structure. The center, if it had a clearly defined one when over New England, and the path of destruction were unusually narrow. Estimates of the latter's width ran from 40 to 50 miles, all to the east of the apparent path of the center. Lowest pressure over land ran close to 29.00″. Five to ten miles west of the center there was not only no destruction, but no winds of storm strength were noted. And the circulation in the western sector hardly conformed with the classical conception of a tropical storm structure. The highest winds reported anywhere in the western semicircle were only force five, and wind flow throughout much of central New England continued at southwest apparently unaffected by the great disturbance a few miles to the east.

It is probable that the friction of passing over a land mass greatly distorted the western sector of the storm which had already started to transform to an extratropical type. For instance, at the 1400 observation time the Smithsonian weatherman at the Massachusetts town of Mendon, close to the Rhode Island border, had a southeast gale force 8; while Worcester, only 18 miles to the northwest, had southwest force 4.[1] All afternoon Worcester reported "steady rain with fresh southwest breeze," while a full gale was raging only 20 miles to the east and a hurricane 40 miles away.[2] The observer at Lunenburg, just east of Fitchburg, reported little damage at his place, but great destruction 20 miles to the south.[3] The same pattern continued farther north where southeast gales caused havoc along the Maine coast, but inland over eastern New Hampshire only light northwesterly winds prevailed. No doubt, the eastern sector continued to be nourished with water vapor energy from the warm ocean, while cool, dry air entering the circulation from the north and west destroyed the former tropical structure of the storm system.

LANDFALL

There were several reports of a severe hurricane at sea during the morning of the 8th, but the exact locations of the ships at the height of the gale were uncertain. One vessel at 72 40W experienced a severe hurricane from 0800 to 1200 with the wind veering from southeast to south-southwest; this is directly south of Moriches on Long Island, our first land station check point.[4] It would indicate that the disturbance was moving east of north.

The Smithsonian observer at Moriches, on east-central Long Island about 50 miles west of the eastern tip, had seen his wind vane back steadily from southeast in the morning, to east by noon, and into northeast in the early afternoon. From 1530 to 1630, for a short time, the heaviest wind came out of the northeast, varying from force 4 to 6. By 1700 it had backed farther, into the northwest, and become quite moderate; sunset witnessed the air flow down to force 2 and coming from the southwest. His notes mentioned at that very time a severe hurricane was raging in Rhode Island and Massachusetts.[5]

Fort Trumbull, on the Connecticut shore of Long Island near New London, also lay in the western side of the storm system. Here the wind at 1400 was coming out of the east at force 5; a heavy rain fell during the afternoon.[6] Other Connecticut points make mention of heavy rains, but no wind damage. Nor do their wind roses show any effect of the disturbance to the east. The *New Haven Journal and Courier* sent a reporter eastward along the shore line to check on the storm:

> Leaving New Haven Thursday morning (9th), the first thing observed while passing on the Shore Line railroad to New London was the corn fields showed signs of a violent storm, and the farther I went east, the stronger the wind had been seemingly.
>
> At New London considerable damage had been done, but not until I neared Providence, R. I., did I comprehend the gale had been so powerful. As we neared the city, I saw at one time seven schooners aground, and so near each other they almost touched . . .
>
> Nearing the city of Providence I could not help noticing the fields where corn had stood. It laid flat on the ground with its leaves all whipped off.[7]

THE EASTERN SECTOR

Over Rhode Island and southeastern Massachusetts it was a different story. Every town and village had dire tales of destruction to report. The editor of the Westerly paper thought: "the storm of Wednesday, although severe, cannot have been as violent here as in Providence." [8]

Our easternmost reporting station, Nantucket, lying about 80 miles to the east of the storm track, had its most violent gale since the coastal hurricane of 2 August 1867. On the morning of September 8th the wind from the southeast increased steadily until 1900 when it veered to the southwest. No wrecks were reported in the vicinity.[9]

On the south shore of Cape Cod the storm on Wednesday afternoon was described in the *Yarmouth Register* as "of great severity, rivaling in violence and destructiveness, the great gale of September 1815. It commenced to blow in this vicinity about 2 o'clock from the S.S.E. and by 4 o'clock the wind had increased to a furious gale, which continued two or three hours." [10] The *Barnstable Patriot* thought "the storm will long be remembered as the *great gale,* and the injury it has caused will not be speedily repaired." [11]

Along Buzzard's Bay, which separates the Cape from the mainland, the southeasterly hurricane coincided with near high tide. The wind-driven waters funneled up the bay to its head at Wareham where it was said they reached the highest point in 234 years, or since the legendary Great Colonial Hurricane of 1635.[12] Though this cannot be confirmed on any scientific basis, the flood tide at its height did enormous damage along the bay shore, moving buildings, undermining trees, and washing out highways and railroad tracks—all communication was seriously disrupted for several days over the narrow isthmus there.

Farther west at New Bedford water damage was less since the gale commenced at low water. If the peak gusts had come three hours later, the destruction would have been infinitely greater; as it was, many boats, cast adrift by the wind and waves, piled up against the main bridge which finally gave way under the pressure. At nearby Fairhaven the beautiful spire of the Congregational Church crashed to the ground about 1630.[13] The gale in lower Bristol County was judged as strong as that of 1815, though it did not last as long. Samuel Rodman, New Bedford's longtime weatherman who had witnessed the gale 54 years before, saw his barometer dip to 29.51″. His rain gage caught only 0.74 inches.[14]

NARRAGANSETT BAY

The main scene of the September gale's destruction came in the complex of bays, inlets, islands, and low-lying ground that make up Narragansett Bay. This tidal estuary, running for 40 miles north to south from Pawtucket to the open Atlantic, is especially susceptible to wind flow from a southerly quadrant. Disaster strikes here whenever a major tropical storm moves inland to the west over Long Island and Connecticut. In 1769, 1815, 1869, 1938, and 1954 the waters of the bay have been driven up the narrowing estuary in tidal bore fashion.

The historic city of Newport lies in a protected harbor almost at the entrance of Narragansett Bay and serves as a good check point for all passing weather in the area. Meteorological observations there commenced in the 1750's and have continued with few intervals to the present. The local observer in 1869 recorded the commencement of a rainstorm with wind from the southeast on the early afternoon of September 8th. By 1500 the wind had increased to a "violent gale," reaching a peak about 1600 when considerable structural damage occurred. The steeples of the Emanuel and Unitarian churches were toppled, the roofs of the railroad car house and depot torn off, and forty-eight vessels cast high and dry on the shore. The tide at Newport would not have been full until 2200, or five hours after the peak gusts subsided, greatly lessening the destructive surge of the waters.[15]

At Bristol, halfway up the bay, the tide rose six feet above normal high water mark at 1730 and "had the wind held southeast two hours longer the damage by water would have been immense." As it was, in addition to much structural damage from wind there, some shade trees known to be a century old were uprooted, and at nearby Poppasquash orchard trees were blown over which had been subjected to the same indignity in 1815 but righted after that storm to bear fruit through the intervening years![16]

The city of Providence, the largest in Rhode Island, is always the principal sufferer in hurricane situations due to its peculiar geographical location. The Narragansett estuary narrows to the size of a small river there; being easily bridged, it was chosen as the site for the first settlement in 1635. Much of the city occupies a low plain on the western side of the river.

On the afternoon of Wednesday, 8 September 1869, President Alexis Caswell of Brown College was at his home on the campus which occupies a point of vantage on a hill overlooking the harbor and city. Caswell had been taking complete meteorological observations since 1832, and on this afternoon fortunately recorded some of the salient facts. He sent these off to the press next day so that we now have a firsthand account from a qualified observer as to the behavior of the elements:

> The 8th, Wednesday, opened with a very heavy fog, wind very light, southerly, barometer 29.900, with appearance of rain. But about 9 a.m. the fog cleared away and the sun came out. Before noon the wind hauled toward the southeast, and increased in force, with occasional heavy clouds. At 1 p.m. it began to rain, the wind still increasing. At 2 p.m. the barometer had fallen to 29.660. At 3 p.m. the wind had become heavy at southeast, with copious rain: at 3:30 it was blowing a gale, and from that time to 5:30 the violence of the storm was fearful, uprooting trees, prostrating fences and buildings, as if they were nothing and less than nothing before it. The extreme violence of the wind was, I think, from 5 to 5:30. At 4 p.m. I observed that the barometer had fallen a good deal, but did not take the reading; at 5 it read 29.106, thermometer attached, 74°; at 5:30 it read 29.026. A few minutes later I looked at it, and it seemed to be rising; at 5:40 it had risen to 29.1000; at 6 to 29.206; at 8 to 29.550; and at 10 to 29.608, the thermometer continuing at 74°. This morning at 7 it stood at 29.712, wind moderate at southwest. On noticing the rise at 5:40 p.m. I at once inferred that the wind had changed. On examination, I found it had hauled to southwest. The violence soon abated, and every one felt a sense of relief. It is many years, I know not how many, since we have had a storm of equal violence; very probably not since the "Great Gale" of September, 1815. My rain guage showed the quantity of rain to be 1.38 inches. But from the violence of the wind, taking in connection with the position of the 47 College street guage, I suspect the quantity to have been somewhat greater.[17]

The September Gale was given historical documentation by the publication of a pamphlet of 24 pages describing its destructive effect in the Providence area. This appears to have been one of a series of souvenirs which have been frequently put together following the occurrence of a New England hurricane. The 1869 material, no doubt, was taken from local press accounts though considerably polished by the unknown author who was well schooled in the classics:

> SEPTEMBER GALE, 1869
>
> Our city has again been visited by a flood and gale, outrivaling in fury and destructiveness the terrible storm of September, 1815. On Wednesday morning, September 8th, the sky was overcast, and occasionally a slight shower fell over the city; in the forenoon the clouds were dispelled somewhat and the sun came out for a short time. About noon the wind sprung up quite fresh from the southeast, blowing up large masses of dark clouds. Between two and three o'clock, p. m., it commenced to rain quite freely, the wind, in the meantime, blowing still heavier. At four p. m., the wind was blowing a perfect hurricane and the rain coming down in torrents.

The combined power and fury of the elements were beyond all description. It seemed as if nothing could withstand them. Our peaceful shores were under the wand of some mighty Ariel, and

"Not a soul
But felt a fever of the mad, and play'd
Some tricks of desperation."

The water in the harbor rose to a great height, and poured over the wharves and into the streets, in the lower portion of the city, with appalling swiftness . . . at one time rising two feet in twenty minutes. Mighty trees bent and bowed before the tempest, some of them being torn up by the roots, while others were snapped off like rotten twigs. Boards, bricks, shingles, broken boughs, portions of gates and fences, shutters, signs, and fragments of all kinds filled the air. Massive buildings rocked like toys, roofs of tons in weight were lifted and carried rods away, or torn into minute pieces. Huge strips of tin and metal were torn from places where they had been securely nailed, and blown like sheets of paper, for long distances. Steeples rocked and fell; huge buildings were crushed in like egg shells; vessels were swept like chips upon the shore; dwellings were overturned and carriages blown along the street like feathers. The spire of the Mathewson street Methodist church, at one time, swayed so violently that its fall was momentarily expected and those watching it were glad to betake themselves to a place of safety. The tower upon which it stands cracked, and the masonry was considerably broken. The Chestnut street steeple went over entirely. The roofs of the towers of the Stewart street Baptist church were blown off. The First Baptist steeple swayed to and fro at a rate that was fearful to look upon, though its stability was not perhaps imperilled. The stone ornamental work above the Mathewson street clock of Grace Church spire, came down with a terrible crash, and the clock face on the south side of the church blown in. The clocks in the different towers stopped at the moment the storm was felt strongest in the various localities. The clock on Grace Church stopped at ten minutes past five; on the Chestnut Street Church at five; first Universalist, half-past five. The jets, balustrades and ornamental carpentry upon the roofs of dwellings was broken and scattered in every direction. Some immense pieces were hurled to a great distance. Windows and doors were forced in, and blinds and scuttles wrenched off, and strewn far and near. For the first time since the advent of telegraphy in this city we were without a single "tap" from outside "barbarians," not a wire of either the Western Union or Franklin Lines being in working order. Verily it seemed as if "Hell were empty, and all the devils were here." If the violence of the wind had continued for half an hour longer, it is probable that the waters of the harbor would have united with those of the Cove, in the very busiest portions of the city. The rise was at the rate of a foot every ten minutes. The hurricane abated somewhat in its fury about 5.45, and very soon afterwards the water rapidly receded, leaving South Water and Dyer Streets completely covered with the wrecks of the gale. The water poured into the PRESS Office in great volumes, putting out the fires in the engine room and submerging the press room to the depth of eighteen inches. Mr. H. M. Coombs, whose book bindery occupied the third story of the PRESS Building was rowed in a boat from the Custom House Street entrance of the building to the steps of the Post Office, as were quite a number of the young women at work in the bindery, and in the box manufactory of Jerauld & Holmes; some of them were rescued, however, by gallant and daring young fellows in the same manner that Aeneas conveyed the venerable Anchises from capitulated and forsaken Troy. An editorial in the PRESS of Thursday, says: "The water mark in the room where we write is eighteen inches from the floor and all around are indications of a great flood, beaten in history only by Noah's celebrated deluge and the Great Gale of 1815."

The Steam Fire Engines of the city were busy all night in pumping out the cellars near the wharves, but several days elapsed before the water was entirely cleared out.

THE BOSTON AREA

After its devastation in Rhode Island, the hurricane rushed on into east-central and eastern Massachusetts, with the axis of the center close to the Milford-Framingham-West Concord-Lawrence line. At both Mendon and Milton, the wind was blowing at force 8 as early as 1400 when the Smithsonian observers took their regular afternoon observation. Some damage occurred to buildings, and trees were uprooted by the southeasterly near-hurricane winds at both places. At Milton, immediately south of Boston, the barometer fell from 29.50″ at 0700 to a low of 28.75″ at 1800—a rise commenced at 1810 and the wind began to abate immediately. Rainfall amounted to 1.40 inches.[18]

At Boston the weatherman for the *Evening Traveller* kept a careful log of the elements that day, as he had for a quarter of a century. There had been some light rain in the morning with a southerly flow. At 1400 the temperature stood at 85° with a light breeze from the south-southwest still prevailing. By 1500, however, it had become overcast with a shift of wind to an easterly quarter, and heavy rain squalls soon struck the city. By 1645 the gusts of wind had become "severe" from the east-southeast; the mercury tumbled to 70°. A shift to south occurred at 1750 as the center moved to the west. The observer thought the heaviest blast of wind came at 1812, the time of the lowest barometer—29.02″ corrected to sea level. From a sunrise pressure of 29.94″, the trace had dropped to 29.14″ at 1800, bottomed at 29.02″ twelve minutes later, and had risen to 29.25″ by 1900—all pointing to a very small, yet extremely intense pressure center. The heavy gusts at Boston ended by 1900, and by 1930 the wind had veered around to south-southwest with heavy rain squalls still continuing. Rainfall that afternoon was measured at 1.22 inches. The tide in Boston harbor had been full at 1400. By the time of the greatest wind force four hours later it was at half ebb, a fact which greatly lessened damage to shipping and harbor installations.[19]

The hurricane winds caused major structural damage throughout the Boston area. The large Coliseum proved the principal victim. The raging gale blew out the east gable exposed to the wind, and then rushed through the structure to force out the west gable; the roof at each end then collapsed, leaving only a small section of the center part still standing. A large organ was completely demolished, as was a special big drum—both were showpieces of the exhibition hall.

Many churches through the city also suffered. The Hanover Street Methodist Church lost most of its lovely spire, and the roof of the Central Church on Berkeley Street was shorn of its wooden roof. In the Public Garden several trees were torn out by the roots. The winds flattened a large tenement house in Chelsea.

In the neighboring towns similar damage occurred. The steeple of the First Baptist Church in Lynn, said to be 160 feet tall, crashed onto the roof of the western wing of the building. In Peabody a large building two hundred feet long collapsed under the pressure, as did a large barn one hundred feet long in Hamilton.[20]

NORTH OF BOSTON

The path of the hurricane continued overland into New Hampshire and western Maine approximately on a Lawrence-Dover-Lewiston line. The blasts of the gale continued to spread terror and injury over Essex County in the northeastern corner of Massachusetts. Its many bays and promontories were open to blasts from a southeasterly quarter. The historically-minded *Salem Register* judged the storm there: "for severity at its height, although shorter in duration, its destructiveness was probably greater than in 1815." [21] The same type of havoc as had been meted out to the southward occurred in most towns and villages close to the sea. In addition, many vessels in the small harbors dragged their anchors and piled up onshore. Several boat crews were saved by daring rescues in lifeboats. The schooner *Helen Eliza* out of Rockport, Mass., enacted the principal tragedy when it was wrecked near the entrance to Portland harbor with the loss of all but one of its crew of twelve.[22]

A short distance inland the cities of Worcester, Fitchburg, Lowell, and Lawrence reported only minor or no damage. At Nashua, the largest city in southeastern New Hampshire, it was considered an ordinary rain and wind storm. The Smithsonian observer at Lunenburg, just east of Fitchburg and 40 miles northwest of downtown Boston, was located along the western fringe of the track of devastation. He experienced strong winds, but they failed to attain the strength to cause trouble. The peculiar structure of the eye of the September 1869 gale is well brought out by the Lunenburg observations. The barometer there fell rapidly from 29.31″ (uncorrected) at 1100 to a low of 28.60″ (uncorrected) at 1815. Heavy rain commenced at 1545, with a rising wind fluctuating between northeast and southeast. "The heaviest puffs came from 1730 to 1930 from the east southeast . . . at 1930 wind very suddenly shifted to northwest and continued fresh," he noted.[23] Lunenburg, no doubt, was on the western side of the path of lowest pressure. The prevalence of east-southeast winds at the peak of the blow again illustrates the unusual transformation that was taking place in the physical character of the western portion of the storm.

Down East in Maine winds at all reporting stations were southeast. Coastal points suffered severely. Portland had major structural damage. When the wind was at its peak at 2030 (two hours and 15 minutes after the maximum in Boston), the spire of a newly-built Catholic church crashed to the street. Many awnings and chimneys were demolished, and the streets of the city littered with fallen trees and limbs.[24]

At the inland towns of Lisbon and West Waterville, the Smithsonian observers reported a severe gale (southeast force 9 at Lisbon at 2100) with trees blown down, but no major damage to buildings. At Gardiner, 50 miles northeast of Portland, the rising east-southeast wind became a gale at 1900 and "almost a hurricane at 2000." Soon after 2200 the wind lulled, and by 2300 was blowing strong from the southwest as the center moved into central Maine. The Gardiner barometer had fallen 0.60″ in the evening, reading 29.14" at 2100—a rise of 0.30″ took place in the two hours following 2200.[25]

In Washington County, in extreme eastern Maine, the gale at Steuben commenced at 2200 and continued until 0100/9th. Trees and fences were prostrated, but little major structural injury occurred. The weather observer there thought that the heaviest part of the blow passed well to the west of his location.[26] Farther Down East the storm was believed to have been "of unprecedented severity" by the *Eastport Sentinel,* but again little material damage resulted on land. At sea over thirty wrecks were reported as the southeasterly gales drove many small vessels onto the rocky stretches of the Maine coast.[27]

The storm system maintained its energy as it moved into Canada. A final tally on its activities came from the Gulf of St. Lawrence where two vessels were sunk and ships attempting to sail up to Quebec were seriously hindered by the varying wind pattern of the unusual storm.[28]

[1] SI.
[2] SI.
[3] SI.
[4] Press clipping in 1869 Folder. (National Archives.)
[5] SI.
[6] SG.
[7] *New Haven Daily Morning Journal and Courier,* 13 Sept 1869.
[8] *New Haven Register,* 11 Sept 1869.
[9] *Nantucket Inquirer,* 11 Sept 1869.
[10] *Yarmouth Register* (Mass.), 10 Sept 1869.
[11] *Barnstable Patriot,* 14 Sept 1869.
[12] C. F. Swift. *Cape Cod.* 291.
[13] *New Haven Register,* 11 Sept 1869.
[14] SI.
[15] SI.
[16] *The Great September Gale of 1869, in Providence and Vicinity.* Providence, 1869. 20–21.
[17] *Idem.*
[18] SI.
[19] *Evening Traveller* (Boston), 9 Sept 1869.
[20] *Idem.*
[21] *Salem Register,* 13 Sept 1869.
[22] *N. Y. Tribune,* 10 Sept 1869.
[23] SI.
[24] *N. Y. Trib.,* 10 Sept 1869.
[25] SI.
[26] *Idem.*
[27] *Eastport Sentinel* clipping in 1869 Folder. (National Archives.)
[28] *New Haven Register,* 11 Sept 1869.

SAXBY'S GALE AND THE GREAT NORTHEASTERN RAINSTORM AND FLOOD OF OCTOBER 1869

In November of 1868, Lt. S. M. Saxby of the Royal Navy forwarded a message to the London press predicting that the Earth would be visited by a storm of unusual violence, attended by an extraordinary rise in tide, at 0700 on the morning of 5 October 1869, or some eleven months hence. He based his prophesy on a lunar coincidence which would place the Moon directly over the Earth's equator at the very time that the satellite's orbit would be at its closest approach. Since the Sun and Moon would be exerting their maximum forces in unison, Saxby reasoned that a tide of record proportions should be expected. This dire prediction caused considerable comment both in scientific and popular circles since Saxby had a previous success with this sort of thing.[1]

The subject was subsequently taken up by Frederick Allison of Halifax and narrowly defined. Though Saxby had allotted himself the entire world for confirmation of his forecast, Allison limited his prognostication to one locality—he pinpointed the event for Halifax Harbor on the Atlantic Coast of Nova Scotia. How closely he came to the fulfillment of his forecast comprises one of the best "near-misses" in the entire history of prediction.[2]

As the fateful day approached, a feeling of serious apprehension became manifest among the more excitable portion of the population in the city and in surrounding areas. Outward preparations were undertaken by some to protect person and property from the coming storm. An atmosphere much like that preceding the Second Advent fervor of the 1840's, when the coming of Christ was daily anticipated, pervaded susceptible minds. Fortunately for those subject to such foreboding, there were no means of rapid communication of weather information over ocean areas that would have warned the people that an intense hurricane was racing northward in the open Atlantic that afternoon and evening of October 4th and might reasonably be expected to burst into their area before midnight. As it turned out, the path of full hurricane blasts bypassed the Halifax region, but localities on the western shore of Nova Scotia, bordering the Bay of Fundy, as well as points in Maine and New Brunswick, suffered severely on the evening of the 4th from the combined onslaught of hurricane force winds and unprecedented tides.

Locations on the western shore of the Bay of Fundy in the Province of New Brunswick and State-of-Maine points in the Passamaquoddy Bay complex were the principal victims when the tides in those areas, normally so susceptible to huge rises and falls, swelled to unprecedented heights. Moncton, near the northern head of the Bay of Fundy, experienced a tide six feet above any previous record as the southeast full gale produced a gigantic tidal flood and piled up the waters to enormous heights. The rivers and streams feeding into the bay backed up many miles as was demonstrated at Fredericton, 55 miles inland on the St. John River, where the tide rose three feet above normal. The largest city in the entire region is St. John where the estuary of the St. John River meets the Bay of Fundy. The timetable of events there was noted by the local weather observer:

1700—wind increasing to gale
1800—rain began to fall
2030—blowing hurricane from south by east
2100—reached maximum force, rain almost ceased
2200—wind began to subside and shift to southwest[3]

A weather map analysis of the Great New England Rainstorm and Down East Hurricane of 1869 presents many problems to the meteorologist who attempts to unravel its pattern. Unfortunately, we do not have hourly observations of wind direction and change. The Smithsonian observers took only three observations per day: at 0700, 1400, and 2100.

Two facts stand out: (1) During the early afternoon a wind shift line moved eastward from the Hudson Valley of New York to the eastern New England coast. Presumably this represented an advancing cold front occupying a trough of low pressure that possessed a fairly steep gradient both to the west and east. The strong southerly flow preceding the trough had transported large amounts of very moist air from either the Atlantic or the Gulf over the Middle Atlantic States, New England, and the Maritime Provinces; the advance of the cold front provided the triggering mechanism to set off intense downpours.

We know that the cold front with attendant wind shift line passed New York City before noon, reached Moriches on Long Island 65 miles to the east by 1400, and arrived at Boston about 1530.[4] Despite the eastward movement of this pressure impulse, the barometer over New England continued to fall for about two hours after the wind shift—probably the reflection of conditions aloft over those stations as the influence of events taking place over the ocean to the eastward began to be felt. (2) At the approximate time the trough reached the eastern Massachusetts shoreline, a tropical storm with full hurricane blasts in its eastern sector was moving north-northeastward at a rapid pace.

Our best checkpoint for the hurricane's movement was provided by Nantucket Island. The always weather-conscious *Nantucket Inquirer* informed us that the barometer had fallen to a low point of 28.70″ (uncorrected) at 1500/4th when the wind veered around from southeast to southwest.[5] Shortly thereafter the weathercock at Boston backed from northeast to northwest.[6] Nantucket is 50 miles longitudinally east of Boston. This wind behavior placed the center track of the tropical disturbance between the two localities—probably close to Martha's Vineyard and across outer Cape Cod. Though southeasterly winds prevailed in early afternoon at New Bedford and the barometer dropped to a sea level pressure of 29.03″, the local observer reported: "the gale was not very severe in this locality." [7] The minimum pressure at New Bedford came at 1515, while at Boston, fifty miles to the north, the barometric low of 29.34″ (sea level) was noted at 1700.[8]

We do not have any meteorological estimation of the force of the "gale" at Nantucket. The editor of the *Inquirer* thought "the gale of Monday last (the 4th) was much more violent here than that of Sept. 8th, but did not do material damage." [9] The main forces of the hurricane were expended seaward. The shipping lists of the day were filled with accounts of the devastation this October gale caused.[10] There were many reports in the vicinity of Nantucket Shoals and off Cape Cod of full hurricane blasts from the southeast which later veered to the south and southwest. There were no reports from any quarter of a northeast hurricane, evidence which again leads to the belief that the deep eastward-moving trough had caused the western part of the tropical storm to become highly disorganized, with its former eye dissolved as cold air entered the circulation. It now presented, in its western part at least, a rather broad area of very low pressure but relatively light winds. The disaster account of one ship in the *New York Shipping List* well illustrated the predicament of vessels which were first lashed by the heavy blows of the southeasterly hurricane and then were caught in the cross seas following the wind shift:

> Ship *Graham Polly,* Burgess, hence 1st instant for Glasgow, returned this port 10th instant, having on Oct. 4 in latitude 41° longitude 68° 18′ (about 100 miles east of Nantucket) encountered a hurricane from SE to SW with a tremendous sea; blew away almost an entire suit of sails. At 8:30 P.M. was boarded by two immense seas, entirely submerging the ship and carrying away bulwarks fore and aft on both sides, broke stanchions, and mainsail, split covering planks, stove in forward house, washed everything out of the galley and forecastle, carried overboard long and port quarter boats and everything movable on deck.[11]

Back on the mainland wind forces appear to have reached their peak in the southerly flow preceding the cold front passage. The highest wind mentioned in the available Smithsonian reports, force 6, occurred well inland in southeastern Worcester County.[12] At New Bedford, where "the gale was not severe," there are no press reports of damage on land, nor at Boston or elsewhere in eastern Massachusetts.[13] Such mishaps as occurred were of a marine nature. Small vessels were driven ashore in Narragansett Bay and Buzzard's Bay where a southerly wind is particularly feared. One ship at Fall River capsized.

Farther north, the wind behavior at Lunenburg, just east of Fitchburg and about 40 miles northwest of Boston, is instructive: fresh gale from southeast in morning; at 1200 shifted to south; at 1430 went to northwest and the storm rapidly abated. The rainfall, totaling 7.50 inches, stopped an hour after the wind shift, the same time the barometer reached its lowest point.[14] At Boston a "short squall" came out of the northwest at 1535, but from 1600 onwards only light west-northwest winds prevailed, according to the *Evening Traveller's* weatherman.[15] There was no strength to the following winds, even though a hurricane was raging just a short distance offshore.

The main story of this storm period for most land stations in the Middle Atlantic and New England States concerned rain, more rain than had fallen over this area in a single storm in recorded meteorological history, before or since. Think of the mighty physical forces at work that could precipitate six inches of water continually in a progressive storm all the way from northern Virginia northeastward into Canada: in each of the following states one or more stations reported at least a five-inch catch: Virginia, Maryland, Pennsylvania, New Jersey, New York, Connecticut, Massachusetts, Vermont, New Hampshire, and Maine, Only coastal Delaware and Rhode Island did not give the rain-bearing clouds sufficient surface lifting to cause such downpours.[16]

Wherein lies the explanation of this greatest of New England deluges? The stage was probably set by the trough of low pressure that produced rain in Virginia and western Pennsylvania as early as the morning of the 2d. In eastern Pennsylvania and New Jersey the first falls did not descend until early on the morning of the 3d. Amounts appear to have been heavy, but not excessive, prior to the cold front's approach from the west; with its arrival, however, there are accounts of some phenomenal falls in brief periods for non-thunderstorm rains at these temperate latitudes. The long-continued southerly flow in advance of the trough had brought very moist tropical air to overlie the whole area. This could have produced the heavy rains that gradually moved eastward on the 2d and 3d, but the excessive precipitation on the 4th must have resulted from the reinforcements of tropical air brought northward by the advance of the hurricane circulation offshore.

Very heavy rains progressed eastward across southern New England on the 4th. At Richmond in the Massachusetts Berkshires about 2.00 inches fell from 0800 to 1100. At Middletown, Connecticut, in a position to benefit from both rain-producing influences, a total of 7.15 inches fell in 24 hours, most of it on the morning of the 4th. At Fitchburg, Massachusetts, three inches fell in the 2½ hours preceding 1430/4th—at Goffstown, New Hampshire, 4.27 inches descended in two hours ending at 1400—at Concord in the same state four inches came in the two hours before 1430. At Canton in Connecticut northwest of Hartford, a hill location noted as a perennial great rain-producer, the observer wrote: "I measured 12.25 inches, but I do not think it could possibly have been so much. I think that there was an unaccountable mistake somewhere." [17] The rainfall attending Diane in 1955 in the same area would confirm the possibility of such rainfalls in this area; so the verity of the Canton figure for 1869 can now be accepted.

Over eastern Massachusetts the rainfall proved much lighter: Milton 1.15, New Bedford 1.51, and Boston 1.52 inches. Farther inland the rain gage catches increased: Mendon 2.80 and Lawrence 3.56 inches. The greatest amounts seem to have been concentrated in a broad band extending northwestward from eastern Connecticut to the Springfield area of Massachusetts and westward into the central and upper Hudson Valley of New York. The hill country of southern New Hampshire also had some remarkable amounts.[18]

Northward of Masachusetts the same deluge story was repeated: excessive rains to the west, and high southerly winds to the east. Dunbarton, in New Hampshire southwest of Concord, had a windshift about noon from southeast to northwest accompanied by a cloudburst that dropped 4.27 inches in three hours. The rain finally ended there at 1645 after a storm total of 7.28 inches had descended.

All day the 4th high winds from a southerly quadrant swept southern and central Maine.[19] At Lisbon on the Androscoggin River north-northeast of Portland, force 6 out of the south was registered at 0700 and force 8 from the southeast at 1400. Though only 2.70 inches fell there, the rains higher up the valley in the foothills of the White Mountains raised the river to its highest stage since the famous floods of 1832.[20] No unusual wind damage was noted in the Lisbon vicinity. Farther northeast at Gardiner near Augusta we have the excellent meteorological reports of R. H. Gardiner who had faithfully maintained his records since 1836. His barometer stood at 29.70″ at 0700/4th, but dropped rapidly thereafter to reach a minimum of 28.99″ sometime around 1900 (the exact time was not given, but by 2100 it had rebounded to 29.21″ with the wind veering from southeast to south).[21] Again wind speeds along the southern Maine coast, though higher than in Massachusetts mainland points, were not of sufficient strength to cause material damage to buildings and structures.

The farther Down East one progressed, the higher the wind speeds. Washington County occupied the easternmost corner of Maine with a seacoast facing south-southeast with many bays and inlets nakedly exposed directly to the south. Here the combined force of tide and gale caused unprecedented havoc. We have good Smithsonian reports from both the western and eastern extremities of the county. The observer at Steuben just east of Mt. Desert Island noted in his weather log: "About 6 P.M. wind increased till it blew a perfect gale from about 7 to 8 P.M. . . . cutting off chimnies, blowing down barns, moving buildings from their foundations, driving vessels on shore . . . There never was such a gale hereabouts as this since the country was settled . . . the tide rose beyond all precedent." By 2100 his wind had veered around to southwest as the hurricane's main force was being expended in the Province of New Brunswick directly north. Observer Parker further noted: "we had a gale the 8th of Sept., went mostly to the west of here, but the best of this gale went east of here." [22]

Northeastern Maine experienced the full brunt of the hurricane with the southerly and southeasterly wind flow producing tremendous tides. There is no evidence of any center or eye of the hurricane as it swept over Maine and New Brunswick—all the damaging winds came from southeast to southwest with no mention of any sustained northerly or westerly flow of any strength.

The complex of bays, channels, and islands making up Passamoquoddy Bay at the western entrance to the Bay of Fundy is particularly susceptible to the intrusive influence of southerly gales. Grand Manan, Campobello, and Deer islands all suffered severely from wind and tide, and the settlements of Eastport, Calais, and St. Andrews shared the same fate that evening when the still dangerous sector of the hurricane, the same that had buffeted the ships off Nantucket and Cape Cod in mid-afternoon, slammed ashore in this important commercial region. The *Eastport Sentinel* summarized the destruction:

> This town was visited by a fearful hurricane last night, vessels, wharves, stores, and fish houses were smashed to atoms. Great quantities of fish and oil were destroyed. The steamer *New York* narrowly escaped loss with all aboard. She was driven ashore and lost both anchors and her rudder, but got off safely and returned to Eastport. Twenty-seven vessels are ashore in Rumney Bay. The schooners *Pomp* and *Percy* were badly damaged. The schooner *Rio* was lost in St. Andrews Bay with all on board—17 in number. Grand Menan is swept with all its weirs and smoke houses. The towns of Lubec, Pembroke, and Perry lost heavily . . . The loss cannot be less than $500,000.[23]

On the day following the storm a man driving from Eastport to Calais, a distance of 30 rural miles, counted ninety houses either blown down or damaged seriously. The relentless force of the wind and tide may be judged from the crewless voyage of a ship which broke loose from its moorings at Eastport, drifted across the Bay, drove up the St. Croix River to the head of navigation at Calais, and grounded at a turn in the river after its 30-mile ghost trip.[24]

The hurricane scene was witnessed by George A. Boardman, of Calais, the naturalist of the St. Croix River area:

> Nothing like it ever took place here. It appeared like a whirlwind. It took the roof off my long woodshed, my old store and part of the roof from the barn on the hill. The Universalist church was a perfect wreck; the railroad bridge over the falls in front of my house fell into the river; also the covered bridge at Baring. More than one hundred buildings in St. Stephen were ruined, and in our cemetery more than one thousand trees were uprooted and broken. At Eastport about forty buildings were destroyed or unroofed, several lives lost, and most all of the fishing craft were wrecked. At Eastport and St. Andrew and about the islands the tide was very high and damaged the wharves much. Sixty-seven vessels were ashore—those that went onto soft places came off, many went onto the rocks and were ruined. The blow did not last but about an hour and was heaviest at eight o'clock in the evening. There was very little wind at Bangor and not much at St. John.
>
> * * * * *
>
> The great loss to this country from the Saxby gale will be to the woods. We have had some of our men up exploring and they say they can walk ten miles at a time on the trees that are down without stepping on the ground. In some places for half a mile about every tree is down. The bridges and buildings can easily be put back, but the woods all down will soon get on fire and burn all over the down district. The wind did not reach very far up the river, only about thirty or forty miles—it was the heaviest about the shores.[25]

1 D. L. Hutchinson. The Saxby Gale. *Transactions of the Royal Canadian Institute.* 9, 1913. 256.
2 *Idem.*
3 *Idem.*
4 *New York Tribune,* 5 Oct 1869; *Evening Traveller* (Boston), 6 Oct 1869.
5 *Nantucket Inquirer,* 9 Oct 1869.
6 *Eve. Traveller,* 6 Oct 1869.
7 SI.
8 *Eve. Traveller,* 6 Oct 1869.
9 *Nantucket Inquirer,* 9 Oct 1869.
10 *N. Y. Ship List,* 9 & 13 Oct 1869.
11 *Ibid.,* 13 Oct 1869.
12 SI.
13 SI.
14 SI.
15 *Eve. Traveller,* 6 Oct 1869.
16 Proprietors of the Locks and Canals on Merrimack River, Lowell, Massachusetts. *Amount of Rain collected in Rain Guages at the several stations named, during the Great Storm which ended in Massachusetts on Monday, Oct. 4, 1869, and produced destructive floods in many Rivers in the Eastern part of the United States.* Broadside, n.d., 1869?. (National Archives.)
17 *Idem.*
18 *Idem.*
19 *Idem.*
20 SI.
21 SI.
22 SI.
23 *Eastport Sentinel,* 5 Oct 1869, in *N. Y. Trib.,* 11 Oct 1869.
24 *N. Y. Ship List,* 13 Oct 1869.
25 S. L. Boardman, ed. *The Naturalist of the Saint Croix: A Memoir of George A. Boardman.* Bangor, 1903. 57–58.

HATTERAS SOUTH: 1815–1870

THE NORTH CAROLINA HURRICANE OF 1815

A major hurricane cut across extreme eastern North Carolina in early September 1815. In the New Bern area it was judged the worst since 1795, though to the northeast at Edenton the destruction this time was thought not as great as occurred in August 1806. Probably a difference in tidal conditions and exposure of the harbors at the two places accounted for the varying degree of reported damage as the storm under consideration was of *great* dimensions. The North Carolina Hurricane of 1815 deserves more prominence than it has received. Neither Tannehill nor Dunn and Miller list its occurrence, and it was allotted only one paragraph in a contemporary article discussing the hurricane season of 1815. Probably the great publicity attending the famous storm just 20 days later, the Great September Gale in New England, has all but obscured its true stature.

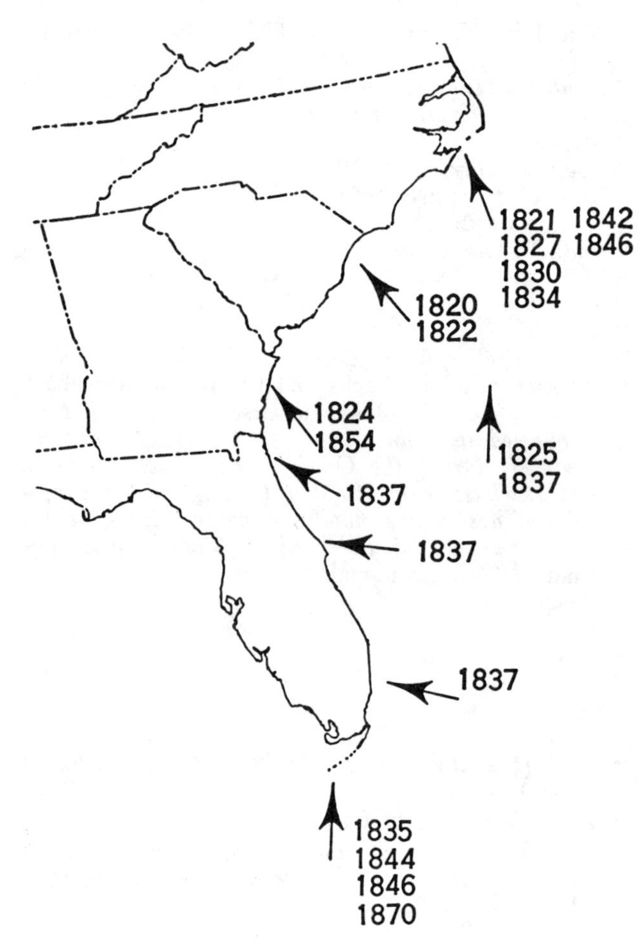

The West Indian antecedents of the Carolina hurricane have been spotted only in the northwesternmost of the Leeward Islands. At St. Barthelemy a severe gale put all the vessels in the harbor on shore as a hurricane moved just to the northeast on Tuesday, 29 August.[1] This could conceivably have been the fringe of the gale that reached the Charleston area on Friday, 1 September, and struck the shoreline near Cape Lookout on Sunday morning. A continuous gale out of the northeast whistled through Charleston from Friday until late Sunday; it then commenced to back to the north as the disturbance drew abreast of Charleston's latitude. By Monday morning the air flow had backed almost to the west as the storm center was then passing over eastern North Carolina.[2]

The sweep of the winds did considerable damage to crops throughout the South Carolina coastal area. Rice suffered heavily along the Cooper River, and cotton on the uplands received serious injury. The accompanying rains, however, put an end to a very troublesome drought situation that had developed during the summer months of 1815.

Our best account of the 1815 hurricane came from the press of Wilmington, North Carolina. The wind there had been out of the northeast for several days, and many sensed that a tempest might be in the offing. On Sunday the wind increased until by evening a full storm raged:

> During Sunday night the wind gradually changed to North, then early Monday morning to North West, increasing in violence with every change; from 8 to 10 o'clock it increased to its utmost violence, during its change from NW to W—about 10 o'clock it gradually came around to the South West, when the rains began to cease and the wind to come in puffs with intervals of 4 or 5 minutes & by 3 o'clock P.M. the rain ceased altogether. The effects of the storm have been visibly marked by a general prostration of fences and trees.[3]

The small tidewater port of New Bern on the Neuse River, northeast of Wilmington, lay in the path of the hurricane's sweep:

> On Sunday night and Monday morning, this town was visited by one of the most destructive gales from the North East, ever experienced in this part of the country. The wind was most violent about three o'clock on Monday morning, and at day light the streets and wharves presented a scene which beggers all description. The streets were rendered almost impassable by trees lying in every direction. Most of the

small wooden buildings in the North East part of town were carried away by the violence of wind and waves—the tide having risen nearly 12 feet above common high water mark.[4]

By 1000 on Monday morning, at the same time that winds at Wilmington were backing into the west, the gale at New Bern, which had been driving hard from the northeast, veered into the south and abated. This timely shift checked the rising waters, though they had already climbed a foot higher than in any storm since 1795. Some streets in New Bern were under six feet of water at the peak flood. The waves were described as breaking against the sides of houses just as if they had been on the seashore itself. Most of the lovely shade trees which embellished this Colonial cultural center of North Carolina were destroyed by the violence of the gale. Along the Neuse River on the south side there was great flood damage, and along the north side the wind proved the chief culprit.[5]

At Beaufort close to Cape Lookout the tide flowed four feet higher than ever known. It burst in doors of houses along the shore even though they were bolted securely. Every vessel at Ocracoke Inlet, some twenty in number, was driven ashore by the shifting gale. Farther north at Edenton on Albemarle Sound the storm raged from 2300 on Sunday night until 1300 the next afternoon. The direction was given as northeast. Destruction there was not thought to have been as extensive as in the great coastal hurricane of 1806.[6]

Inland the sweep of this large hurricane also brought widespread devastation. Fayetteville reported many trees downed and the roads in all directions blocked.[7] Raleigh had the worst storm in years with many trees uprooted in the area.[8] Westward near Winston-Salem the *Moravian Diary* recounted "a frightful storm with hard rain. The wind did much damage to fences, fields of corn, orchards, and in the woods." [9]

An analysis of our few reports on the wind pattern of the North Carolina Hurricane would indicate that the center came ashore close to Cape Lookout. It had passed east of Wilmington and moved very close, but to the east of New Bern. A recurvature to the northeast is then manifest as Edenton, with northeast winds prevailing, lay to the west of the storm track.

The Norfolk *Public Ledger* made several references to the storm of the 5th. Four British ships among others had been forced into that port for repairs.[10] Vessels off Long Island on the 5th also met heavy gales lasting 12 hours. Points along the New England coast, such as New Haven and Salem, endured a hard northeast rainstorm on the 5th and morning of the 6th, though the dangerous part of the hurricane was seaward and caused no material damage on the New England shore.[11]

[1] Gustavia, 2 Sept, in *N. Y. Spectator,* 22 Sept 1815.
[2] *Courier* (Charleston), 4, 5 Sept 1815.
[3] Wilmington, 9 Sept, in *Courier,* 16 Sept 1815.
[4] New Bern, 9 Sept, in *Courier,* 20 Sept 1815.
[5] *Idem.*
[6] *Edenton Gazette,* 12 Sept, in *N. Y. Spectator,* 18 Sept 1815.
[7] Fayetteville, 5 Sept, in *Courier,* 8 Sept 1815.
[8] *Raleigh Minerva* in *Courier,* 13 Sept 1815.
[9] *Moravian Diary,* 7, 3278.
[10] *Public Ledger* (Norfolk), 5 Sept 1815.
[11] Jeremiah Day. Ms. Met. Reg. (Yale); Edward A. Holyoke. Ms. Met. Reg. (Harvard).

THE WINYAW HURRICANE OF 1820

The concave curve of the shoreline of Long Bay, reaching from Cape Fear in North Carolina to Santee Point in South Carolina, lies entirely open to the southeast from whence tropical storms often strike direct blows at the unprotected coast. The modern resort of Myrtle Beach, lying in the center of this arc, has received much publicity in recent years when a seeming concentration of hurricanes have struck with devastating effect. But in earlier times the principal settlement of the area was the port of Georgetown on Winyaw Bay, into which drain the Black, Pee Dee, and Waccamaw rivers. In the early 1800's many thriving plantations with large numbers of slaves clustered about the water courses that comprise much of northeastern South Carolina. During the first quarter of the Nineteenth Century, as in recent years, another cluster of hurricanes struck in 1804, 1811, 1813, 1820 and 1822—so that the phrase "September Gale" became a commonplace and was looked for and feared every year.

The *Winyaw Intellingencer* of 13 September 1820 took note of the September gale of that year:

> "On 10th wind blew tempestuously all day fluctuating between points ENE and NE, but more generally blowing from NE. About sunset the scene became truly awful, the wind increasing in violence, and the tide running with frightful impetuosity. About this period, the church was blown from its foundations, and many of the inhabitants were seen removing from such houses as appeared most exposed to the dangerous tide and wind. After dark the gale continued to increase, and about 10 or 11 o'clock, there raged one of the most violent hurricanes that has ever been experienced here. At this hour the wind began to back (as it is called) to the N, blowing at times in squalls of incredible violence, bringing with them such floods of rain, that there was not a house in the village could entirely resist their fury. The wind about 1 o'clock appeared to have backed as far as NW from which quarter it continued to blow, but with decreasing violence until morning." [1]

The tide in the bay at Georgetown rose four feet above a normal spring tide, to about the height reached in the Great Gale of 1804. The temperature during the storm was reported to read 77.5°.[2]

The maneuvering of the wind at Georgetown would indicate that the center passed inland north of that place. Southward at Charleston the disturbance was not severe, being described as "a smart gale from the north-northeast" accompanied by heavy rain. The wind at Charleston had commenced to blow about midnight and backed into the west by dawn. Fruit and shade trees were stripped of branches and in some instances prostrated, and some lesser buildings were blown down.[3]

To the north we have two significant reports. Fort Johnston at the mouth of the Cape Fear River had "very high wind & rain—a gale most from the southeast" on the 10–11th.[4] A ship off Ocracoke Bar met a heavy southeast gale on the 10th and then experienced an almost dead calm at 0630 of the 11th when a waterspout passed over the ship. Another gale commenced in the afternoon.[5] Thus, the center of the storm must have passed inland north of Georgetown and south of Cape Fear, then recurved to the northeast, to pass close to Ocracoke and Hatteras. The long interval between the two gales off Ocracoke could be accounted for by a slowing-down of the forward movement of the center to an almost stationary phase such as often occurs in this latitude when the eye of a hurricane is in the recurving stage.

[1] *Winyaw Intelligencer* (Georgetown) in *Nat. Int.* (Wash.), 25 Sept 1820.
[2] *Idem.*
[3] *City Gaz.* (Charleston), 12 Sept 1820.
[4] SG.
[5] *Amer. Beacon* (Norfolk), 19 Sept 1820.

THE MEMORABLE CAROLINA HURRICANE OF 1822

A hurricane of fairly small size, but packing tremendous energy struck the South Carolina coast between Charleston and Georgetown close to midnight of 27–28 September 1822. The tight whirl of winds moved forward at an unusual rate so that areas experienced full hurricane blasts for less than four hours. Nevertheless, it was the most devastating storm to hit the South Carolina coast since 1804, and 32 more years would pass until its equal put in an appearance.

From the detailed wind descriptions at Charleston and Georgetown, it is certain that the center passed between the two points, and probably closer to the latter which lay in the dangerous right semicircle of the storm. An unprecedented storm tide accompanied the peak of the wind in this sector, resulting in several hundred deaths by drowning as the rising waters engulfed the extensive lowlands surrounding Winyaw Bay.

At Charleston it was mainly a wind storm that "raged with most ungovernable fury" from north and northwest. The editor of the *Courier* supplied some of the meteorological facts:

> About 10 o'clock a breeze sprang up at N.E. which had increased by 11, to a pretty heavy blow. At 12 it had assumed the desolating power of a West Indian Hurricane, and at 1 o'clock was at its extreme height—having come around by N., from N.E. to N.W. at which later point we encountered its greatest fury—Shortly after two o'clock, it began to abate in violence, and by 3, was again perfectly calm.
>
> The havoc occasioned by this tremendous visitation in the city, is without parallel in the memory of our oldest inhabitants. The Tornado, which passed over a part of it, in the year 1811, was perhaps of equal or even greater violence; but its effects were then confined to a very narrow limit, while the desolation on this occasion, is extended to every part of the city and suburbs.[1]

The temperature at 2200 stood at 70° and the barometer at 30.15″ just before the onset of the gale. Three and a half hours later at the peak of the storm the mercury had risen in the tropical air to 77.5° and the barometer had tumbled rapidly to 29.50″. The areas of the city in the western and northern districts suffered the greatest damage. Hardly a house did not receive some minor structural damage, to attest the strength of the gale. But the harbor area escaped the extensive losses to shipping and wharves that occur when the storm winds come out of an easterly quarter. The tide rose and fell six feet in the short period of 45 minutes around 0100.[2] Eight persons were reported drowned at Charleston and four at Sullivan's Island.[3]

According to press reports the gale did not extend more than 25 miles inland from the coastal area. At Pineville its strength was not great, and advices received from Combahee, Parker's Ferry, and Monk's Corner said the "wind by no means violent there."[4] To the south, Savannah had a brief gale of only 20 minutes duration with no damage being sustained.[5]

The northern sector of the storm appears to have been of much greater areal extent and to have carried far superior wind energy. The onshore slant of the winds caused a high storm tide from Winyaw Bay northward at least as far as Cape Fear where Fort Johnston had "rain with very high wind from east" on the 27th.[6]

A full account of the meteorological circumstances on the 27–28th accompanying the hurricane at Georgetown, along with a description of the storm tide at

North Island astride the entrance to the bay, was carried by the *Winyaw Intelligencer:*

> The weather had been for a week or ten days very unpleasant, the wind blowing occasionally fresh from the E and SE, but as there appeared none of those indications which usually precede a hurricane, and as the mercury in the thermometer continued low, very little apprehension was entertained; even at sunset Friday evening the 27th inst. although the weather was bad, yet there appeared no cause to apprehend a gale; at the close of the day, there was a heavy shower from the S.E. accompanied by some wind, after which the weather appeared better; between ten and eleven o'clock, however, we had a squall from the north-east, from which quarter the wind continued to blow high till about 12, when we experienced another more violent squall from about E.
>
> The mercury about this hour had risen to 79, and continued to rise for some time after. From 12, the wind continued gradually to change to S.E. and S., increasing in violence as it shifted; from S.E. it blew with frightful and unprecedented violence; most of the injury caused by the wind must have occurred about 2 o'clock in the morning, and whilst it blew from this quarter.
>
> As the time of high water had been about seven o'clock in the evening, the inhabitants apprehended no danger from the tide, as, from the violence of the gale, it was presumed that it could not continue until the period of the succeeding high water. In this expectation, however, it pleased the Almighty to disappoint them, and, by the awful result, to prove how fallacious are all human calculations.
>
> The tide could have ebbed little at all, when the waters returned with irresistible violence, and between 3 and 4 o'clock in the morning had reached a height far exceeding that of the gale of 1804, and we believe of any other tide within the memory of the oldest inhabitants—a very small portion of the Island remained above the ocean.
>
> The gale began to subside, we believe, about half past 3 o'clock, the wind then blowing from S.W. It was oppressively warm during the gale, and many of those luminous bodies, or meteors, unusual in our fall gales, passed near the surface of the earth. The gale was of shorter duration and accompanied by less rain than usual.[7]

The death toll ran very high as many plantations on North Island and the shores of the bay were completely overflown by the rampaging tide. The editor estimated that more than 120 negroes and five whites were drowned on North Island alone! A total loss of about 300 seemed a reasonable figure.[8] As a result of the tragedy, a number of stormproof structures resembling small towers were built on North Island to afford protection for the slaves whose flimsy houses were engulfed by the waters of the Atlantic Ocean. Some of these were still standing in 1893 when another major disaster took another enormous toll of lives along the Carolina coastal plain.[9]

The settlement of Georgetown, itself, lying on the western shore of the bay about eight miles inland, also felt the lash of the southeasterly hurricane:

> We must content ourselves with stating that our town exhibits, at this moment, a scene of ruin and desolation never surpassed in this State. The wind appears to have been full as violent as it was at North Inlet—the tide, however, certainly did not rise so high.
>
> The Court-House has sustained very serious injury, and many of the records of the Clerk's Office destroyed; the Sheriff's Office had every door and window blow in and the records and papers destroyed; the four chimnies of the Jail have been blown down and the building in other respects much injured; many of the tiles have been blown from the roof of the Bank; the building over the Market, occupied by the Town Council, is nearly down, every pillar which supports it, being fractured.
>
> We have no particular accounts of the injury sustained in the crops; but it must necessarily be great, as much of the rice which was harvested has been blown out of the barn yards and dispersed—many negroes have been killed, and most of the barns and mills have been unroofed and otherwise injured, and the banks and trunks torn to pieces.
>
> The schooner *Little Jack,* captain Thomas Davis, which was up the Waccamaw River taking in a load of rice, nearly foundered at her anchors, and when she parted her cables was driven on shore and bilged. From the number of trees which have been thrown across the roads they are rendered impassable. Planters who have visited their plantations eight or nine miles from Town have been three or four hours in meeting them, being obliged to pursue their way, through the woods, the roads being literally blocked up.[10]

From its landfall near Cape Romain, the eye of the hurricane continued on a course slightly west of north, on an approximate Florence-High Point-Roanoke axis. The mail driver reported that roads were completely blocked north of the Santee River for many miles, but south of that stream toward Charleston they were in a much less tangled condition.[11] One planter, "a stout man with an excellent horse," was able to make only 14 miles northward progress from Georgetown in two days time following the storm.[12] Fayetteville, 75 miles inland and directly north of Georgetown, experienced the same meteorological conditions about three hours later. At 0300 the wind shifted into the southeast accompanied by "a heavy fall of rain" and "blew for two hours with uncommon violence." At daybreak devastation lay all about.[13]

The next checkpoint was Hillsboro, the old provincal capital lying about 13 miles northwest of Durham in the central part of the state. There had been a period of showery weather until "early on Saturday morning, when it commenced blowing with great violence from east-southeast accompanied with rain which continued for about two hours." Houses were unroofed and trees blown down by the force of the gale. Raleigh and

Salisbury noticed the wind, but reported no unusual destruction.[14]

The sweep of the gale force winds reached as far inland as Monticello in west-central Virginia. Thomas Jefferson's granddaughter wrote: "We have had constant rains, & on the 25th of September a violent storm which strewed the whole mountain top with broken boughs of trees, & tore one of our willows completely asunder . . ." Miss Randolph, though writing only four days later, apparently misdated the occurrence of the storm which took place there on the 28th.[15] Lynchburg also experienced winds strong enough to topple chimneys and to tear up trees by the roots.[16]

In the Blue Ridge Mountains of western Virginia, as well as in the middle Alleghanies, the storm brought very heavy rains. The downpours commenced on the evening of the 27th in the Staunton River watershed with an easterly air flow and continued until 0900 on the 28th. In this short time the river rose to "the greatest height ever known" and swept away crops which a few days before had been parching in a major drought.[17]

Our last checkpoint for the 1822 storm came at Washington and Baltimore. "Stormy weather which usually accompanies the equinoctial period" visited the Nation's Capital on the 27–28th with east and southeast winds, but no gales. Rainfall at Baltimore amounted to only 0.15″, indicating that the storm's scene of action and line of maximum rainfall lay well south of that district.[18]

[1] *Courier* (Charleston), 30 Sept 1822.
[2] *Southern Patriot* (Charleston) in *Amer. Adv.* (Phila.), 9 Oct 1922.
[3] *Charleston Mercury,* 1 Oct 1822.
[4] *Idem.*
[5] *Georgia Republican* (Savannah), 1 Oct 1822.
[6] SG.
[7] *Winyaw Intelligencer* (Georgetown), 2 Oct, in *Nat. Int.* (Wash.), 16 Oct 1822.
[8] *Charleston Mercury,* 7 Oct 1822.
[9] *Monthly Weather Review* (Wash.), 24 May 1896, 155.
[10] *Winyaw Int.,* 5 Oct, in *Mercury,* 12 Oct 1822.
[11] *Mercury,* 8 Oct 1822.
[12] *Nat. Int.,* 16 Oct 1822.
[13] Fayetteville, 3 Oct, in *Amer. Adv.,* 9 Oct 1822.
[14] *Idem.*
[15] Virginia Randolph to Nicholas Trist, Monticello, 1 Oct 1822, Nicholas Trist Papers (Library of Congress), quoted by E. M. Betts. *Thomas Jefferson's Garden Book* 1766–1824. 604.
[16] *Nat. Int.,* 9 Oct 1822.
[17] *Norristown Herald* (Pa.), 23 Oct 1822.
[18] *Nat. Int.,* 30 Sept 1822; Brantz. Ms. Met. Reg. (National Archives).

THE GEORGIA COASTAL HURRICANE AND STORM TIDE OF 1824

A memorable storm tide, generated by a small, savage hurricane, overwhelmed the Georgia coastal area on 14–15 September 1824 as the center of the disturbance smashed ashore somewhere between Jacksonville and Savannah. Its previous history is shrouded in silence. Neither Poey nor Tannehill mentions any activity in the West Indies at this time. Some hint of its inexorable course northwestward toward the American mainland can be gleaned from the contemporary disaster reports in the *New York Shipping List;* these tell of a vessel being driven high and dry at Turk's Island on the 11th, and of others ashore in the central Bahamas on Rum Key, Long Island, and Harbor Island at the same time. Turk's Island lies some 600 miles southeast of the Georgia coast, a not unreasonable distance for a tropical storm to cover in two-and-a-half days.[1]

The concave spread of the Georgia coastline is made up of a jumble of irregular islands surrounded by little bays and crisscrossed by innumerable tiny inlets. The central part of the complex comprises the delta of the Altamaha River which is served by the tidal port of Darien and flanked by the guardian islands of Sapelo and St. Simon. These bits of land, though attractive to the planter and the vacationer, were especially subject to storm inundations from the sea since the whole delta area lies only a few feet above sea level. It was here that the Great Gale of 1804 took a heavy toll, and this was repeated to an even greater degree just twenty years later. Sir Charles Lyell, the eminent touring British geologist, when visiting St. Simon Island in the 1840's, learned of the disastrous effects of the 1824 storm flood and gave it wide publicity in his famous work: *A Second Visit to the United States.*[2]

We do not have any satisfactory meteorological report from the coastal strip at this time. The *Savannah Georgian* judged: "the storm at Darien, and in its neighborhood, exceeded that of 1804 both in violence and destruction." [3] On St. Simon Island early reports had nearly all property destroyed and the entire island overflowed by the inrushing waters, driven by the southeast hurricane. A week after the storm the press reported that 83 persons, at least, had been drowned or lost there. An early letter from Darien to the *Savannah Republican,* in confirming the fears that had been expressed in the city for the safety of their compatriots on the coast, painted a doleful picture of the scene:

> To attempt to describe the effects of the gale with us would be undertaking a task which I am bold to say no man can do justice to: I will not pretend to say anything of it further than that it has been with us and has passed, desolating and making bare everything in its passage—The damage to property is

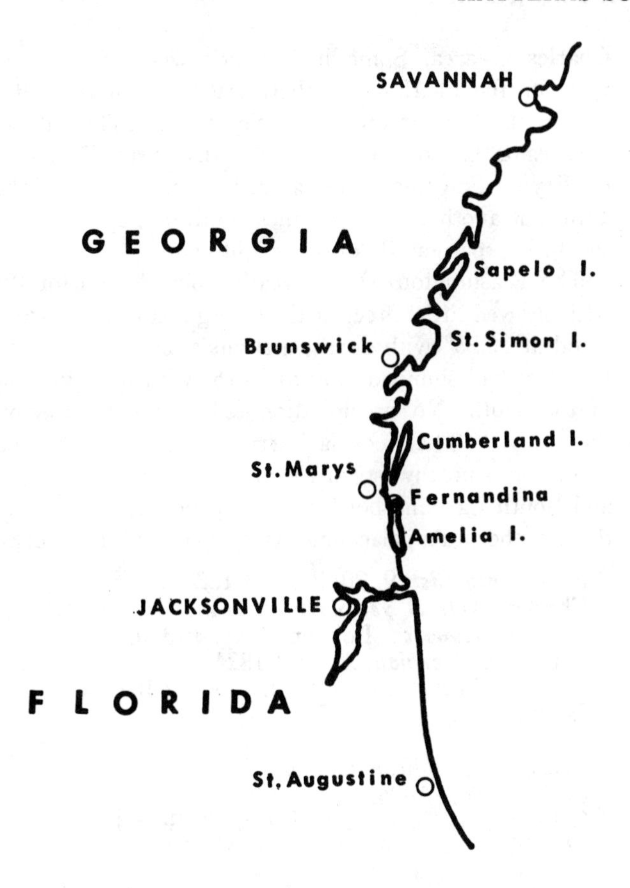

beyond calculation, the loss of lives immense—I suppose in our country not short of one hundred persons have been destroyed, some of the most heart-rendering and melancholy cases—whole families separated and crushed amidst the ruins of buildings, or drowned in the water thrown up by the sea. I look upon it as death to all our prospects. I do not know what is to become of our country.[4]

The hurricane's effect on transportation facilities between Savannah and the shore islands was summed up by the editor who interviewed the first mail carrier to arrive from Darien:

> From Mr. Page, who arrived with the mail last evening, we learn that the Bridge over Aiken's Creek, 7 miles from town, and also both ends of the Ogechee Bridge are gone.
>
> The bridge at South Newport, the two Sapelo bridges, and nearly all the smaller bridges between this (Savannah) and Darien are swept away. The road is nearly covered with trees. Upwards of a week must elapse before the Commissioners will be able to commence repairing damages, owing to the negroes being wanted for that purpose on the plantations.
>
> Mr. Page was three days going to Darien, and three in returning, and performing the journey each way by alternately swimming and using canoes. The water in some places is still rising.
>
> A gentleman arrived from Bullock County reports the crops in that quarter, as also having suffered excessively. Jenk's bridge is entirely gone, and the water of the Ogechee still rising.[5]

On St. Simon Island the heavy losses were summarized by the editor of the *Georgian:*

> On St. Simon's (sic) the loss has been excessive. Mr. Cooper's cotton crop is entirely lost, and a great many of his out buildings demolished, and considerable damage done to his old and new dwelling houses. About 30 acres of his cotton is covered with 2 to 3 feet of thick marsh grass. An effort will be made to remove it. The estate of Butler has lost crop, out buildings, &c. Dr. Grant's crop, out buildings, &c. on the island destroyed, and also the crop &c. on his plantation on the main, with three feet of water on the fields. Mr. Hamilton's on the island has lost crop and several buildings, Capt. Demere has lost his crop, negro houses &c., together with his boat's crew of five negroes in going from Darien to St. Simons. Major Page has lost crop and out buildings &c. The houses and crop of Gould are also destroyed, excepting the light house and dwelling. The Rev. Mr. Matthews and Mr. Hazzard have suffered in a like degree with the others on the island.
>
> The crops on Champney's and Butler's Island, and Brailsford's are all gone; at Brailsford's one man was killed.
>
> On little St. Simon's, 100 sheep lost. Mr. Spaulding, on Sapelo, lost his crop and 250 head of cattle, and 27 horses and mules.
>
> At Hopeton, Mr. Cooper's plantation above Darien, all the cotton, most of the corn, rice, buildings are gone.
>
> The Light House at Doboy, has been injured and lost the lantern, and of course cannot be lighted for some time.[6]

Our only wind direction report for the entire area indicated that the center of the hurricane passed to the south of Savannah, very likely coming ashore not far to the north of Cape Fernandina and the St. Marys River on the Georgia-Florida border. A report from the settlement of St. Mary, which lies on the north side of the river estuary, stated that, though the storm was felt there, extensive damage had not occurred such as had visited St. Simon. Lowlying St. Mary is susceptible to inundations from storm tides when the wind is from an easterly quarter. Assuming that St. Mary lay to the south of the storm track, the air flow there would be mainly from a westerly quadrant which would greatly lessen the danger of serious flooding.[7]

A study of press accounts of the storm's behavior in the Savannah area pointed to the passage of a hurricane of short duration, first approaching from the southeast to south, then moving to the west and northwest of that place:

> On Tuesday night last (14th), our city experienced probably the most severe gale ever visited since that of 1804. We had had threatening weather for some days previous, but no serious apprehensions were entertained until about 3 o'clock P.M. that day at which commenced blowing fresh from N.E. accompanied with heavy rain, and continued to increase until about 2 o'clock in the morning when it suddenly

changed to S.E. and blew with increasing violence for about an hour and a half and then gradually subsided.

The tide came up over the wharves, but not high enough to do any damage to goods in the stores.

The damage done to the city is very considerable. Almost every slate roofed building is more or less injured,—and we should think that at least three-fourths of the ornamental trees in the city are prostrated, or otherwise injured. The old trees on the Bay stood out the gale better than any others, but a large number of them are down, or nearly deprived of their branches.[8]

A freshet of immense proportions followed on all rivers draining from the Georgia and South Carolina hills eastward into the Atlantic as tropical downpours deluged both the coastal plain and the Piedmont. Along the Savannah River it mounted to the worst flood since the memorable high water of January 1796, and damage was reckoned equally severe with that unforgetable event since the flood this year came at a time when the crops were mature and ready for harvest.[9] Reports of immense damage came from all inland areas. Losses on plantations around Columbia, South Carolina, ran to extreme figures: Wade Hampton estimated his damage at $35–40,000.[10] The only rainfall measurement we have from the storm area was made at Fulton in Sumter County in central South Carolina where a September total of 9.20″ would indicate a sizable contribution from the hurricane.[11]

The storm was not of serious proportions in the Charleston area. Some high winds were experienced there on the 14–15th, but there was "no material damage in the city or on Sullivan's Island." The tide in Charleston harbor rose no higher than usual.[12] A ship off Frying Pan Shoals met a hard northeast gale on the 13th, but another only 30 miles south of Cape Fear on the 14th experienced "only a slight gale." [13]

The Washington, D. C., weather observation for the 14th showed little effect of the Georgia storm. Pressure stood at 30.25″ with northwest winds, and on the 15th the glass had slumped only to 30.15″ with the wind out of the south. No precipitation fell at the Capital on these days.[14] The Georgia hurricane apparently passed inland on a northwestward course close to the Georgia and South Carolina border and expended its energy in the Southern Appalachians as a torrential rainstorm.

[1] *N. Y. Ship List*, 9, 20, 23, Oct 1824.
[2] Charles Lyell. *A Second Visit to the United States of North America.* London, 1849. I, 340.
[3] *Savannah Georgian,* 23 Sept 1824.
[4] *Georgian* in *Nat. Int.* (Wash.), 9 Oct 1824.
[5] *Idem.*
[6] *Idem.*
[7] *Georgia Republican* (Savannah), 30 Sept 1824.
[8] *Ibid.*, 16 Sept 1824.
[9] Columbia, 17 Sept, in *Nat. Int.*, 30 Sept 1824.
[10] *State Gazette* (Columbia, S. C.), 25 Sept, in *Republican,* 28 Sept 1824.
[11] Ms. Met. Reg. (National Archives).
[12] *Republican,* 21 Sept 1824.
[13] *Idem.*
[14] *Nat. Int.,* 16 Oct 1824.

THE EARLY JUNE HURRICANE OF 1825—I

Stormy conditions prevailed in the Greater Antilles at the close of May 1825. Santo Domingo had a severe storm on 28 May in which among others four American ships were lost.[1] A gale lashed Cuba on 1 June; many ships in Havana harbor were forced to run to sea to save themselves.[2] No doubt, the hurricane under discussion was spawned in these disturbed conditions. All reports point to this disturbance as having achieved full hurricane force with major damage reported from land stations from Charleston to New York City—the outstanding example of an early season mainland hurricane.

Our first notice on the continent came from St. Augustine, Florida, where the surgeon of the U. S. Army post at Fort Marion kept his eye on the weather as was now required by the new regulations governing the U. S. Medical Corps. With the wind out of the east, rain commenced on the morning of Thursday, 2 June, with a shift later in the day to southeast-by-south. No mention was made of any excessive wind speeds. By next morning the wind had backed into the northeast with the rain continuing until 1400.[3] The wind behavior would place the storm center as first approaching from the south or southwest, but finally passing to the east on the morning of the 3rd.

The disturbance reached the Charleston area about 1000, Friday morning, and continued throughout the day. Trees and fences were reported leveled at the South Carolina city.[4] Farther north in eastern North Carolina the *Star* of Elizabeth City clocked the storm as reaching that area on Friday afternoon with fresh winds from the northeast which increased to "a heavy gale" by midnight. The June tempest raged all night and until 1000, Saturday morning the 4th, causing great destruction to trees, some of which were pulled out of the ground by their roots, to attest the whole gale or hurricane strength of the wind. [5]

Along the Outer Banks of North Carolina the hurricane lashed at shipping and settlements. The post surgeon at Fort Johnston at Cape Fear reported "a very high wind & storm which lasted 30 hours. Wind south." [6] A press dispatch from Adams Creek told of very heavy losses with crops destroyed and cattle drowned as the storm tide rose 14 feet above low water to engulf fields and barns.[7] Along the shore of the

Outer Banks near Ocracoke Inlet, 25 sail were driven ashore by the unexpected severity of the early season blow.[8]

The evidence from North Carolina—the height of the tide in Albemarle and Pamlico Sounds and the strength of the wind—would place the track of the June hurricane as cutting a traverse across the coastal sections. If the Elizabeth City report of northeast winds is correct, the center recurved seaward south of that place.

The occurrence of a severe storm of tropical origin in June along the South Atlantic coast is not unique. In the period under consideration, others swept up the coast on 10 June 1831 and 3 June 1838.

[1] *N. Y. Ship List,* 9 July 1825.
[2] *Ibid.,* 25 June 1825.
[3] SG.
[4] *National Journal* (Wash.), 14 June 1825.
[5] *Star* (Elizabeth City) in *Nat. Jour.,* 9 June 1825.
[6] SG.
[7] *Sat. Eve. Post* (Phila.), 25 June 1825.
[8] *N. Y. Ship List,* 15 June 1825.

THE GREAT NORTH CAROLINA HURRICANE OF 1827

The season of 1827 proved prolific in tropical storm production. At least two intense hurricanes roared up the Atlantic Coast and another pushed westward across the Gulf of Mexico to strike a savage blow at the coast of Mexico near Vera Cruz.

William C. Redfield, who had first subjected the Long Island Hurricane of 1821 to scholarly dissection, took up the storm of late August 1827 in his second study. With data gathered from ship logs, newspapers, and conversations with mariners, he established the following timetable:

Windward Islands August 17th
St. Martin and St. Thomas 18th
Passed northeast coast of Haiti 19th
Turks Island 20th
Bahamas 20–21st
Off Florida and South Carolina 23–24th
Off Cape Hatteras 25th
Off Delaware 26th
Off Nantucket 27th
Off Sable Island and Porpoise Bank 28th [1]

We have no reports at hand to confirm the storm's presence in the Windward Islands, but its activity in the Leeward group on the 17th received wide press coverage. Antigua and St. Croix were hit by the already severe disturbance on the 17th; [2] at the latter the wind was considered "the worst in forty years" and continued at hurricane strength for ten hours. At St. Kitts it was the "most violent in more than 50 years." [3] Moving westward, the storm's encircling winds encompassed Puerto Rico where enormous crop damage resulted. It appeared probable that the center passed along the southern shores of Puerto Rico and part of Hispaniola as Port au Prince had "a severe hurricane" on the 18th.[4] A traverse across the Dominican Republic is indicated as this would give Port au Prince west or northwest gales, the only quarter from which its harbor is not protected.

Like many others before and after, this hurricane must have hesitated and churned around the Bahamas region for five days before a steering current decided on a definite course toward the American mainland. Redfield placed the center near Turks Island in the southeastern Bahamas on the 20th, yet it was not until early afternoon on the 25th that the eye was opposite Charleston and about to move inland somewhere along the stretch of sandy barrier beaches on the North Carolina coast between Cape Fear and Cape Hatteras. At Charleston, though well to the west of the storm track, the wind blew "with great violence" for six hours on Saturday afternoon the 25th with the strongest blow coming from north and northwest between 1200 and 1800. The peak wind period coincided with low tide so little damage occurred at the South Carolina port.[5]

The exact landfall of this large storm cannot be determined with the data at hand, nor is there sufficient meteorological information available to chart its course over the land. William C. Redfield in his 1837 study showed the track of the center as passing close, but to the east of Cape Hatteras. The data assembled below, however, would indicate that the eye of the hurricane did pass over land while making its recurvature to the northeast. There is little doubt that this was a *great* hurricane with destructive winds, extraordinary tides, and very heavy rainfall, and that all of these elements covered an unusually large geographical area. It must be accorded a place in top rank among North Carolina storms.

The North Carolina port of Wilmington on the Cape Fear River was the only city possessing a press that lay near the landfall area. The *Recorder's* account, as copied by exchange papers, is not completely satisfying from the meteorological point of view. It mentioned the wind at 1400, without giving its direction, as "increasing with redoubled fury" and "the water rising with astonishing rapidity." This apparently referred to a shift of wind from northeast to southeast as the center passed to the southwest. Waves rolled over the tops of garden fences as far as 600 feet from the beach; at its peak the water was estimated at ten feet above normal high water mark. Only a southeast wind can create such a condition on the Cape Fear River.[6]

The towns of New Bern and Washington, both heads of navigation for tidal rivers emptying into Pamlico Sound, suffered severely from high tides, and such high waters here, too, are always caused by wind with an easterly component. At Washington the tide was "12 to 15 feet above ordinary tides" and houses on Water Street found the river five or six feet deep in their first floors during the height of the storm tide. At New Bern all communication for awhile was by canoe.[7]

Near Cape Hatteras two New York-to-North Carolina packets were driven ashore and smashed to pieces by the tremendous breakers. The new Cape Hatteras Lightship off Ocracoke Inlet, only two years on the job, broke loose and piled up on the south side of Ocracoke Island where the captain, his wife, and three daughters were rescued from the waves. The inlet opposite Masonborough Sound was widened by half-a-mile. "It was more violent than the memorable one in 1815," concluded the editor of the Washington *Recorder.* He noticed that out of 20 sail in Portsmouth Roads near Hatteras, only two rode out the gale without beaching or going to sea, and that but one mill was left standing within 10 miles after the combined onslaught of wind and tide.[8]

The storm area spread its influences widely over North Carolina. Fayetteville in the southeast had a 30-hour period of gales and rain out of the northeast commencing early on Saturday morning, the 25th. The editor of the local *Observer* considered the storm "little inferior to that of 1822."[9]

In the west-central part of the state near Winston-Salem, the gales together with flood waters caused considerable damage. The *Moravian Diary* at Bethabara related: "Aug 25—From morning on we had heavy rain accompanied by hurricane wind that broke or uprooted fruit and forest trees, overturned fences and bridges, even damaged roofs here and there. This frightful storm continued until during the morning of the following day, the 26th."[10] Wind direction was given as northeast, the same as at Fayetteville and Raleigh, putting these places in the western sector of the cyclonic system.

The violence of the storm was largely spent over North Carolina as far as the mainland was concerned. Unfortunately, we do not have reliable wind reports to trace its course northward and northeastward. The wind activity at Norfolk proved worthy of comment by the editor of the *American Beacon:* "that were it not that there is much less destruction of property, he should not pronounce it nearly equal in violence to the memorable tempest which visited there in 1821." There was little tidal damage at Norfolk, due to a happy combination of wind direction and ebb conditions.[11]

Farther north there were high tides at Baltimore as a northeast rain storm swept the area. Washington and New York papers also mentioned the northeaster.[12] At East Windsor, Connecticut, the 27th brought a hard rain during the previous night and forenoon.[13] Out at Nantucket the *Inquirer's* weather reporter saw his barometer near its low at 1400 on the 27th with wind from the south—by 2200 the glass had commenced to rise with a northerly air flow prevailing.[14] A recurvature of the storm seaward over North Carolina is indicated, very probably passing along the familiar track southeast of Norfolk and then northeastward. The effects of this vast disturbance reached well inland over New England and the Nantucket report is indicative of a passage of the center inside that island and Cape Cod.

[1] W. C. Redfield. *Amer. Jour. Science.* 31–1. Jan 1837. 123–24.
[2] *N. Y. Ship List,* 5, 8, 15 Sept 1827.
[3] *Ibid.,* 26 Sept 1827.
[4] *Nat. Int.* (Wash.), 15 Sept 1827.
[5] *City Gaz.* (Charleston), 27 Aug 1827.
[6] *Wilmington Recorder* in *Nat. Int.,* 5 Sept 1827.
[7] Newbern and Washington, N. C. in *American Beacon* (Norfolk), 2 Sept 1827.
[8] Washington, N. C. in *Nat. Int.,* 5 Sept 1827.
[9] *Fayetteville Observer* in *City Gaz.,* 4 Sept 1827.
[10] *Moravian Diary.* 8, 3823.
[11] *Amer. Beacon,* 2 Sept 1827.
[12] *Nat. Int.,* 1 Sept 1827.
[13] Thomas Robbins. *Diary.* 2, 64.
[14] *Nantucket Inquirer,* 1 Sept 1827.

THE TWIN ATLANTIC COAST HURRICANES OF AUGUST 1830—I

Two destructive hurricanes swept up the Atlantic seaboard states within a week of each other in late August 1830. The first was particularly destructive along the South Atlantic coast from Georgia to North Carolina, and the second struck its heaviest blows at the Cape Cod area of southeastern Massachusetts. Shipping in the sea lanes offshore took a severe beating from each storm, some vessels being caught in both. The occurrence of these hurricanes coincided with William C. Redfield's first intensive researches into the subject of storm behavior. We are fortunate, in this case, that he collected much contemporary data about each disturbance; the first, especially, afforded him considerable original material about coastal hurricanes from which he was able to confirm his beliefs in the rotary circulation and progressive movement of such storms.[1]

The first disturbance appeared to have originated somewhere to the east or northeast of the Leeward Islands. Its first identification came from the Virgin Islands where a hurricane was experienced at St.

Thomas on the night of August 12–13th. The Bahama Islands lay in its path on the 14–15th, and late on the 15th the storm was "very severe" when abreast of St. Augustine, Florida, where the weather observer at Fort Marion recorded "a violent gale" from the northeast on the 15th with rain and temperature close to 80°.[2]

From the Revenue Cutter *South Carolina* stationed in St. Andrews Sound (near Brunswick in southeastern Georgia), the captain later reported that the wind commenced to blow strong from the northeast about 2000 on Sunday evening, the 15th, remaining high until 0200 when a shift to southwest occurred. The gale raged through early morning until 0800 when it commenced to abate. The hour of wind shift at 0200 can be taken as the approximate time that the center was abreast of the southeastern Georgia point.[3]

At Savannah the wind picked up Sunday evening as the storm approached from the south. Early on the 16th, from 0100 to 0900, it blew violently from the northeast. A shift took place between 0900 and 1000, with speeds remaining high until noon; then it moderated and veered more and more to the west. A total of five inches of rain was measured in the local rain gage.[4]

Northward in the Charleston area, while the hurricane was raging off Florida's coast on Sunday, the wind had been from the east and southeast. Increasing during the night, it blew violently between 0700 and 0800 from the southeast. After a brief period of moderation, a renewal came at 1000 with hurricane force speeds until 1600. At that hour the direction suddenly shifted to northeast as the center apparently passed a short distance seaward. During the evening the wind worked around to northwest, dropped below gale force by 2100, and the storm entirely ceased by 2230. At Sullivan's Island at the mouth of Charleston Harbor the shift from southeast to northeast is credited with saving the entire island from the threat of total inundation. The press carried a long list of shipping mishaps along the Carolina and Georgia coasts.[5]

Farther north on the coast at Georgetown the same wind shift characteristics were repeated. At first southeast gales swept Winyaw Bay, doing great harm to the dikes that protected the rice fields along the northern shore; but the subsequent shift to a strong northerly flow, while the tide was still three feet above normal high water, caused the southern banks to crumble also. Salt water spilled into the rice fields. The Georgetown *Intelligencer* estimated that at least 20 plantations on Winyaw Bay would make little more than seed as a consequence of the salt water intrusions.[6] A writer in the *Southern Agriculturalist* thought "the destruction of the rice crop would exceed anything of the kind within the recollection of the oldest inhabitant."[7] It was the coincidence of high tide at the period of peak winds that spelled great misfortune for the often suffering Winyaw Bay area.

Though our wind reports from eastern North Carolina are not satisfactory, it appeared that the center of the hurricane, after skirting both Savannah and Charleston, cut across the extreme eastern part of the Tarheel State, probably passing inside Cape Lookout and Cape Hatteras and then veering toward the northeast with a seaward movement south of Norfolk. Ships in the vicinity of Cape Hatteras experienced gales first from the southeast and then from the northwest. At Norfolk the only direction mentioned was northeast, to confirm the above surmise as to the path of the eye.[8]

All settlements in eastern North Carolina underwent the lashing of severe gales with widespread structural damage to buildings and extensive losses to shipping. Here the storm raged all day Tuesday, the 17th. At exposed Smithville, where the wind first came from the east, two houses were blown down before the female residents of the community sought the shelter of the U.S. Army garrison for safety.[9] From Fayetteville, New Bern, Washington, and Elizabeth City storm accounts were featured in the press. Many ships rested high and dry near New Bern and also on Beacon Island.[10] In the Norfolk area corn crops throughout the lower Chesapeake Bay farms suffered grievously.[11] The storm there commenced on Tuesday morning, or more than 24 hours after it had first struck Savannah.[12]

[1] William C. Redfield. Remarks on the prevailing storms of the Atlantic coast, of the North American States. *Amer. Jour. Science.* 20, July 1831, 34–38.
[2] SG.
[3] *Savannah Georgian* in *N. Y. Mercury,* 1 Sept 1830.
[4] *Idem.*
[5] Charleston dispatch in *N. Y. Mercury,* 1 Sept 1830.
[6] *Idem.*
[7] *Southern Agriculturalist* (Charleston), Dec 1830, 640.
[8] *Norfolk Beacon,* 20 Aug 1830.
[9] *N. Y. Mercury,* 1 Sept 1830.
[10] *National Gaz.* (Phila.), 26 Aug 1830.
[11] *Norfolk Beacon* in *N. Y. Mercury,* 25 Aug 1830.
[12] *New Jersey Mirror* (Mount Holly), 1 Sept 1830.

THE CAROLINA HURRICANE OF 1834

A small hurricane came inland on the Carolina shoreline close to the two-state border on September 4th. The South Carolina port of Georgetown received the brunt of the storm tide and wind, suffering its most serious inundation since at least 1804.

The first account of the disturbance came from a

ship report of a gale off Tybee Island, Georgia, on the 3d.[1] The small storm center kept well at sea, moving east and north of Charleston where a northeast gale prevailed throughout Thursday, the 4th. The *Courier* noticed the storm:

About 11 o'clock on Wednesday night it commenced raining very heavily, which continued accompanied by much wind, until 8 o'clock yesterday morning—the wind then increased to a gale, blowing from the North West. In a short time it backed around to West South West, and blew very heavily until about 3 o'clock P.M. at which time it abated considerably, although the rain continued to fall near sun set, when it ceased and there was some appearance of the clouds breaking away.

The wind being from a quarter favorable to the safety of shipping, we understand that they sustained no damage—but in the city, a great number of trees were upturned and broken off; some of the slates on the houses broken off, and in several instances the roofs injured; and many fences in the out skirts of the city were prostrated.[2]

The gale merely sideswiped the central South Carolina coast as it was not felt 40 to 50 miles inland on the railroad to Columbia. Darien, S. C., had heavy rain, but no wind; Aiken had neither wind nor rain.[3]

The storm flood at Georgetown was of record proportions. The rice in the fields was under water for a period of about 12 hours. The water rose 12 to 15 inches above the floors of the warehouses along the river, and all of the wharves were flooded and suffered some injury with the exception of those which had recently been repaired.[4] The editor of the *Georgetown Union* compared the event with past occurrences:

In fact the tide did not fall perceptibly before one o'clock. It is said by one of our respectable and oldest citizens, to have risen higher than in the gale of 1804. The fatal storm of 1822 was of short duration; and by a sudden change in the wind the water was driven back and did not rise near as high as in 1804—while at North Inlet the tide, impelled by a tornado, rose to unparalleled heights and destroyed 7 dwelling houses, &c. one church and 37 lives. The storm of 1822, for strength and mischief done while it lasted, certainly claims preeminence above any known or recorded to have occurred in this neighborhood; but for duration and loss of every sort, except life, this gale of 1834 is unequalled. We have already said no wharves were to be seen. This cannot be said of any gale since 1804, and then the water was not so high. Far as the eye could reach, the fields were covered, and but for the appearance here and there of a tree or cluster of bulrushes, vines, &c. we should not have known that valuable plantations lay under the overwhelming waters. It seemed much as if Old Chaos was about to come again.[5]

Wilmington, North Carolina, though more protected than Georgetown, also received the full brunt of the hurricane. The gale was "very severe" there—three vessels were noticed in distress off the bar and others were thought to have beached in the vicinity.[6] Inland the storm created a great freshet. The Cape Frear River rose an incredible 46 feet, according to press reports, and great damage occurred in the Fayetteville area.[7] Raleigh also experienced a "tremendous fresh" with hardly a mill or bridge in the vicinity left standing when the waters subsided. The wind also did considerable damage to trees and fences in the central North Carolina farm country.[8]

[1] *N. Y. Ship List,* 10 Sept 1834.
[2] *Courier* (Charleston), 5 Sept 1834.
[3] *Ibid.,* 8 Sept 1834.
[4] *Georgetown Union,* 4 Sept, in *Nat. Int.* (Wash.), 9 Sept 1834.
[5] *Idem.*
[6] Wilmington, 6 Sept, in *Courier,* 12 Sept 1834.
[7] *Camden Journal,* 13 Sept, in *Nat. Int.,* 27 Sept 1834.
[8] *Raleigh Register,* 9 Sept, in *Nat. Int.,* 16 Sept 1834.

THE SOUTH FLORIDA HURRICANE OF 1835

The season of 1835 produced a number of hurricanes of widely varying character, diverse geographical origin, and divergent tracks. The outstanding August storm had traveled almost due west from the West Indies, across the Gulf of Mexico, to the Mexican coast below Matamoras. A shift in the upper-air steering currents had taken place by mid-September when the principal hurricane of the month crisscrossed the path of the August storm to the east of Key West. From a point of origin near Jamaica, the September storm moved almost directly northward to strike a mighty blow at South Florida, and then continued along the Atlantic coastal plain causing severe southeasterly storm conditions from northern Florida to New England.

The South Florida hurricane of 1835 was first detected in the vicinity of Jamaica on the 12th.[1] As it crossed central Cuba on the 14th, winds commenced to rise in the Florida Keys and were described as high by nightfall. This is the first major hurricane, of which we have adequate documentation, to strike the Key West area—that many times storm-battered outpost. A surgeon weather observer was on hand at Fort Taylor, and the local newspaper at the time was ably edited by William Whitehead, who was later to maintain forty years of excellent weather records at Newark, New Jersey.

The weather observer at Fort Taylor noticed a northerly flow on both the 14th and 15th with rain: "high winds day and night." [2]

Whitehead commented on the recent brush with the hurricane in the 19 September 1835 issue of the *Key West Enquirer*:

> GALE: On the 15th inst. we experienced here a severe gale with the wind from the northward, which continued for 2 days, the wind gradually veering to the southwest. We have as yet heard of no damage sustained by shipping, with the exception of one or two pilot boats injured at their moorings. We remember seeing, sometime since, the prognostications of an officer in the British army or navy, who judged (upon what authority we know not) that the visit of Halley's Comet, now expected, would cause the year 1835 to be remarkable for the frequency of gales of wind and other atmospheric phenomena; and whether it may be considered a "strange coincidence" or not, we cannot say, but there has certainly been an undue number of severe storms, tornadoes, gales, etc. throughout the country for the last few months.[3]

A week later he was able to supply further details of the havoc wrought by the storm to the northward and eastward:

> The brig *Sea-Drift* of New York, Capt. Hoyt, bound for Mobile, with an assorted cargo, reported in our last, as ashore on Carysford Reef, was carried bodily to the beach at Key Largo and now lies high and dry, 15 feet from the water.
>
> Nothing is known of the disasters that may have occurred farther north than 10 or 12 miles beyond the Light Ship on Carysford Reef, where the greatest strength of the gale appears to have been felt.
>
> The gale which commenced here . . . (Key West) . . . from the north, was felt at the Cape . . . (Cape Florida) . . . and its vicinity from the northeast—the consequence was, that the wind acting against the current of the Gulf Stream, created a tremendous sea and banked up water to an astonishing height. Many of the Islands were completely overflowed; and at Key Biscayne the water was four feet deep around the light house premises, carrying away the keepers stock of poultry, etc. The Government boat attached to the station was carried up on the island high and dry.[4]

In another column Whitehead declared: "It was far more violent to windward, and is thought to have exceeded any gale which can be recollected in its disastrous effects as well as its violence . . . We do not believe such a series of disasters have been recorded, as occurring on the Florida coast in such a short space of time." Many vessels were thought to be ashore northward from Cape Florida to Cape Canaveral, and a vessel from Charlotte Harbor brought the news that the storm had raged furiously along the western shore of the peninsula also.[5]

At Fort Brooke on Tampa Bay a north wind prevailed on the 15th and 16th with rain; it became "very stormy" during the night of the 16th. The next day was "very stormy" with southeast winds prevailing—perhaps an indication that the storm center had crossed the southern peninsula and was moving northward in the Gulf a short distance offshore.[6]

The press carried a lengthy list of vessels driven ashore. Loss of life was thought to have run high, but there were no means of determining the exact number of victims. Damage estimates mounted to $200,000, a relatively high figure for an area of light settlement and only occasional shipping.[7]

The farther track of the storm must have carried northward along the West Coast. A ship only 20 miles south of St. Augustine met with a southeast gale on the 18th, and that Florida East Coast city had experienced squalls from the northeast for three days when a sudden shift to the southeast took place on the 17th and a hard blow set in which lasted until noon of the 18th. There was little damage except for a few fences blown down.[8]

The center of the hurricane continued northward through Georgia and the Carolinas remaining well inland. Savannah had a gale from east to southeast on the 18th with some damage to trees.[9] At Charleston there was a "heavy gale" on Friday from the south-southeast which subsided about 2200: "Damage to the harbor was more trifling than anticipated" from the strength of the wind. There was no serious damage on Sullivan's Island though the beacon light was blown down.[10] At Georgetown "a violent gale" prevailed on Friday evening from 1800 to 2400. The tide, under the heels of a southeast gale, surged into the harbor, but failed to reach the 1834 high mark by two full feet. There was no damage in the city. The usual crop destruction reports came in from the plantations.[11]

[1] *N. Y. Ship List,* 17 Oct 1835.
[2] SG.
[3] *Key West Enquirer,* 19 Sept 1835.
[4] *Ibid.,* 26 Sept 1835.
[5] *Idem.*
[6] SG.
[7] J. B. Browne. *Key West. The Old and the New.* 156–57.
[8] *Florida Herald* (St. Augustine), 21 Sept 1835.
[9] *Savannah Georgian* in *Courier* (Charleston), 21 Sept 1835.
[10] *Courier,* 21 Sept 1835.
[11] *Union* (Georgetown) in *Courier,* 21 Sept 1835.

THE GREAT HURRICANE SEASON OF 1837

Lt. Col. William Reid of the Royal Engineers was dispatched to the Lesser Antilles soon after the Great Hurricane of 1831 to supervise the rehabilitation of the islands which had been cruelly devastated by the storm. Much impressed by its physical force and destructiveness, he sought to obtain a better knowledge of the nature and behavior of hurricanes. Upon his return to England after two and a half years in the West Indies, he continued his studies by consulting the historical collections and marine archives available in London.

In the summer of 1837 a packet from the West Indies brought news of two early season hurricanes of disastrous proportions. Reid extended his studies to include these recent occurrences, and in doing so collected much data about additional storms which plagued the West Indies and the American coastline in 1837. We now have accounts of at least eleven tropical storms in that season. In the modern periods for which compilations have been made, 1886 to 1961, there have been only seven seasons with more than a dozen tropical storms mapped.[1]

Reid devoted a chapter in his *Law of Storms,* published in 1838, to the season of 1837.[2] It is owing to his diligent labors that we can now present such a full coverage of this most interesting hurricane season.

1837—No. 1—BARBADOS

The first hurricane of the active season of 1837 moved through the Windward Islands on 9–10 July. The *Port of Spain Gazette* announced that the barque *Trinidad* had "experienced a severe gale of wind, approaching to a hurricane" on July 9th when east of Barbados. The *West Indian Gazette* told of extensive damage to houses and mills on the island during the same day. St. Lucia also had severe gales from the north, then from the south, during the night of the 9–10th. The same report came from St. Vincent. This disturbance was noticed only in the Windward group.[3]

1837—No. 2—BARBADOS–FLORIDA

The second hurricane of this eventful season passed over Barbados in the Windward Islands on the morning of 26 July. A wind pattern working around from

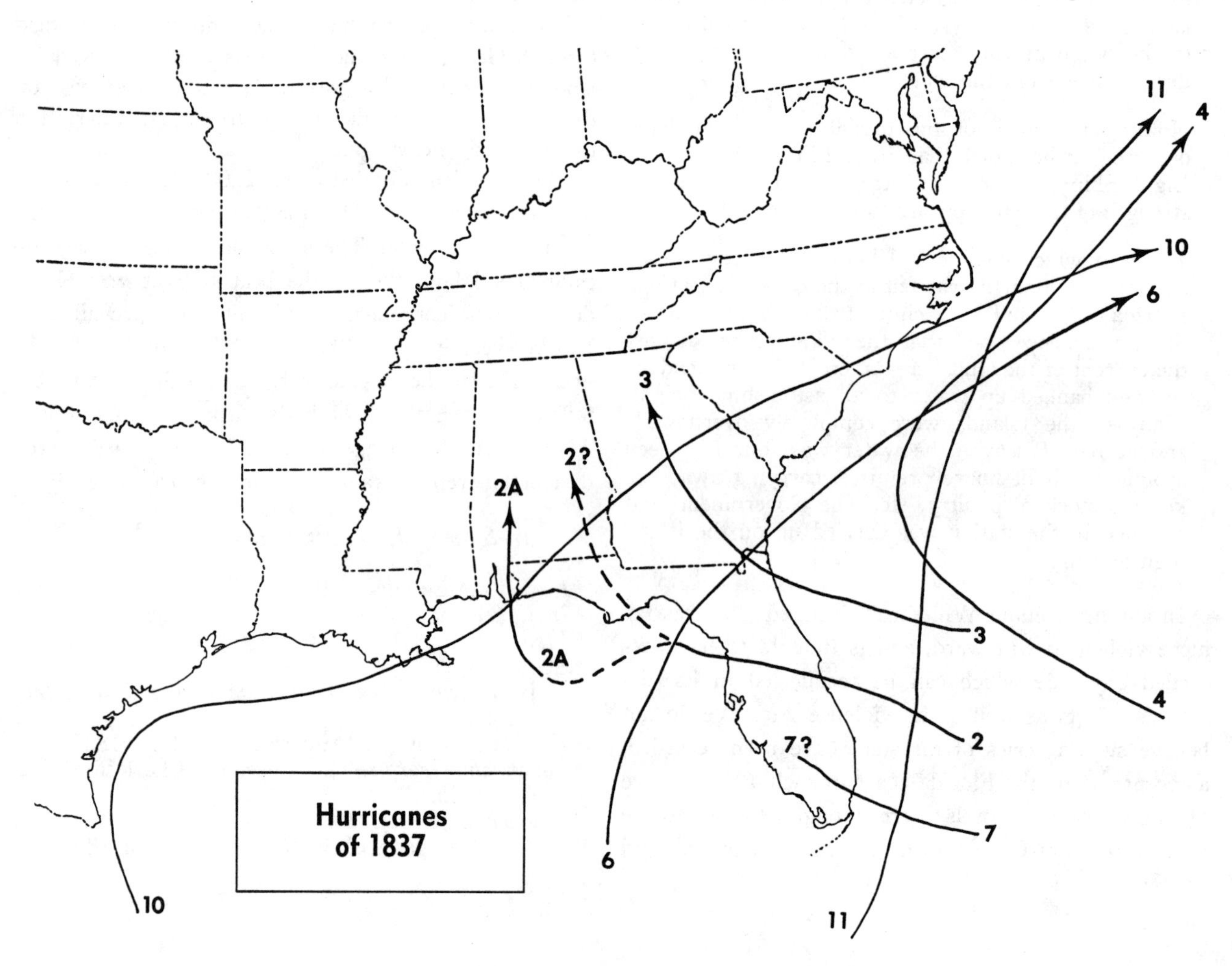

east-southeast to south and eventually to west-southwest drove all but three vessels in the harbor on the shore as the center of the hurricane passed just to the north. The storm was at Martinique in the evening of the 26th, again commencing in the southeast. St. Croix was reached soon after midnight and given a 12-hour lashing. Both Puerto Rico and Santo Domingo reported the gale on the 28th as causing several disasters to shipping around their shores.[4]

The center, passing to the northeast of Cuba, struck Nassau in the central Bahamas on the 29th as a "violent gale" from the east and southeast. It continued there until 1400 on the 31st, a period of 48 hours.[5] H. M. Packet *Sea Gull,* when leaving the Florida Straits south-southeast of Miami, endured a westerly hurricane for four hours on the morning of 1 August.[6] Two ships near St. Augustine also met the gale on the same day. Ashore at the northern Florida settlement the surgeon weather observer at Fort Marion on 2 August observed the wind at northeast in the morning and at southeast in the afternoon "with showers and heavy rain all day." [7] Jacksonville, too, had a hard gale on the 1st and 2nd.[8]

The wind action at northern Florida points, in conjunction with the Florida Straits observation, indicated that the center cut across central Florida with recurvature, if it took place, delayed until the center was in the Gulf of Mexico. The presence of the storm west of Florida was confirmed by the experience of the ship *Ida* in the Gulf where she met a gale on the 3rd, violent enough to sweep twenty of her crew into the sea along with all her small boats. The captain with difficulty brought the *Ida* into port with the remaining five men of the crew.[9]

1837—No. 2A—MIDDLE & WEST FLORIDA

The ultimate path of this destructive hurricane after it passed inland between St. Augustine and Cape Fernandina on the 6th is not clear. No reports from southern Georgia and Alabama are available. Those from northern and western Florida are difficult to interpret from a meteorological view if No. 3 continued on a projected course northwestward from its East Florida landfall. The presence of a southerly gale at Pensacola on the 7th is hard to reconcile with a storm center moving through Georgia or Alabama. William C. Redfield was puzzled by this, too. He surmised that there must have been another storm center in the Gulf, and we are inclined to agree.[21] It could have been No. 2, which moved into the Gulf on the 3d, or it could have been a new center which formed in the trough of No. 3 as the latter moved over Georgia. From the wind behavior at St. Marks and Pensacola, we do know that a marked storm center moved west of both places on the morning of the 7th. Whether it was an old storm center, a new center, or a secondary mattered little to the inhabitants of Middle and West Florida who experienced their first major storm since 1821.

Storm No. 2A commenced at St. Marks on Apalachee Bay at 0100 on the 7th, continued to rage for seven hours, and drove a tide higher than ever known into the harbor. At the lighthouse the waters were estimated at six feet above ordinary levels. In the town, itself, water stood two feet deep; several warehouses were undermined and collapsed.[22]

Westward at Pensacola the wind commenced to mount from the east at breakfast time, but at 1430 it shifted to the south and became very violent. The blasts from the south during the afternoon drove the rain through thick walls and scorched all the leaves of trees exposed to the south. Next morning the editor of the *Pensacola Gazette* declared the harbor a vast scene of wrecks the like of which had never been witnessed there before.[23] The wind flow of the strongest gusts came from the south at Pensacola and east-southeast at St. Marks, indicating that the storm passed to the west of both places.

1837—No. 3—ANTIGUA-FLORIDA

The Leeward Islands were ravaged by a second, even more intense hurricane on 2 August, the same day that Number Two was crossing the Florida peninsula. It struck Antigua about 0230 in the morning and moved its attention on to the other islands in the group later in the day. The Virgin Islands were reached late in the afternoon.[10] The tempest struck savagely at Puerto Rico where it has since been known as "Los Angeles" and has been rated as one of the seven most destructive to hit that island in the period 1825 to 1928.[11]

The harbor master at San Juan, keeping a close eye on the storm, noted his lowest barometer reading at 28.00″ near midnight of the 2nd–3rd. The wind veering from north-northeast to east finally came around to south as the storm center passed south and west of San Juan. The severity of the gale on Puerto Rico may be judged from another press report that 250 buildings were blown down on the small island of St. Barthelemy, just to the southeast of Puerto Rico and apparently very close to the storm track.[12]

The disturbance reached the Bahama Islands on the 4th and 5th where it caused more damage than the first Barbados storm had. "The sea rose on the south side of Grand Bahama, and washed away some low land. At San Salvador the storm was very severe, and several houses were blown down, as well as stock destroyed," Colonel Reid informed us. "At Long Island (more particularly the north part of it) an unusual and destructive rise of the sea took place, and drowned a number of cattle. At Rum Key the loss was great indeed." Off Abaco the brig *William* encountered severe gales shifting from northeast to southwest at 2100 on the 5th.[13]

St. Augustine, already buffeted by Hurricane No. 2, lay close to the projected path of the next:

> On Sunday morning last (6th) we were visited with another gale, much severer than the first. The wind commenced blowing from the N.E. and blew with violence until about 11 A.M. when it suddenly changed to N.W. The tide rose considerably higher than usual, but no material damage was done. A portion of the wall of one house was blown down. The gale ceased about 1700.[14]

At Jacksonville press reports told of a severe gale on the 6th with winds varying from northeast to southeast as the storm approached. Two government warehouses were blown down there, and the cotton crop destroyed at coastal points.[15]

The eye of this great disturbance appeared to have come ashore south of Cape Fernandina, the most northeastern point of Florida, where the St. Marys River empties into the Atlantic Ocean. At the town of Old St. Marys, on the northern or Georgian shore of the tidal estuary of that name, a severe inundation took place such as can occur only with an easterly wind flow. The streets of the village were knee-deep with water and the areas along the bay were waist-deep. "Had the wind continued for two or three hours longer there could not have been a house left standing. The oldest inhabitant does not recollect a similar occurrence, and the buildings are all more or less damaged," concluded a press report. The market house was carried away by the flood tide, many of the Pride of India trees along the river downed by the winds, and the steamboat *Florida* had its side bashed in at a wharf. Out on the coast at Cape Fernandina, nineteen houses were blown down. A letter from St. Marys summarized: "the gale of the 6th was as severe as that in 1813, and has done as much injury to the place." [16]

A dispatch from coastal Georgia described the local effects of the storm there:

> Darien, August 10.—During the last week we have been visited by a storm which has not been equalled since that of the year 1824. The wind on Sunday last, in the morning, blew fresh from the northeast; in the after part of the day, it shifted around to southeast, when the rain began to fall in heavy torrents. The wind then rose very high, and began to blow with fearful violence, tearing up the oldest oaks and mulberry trees in the place by the roots, while limbs and branches of the different trees were flying in all directions. The water of the river then rose, and covered the rice plantations so completely, that they appeared to the eye to form part of the river. The rice, there is no doubt, will be greatly injured by the salt with which the water is impregnated.[17]

Northward at St. Simon Island the wind commenced blowing at noon of the 6th from the northeast, but between 1500 and 1700 shifted into the southeast "and became one of the most furious hurricanes we have had since 1824." The blasts continued until very early on the morning of the 7th when the storm suddenly ceased.[18]

Savannah, too, lay within the circle of maximum destruction from wind and tide. "Our city has suffered in the prostration of trees and fences. The tide yesterday was over our wharves, and no doubt those who had planted on low lands on the river have suffered materially." Another account took a doleful view: "All the goods in front of the stores are damaged, and many of the vessels in the harbor, after having dragged miles up the river, are left high and dry on the marsh . . . I suppose that destruction by the hurricane in this part of the country was never before so universal . . . it is my opinion that we shall scarsely recover in five years." [19]

A tragedy occurred off Jekyl Island during the gale when the Charleston-to-St. Augustine Packet *S. S. Mills* upset with the loss of 14 of the 15 persons aboard. Among them were some of the leading citizens of the Florida city. Lack of ballast in the high winds was given as the cause of the disaster.[20]

1837—No. 4—CALYPSO

The fourth storm originated far to the east of the West Indies. First notice came on 13 August from the barque *Felicity* when upwards of 400 miles from the Leewards, and it probably developed much farther eastward of that. Its track moved north of Puerto Rico and Santo Domingo and into the eastern Bahama group. It was at Turks Island on the 15th and east of the central Bahamas on the 16th.[24] The ship *Calypso* found itself on the 16th in the path of the oncoming storm when about 27 N, 77 W. The wind, having shifted from northeast in the morning to west by midnight, was coming out of the southwest by daybreak with hurricane force, of sufficient strength to stove the fore-scuttle. Quickly the task of cutting away the foremasts was undertaken in an effort to keep the ship from sinking, but before this could be accomplished the *Calypso* heeled over on her beam-ends with the masts in the water. It appeared that she was going down. But with the masts finally cut away, the ship righted herself slowly so that the fifteen crew members could scramble back aboard from their mast perches. A full hurricane blew from noon of the 16th to noon of the 17th, without any semblance of a lull, though gradually backing from east-northeast to southwest.

After pumping night and day until the 21st the weary sailors finally gained on the rising water in the hold. A jury mast was then rigged and course set for the mainland. By the 31st, when only 30 miles south of Cape Fear, another gale rose from the east. This soon backed into the north and into the west-northwest by the morning of 2 September, to further hamper the efforts of the stricken *Calypso* to reach port. Later that day, however, they slipped their battered ship into the mouth of the Cape Fear River and were kindly received by the people of Smithtown and Wilmington. Those on land complained bitterly of the severity of the late storm, as many of their houses had been unroofed and many trees blown down.[25]

The center of this powerful hurricane passed close enough to the mainland to give coastal points a good lashing, but failed to follow the general track of Number One and Two by making the normal recurve when off the Georgia and Carolina coasts.

At St. Augustine, well to the west of the center, the following weather report indicated the passage of the storm seaward: "16th, wind blowing briskly from the north; 17th, A.M. blowing very fresh from northeast—P.M. increased to a gale with moderate rain; 18th, A.M. continued to blow very heavy from northeast."[26] Charleston, also well west of the path, had a three-day gale from the 17th to 19th, but aside from several small boats sunk had no major damage.[27]

Northeastward along the coast the rice plantation country around Georgetown received a lashing. "The Gale of the 17–18th was the most destructive of crops in this neighborhood of any that has occurred in the recollection of our oldest Planter." Rice in mid-August was in blossom, and that is why a "September Gale in August" in the rice country of South Carolina is always much more feared than in its usual month. The northeast gale was thought to be at its height near midnight of the 17–18th at which time the tide had risen within 18 inches of the memorable height reached in September 1834. One-fifth of the shade trees that adorned the town were blown down and nearly all the fences. The storm continued for a period of 24 hours, though damage was not considered as great as in the six-hour blow in 1822.[28]

The southeastern North Carolina coast lay close enough to the storm center to receive both wind and flood damage. From Wilmington the press reported: "On the afternoon of Friday, the 18th, the wind shifted to the northeast, and rain began to pour heavily. Before midnight, the storm increased, threatening ruin; and daylight revealed to us uprooted trees, and our streets washed into gullies, roads obstructed, and bridges carried out . . . The embankments of the sea it is said have given way, and that two new inlets are formed opposite M'Rae's, of Peden Sound. The tide rose six feet higher than usual."[29]

At New Bern, farther northeast, the gale commenced on the 18th at midnight and continued until Sunday the 20th at daybreak.[30] This dangerous hurricane made very slow forward progress while engaged in recurving and expanded into a very large storm upon moving to the northeast on the 20th to 22nd. Reid believed that the recurvature took place near 31 30 N, 78 W, or directly east of St. Simon, Georgia, and directly south of Wilmington. It passed northwest of Bermuda in its drive up the North Atlantic.[31]

1837—No. 5—BERMUDA EAST

Another August hurricane followed close on the heels of Number Four. Its point of origin is in doubt as is much of its course. Two ships encountered the hurricane when east and southeast of Bermuda. The British ship *Castries,* bound from St. Lucia to Liverpool, met a southeast backing to north-northeast hurricane on the night of 24–25 August, when southeast of Bermuda. The ship *Victoria* was upset and dismasted on the 24th at 33 N, 58 W. The barque *Clydesdale* also encountered "a complete hurricane" at noon on the 24th at 32 30 N, 59 30 W.[32] Bermuda lay between the courses of Number Four and Number Five.

1837—No. 6—APALACHEE BAY

See page 143

1837—No. 7—BAHAMAS

A tropical disturbance of undetermined intensity and size was located in the central Bahamas from September 13th to 16th. The *S. S. Pennsylvania* was wrecked about 400 miles off Cape Hatteras on the 14th. St. Augustine, Florida, had a northeast gale on the 13–14th. No other continental land station reported the presence of the storm.[33]

1837—No. 8—EAST FLORIDA

After four days of fresh northeast winds, "heavy winds" prevailed at St. Augustine on 25 September, and on the 26th they were described as "heavy gales," now from the east and southeast. Heavy showers occurred at short intervals during day and night. No other information available.[34]

1837—No. 9—BERMUDA

Tannehill listed a storm at Bermuda on 1–3 October.[35]

1837—No. 10—RACER'S STORM

See pages 144–47

1837—No. 11—CUBA-HATTERAS

Severe hurricane near Trinidad and Cienfuegos, Cuba, on 26 October—violent at Hole-in-the-Wall, Bahamas, on 27th—felt at Cape Hatteras on 29th, at Norfolk on 30–31st.[36]

[1] G. W. Cry, W. H. Haggard, and H. S. White. *North Atlantic Tropical Cyclones. 1886–1958.* Washington, 1959.
[2] William Reid. *The Law of Storms.* 1st ed. London, 1838.
[3] Reid, 44–46.
[4] Reid, 48–51.
[5] *Courier* (Jacksonville) in *Pensacola Gazette,* 26 Aug 1837.
[6] Reid, 51.
[7] SG.
[8] *Nat. Int.* in Reid, 52.
[9] *N. Y. Gen. Adv.* in Reid, 53.
[10] Reid, 58.
[11] Tannehill, 154.
[12] Reid, 60.
[13] Reid, 61.
[14] *Florida Herald* (St. Augustine), 12 Aug 1837.
[15] Jacksonville press in Reid, 62.
[16] *Pensacola Gaz.,* 26 Aug 1837.
[17] Reid, 67–68.
[18] *N. Y. Gaz.* in Reid, 64.
[19] "Times paper" in Reid, 63–64.
[20] *Florida Herald,* 29 Aug 1837.
[21] W. C. Redfield. *Amer. Jour. Science.* 35–2. Apr 1839. 206.
[22] *Floridian* (Tallahassee), 12 Aug, in *Pensacola Gaz.,* 19 Aug 1837.
[23] *Pensacola Gaz.,* 12 Aug 1837.
[24] Reid, 69.
[25] Reid, 74.
[26] SG.
[27] *Courier* (Charleston), 21 Aug 1837.
[28] *Union* (Georgetown), 26 Aug, in *Pensacola Gaz.,* 9 Sept 1837; *Southern Agriculturalist* (Charleston), Aug 1837, 412–16.
[29] Wilmington, 25 Aug, in *Mercury* (Charleston) in Reid, 77.
[30] Reid, 77.
[31] Reid. Chart VII, appendix.
[32] Reid, 117.
[33] *N. Y. Ship List,* 4, 7, 25 Oct 1837.
[34] SG.
[35] Tannehill, 253.
[36] *N. Y. Ship List,* 8 Nov 1837 and 2 Dec 1837.

THE DESTRUCTIVE NORTH CAROLINA HURRICANE OF 1842

"A gale, believed to have been the most violent experienced at Ocracoke Bar for eighty years, commenced in the morning and continued until three o'clock in the evening of the 12th July, blowing from NNW."[1] From the lists of destruction that accompanied the above press dispatch, both to shipping and to land facilities, this early-season hurricane must be accorded a place with the *great* storms of North Carolina history. Approximately eighty years before, in 1761 according to legend, another great hurricane changed much of the coastline of the Outer Banks and cut through the famous New Inlet near Wilmington. David Stick in his *Graveyard of the Atlantic* considered "the severe hurricane of 12 July, one of the worst in the history of coastal Carolina."[2]

Along the sandy banks that make up the outer fringe of North Carolina the damage was immense. The entire village on Portsmouth Island near Ocracoke Inlet, with the exception of one building, was wrecked. A store at the settlement was blown down and floated away at the height of the storm. Fourteen vessels were stranded on the ocean beach near Ocracoke Inlet, and fourteen more were aground on the inside beaches. Six other vessels had been forced to put to sea.[3] Two unknown vessels were dashed to pieces in the breakers on Diamond Shoals, their entire crews drowned, and seven men who later went out to try to salvage some of the wrecked goods were also lost. Numbers of dead horses and cattle were seen drifted down the sound after the blow was over.[4]

A bottle washed ashore at Shelby Bay, Bermuda, 27 October 1842, with the following note inside: "Schooner *Lexington,* off Cape Hatteras, July 15, 1842. This morning at half past two o'clock A.M., it commenced blowing a strong North Wester, which increased to such a degree that it was certain that my vessel could not stand it. At 5 I tried the pumps and found that she made eleven inches. She being an old vessel, worked in her joints. At half past eleven, I determined to leave her with my crew (three men and myself) in our launch; but before leaving sounded the pumps, and found she had increased the water in her hold three feet. I write this and enclose it in a bottle, so that if we should not be saved and the bottle be found, it may be known what became of the vessel and us. At 1 P.M. got into the boat with provisions and water sufficient for six days, having beforehand offered up our prayers to God to protect and save us. Signed Wm. H. Morgan, Captain; John Rider, Mate." Newspapers of the day make no more mention of the fate of the *Lexington* and her crew.[5]

Inland at Washington on Pamlico River the wind began to increase in intensity on Monday night, the 13th; next day an ever-heightening gale continued to churn the waters of the rivers and sounds of the Outer Banks to reach full fury by nightfall. Streets about the

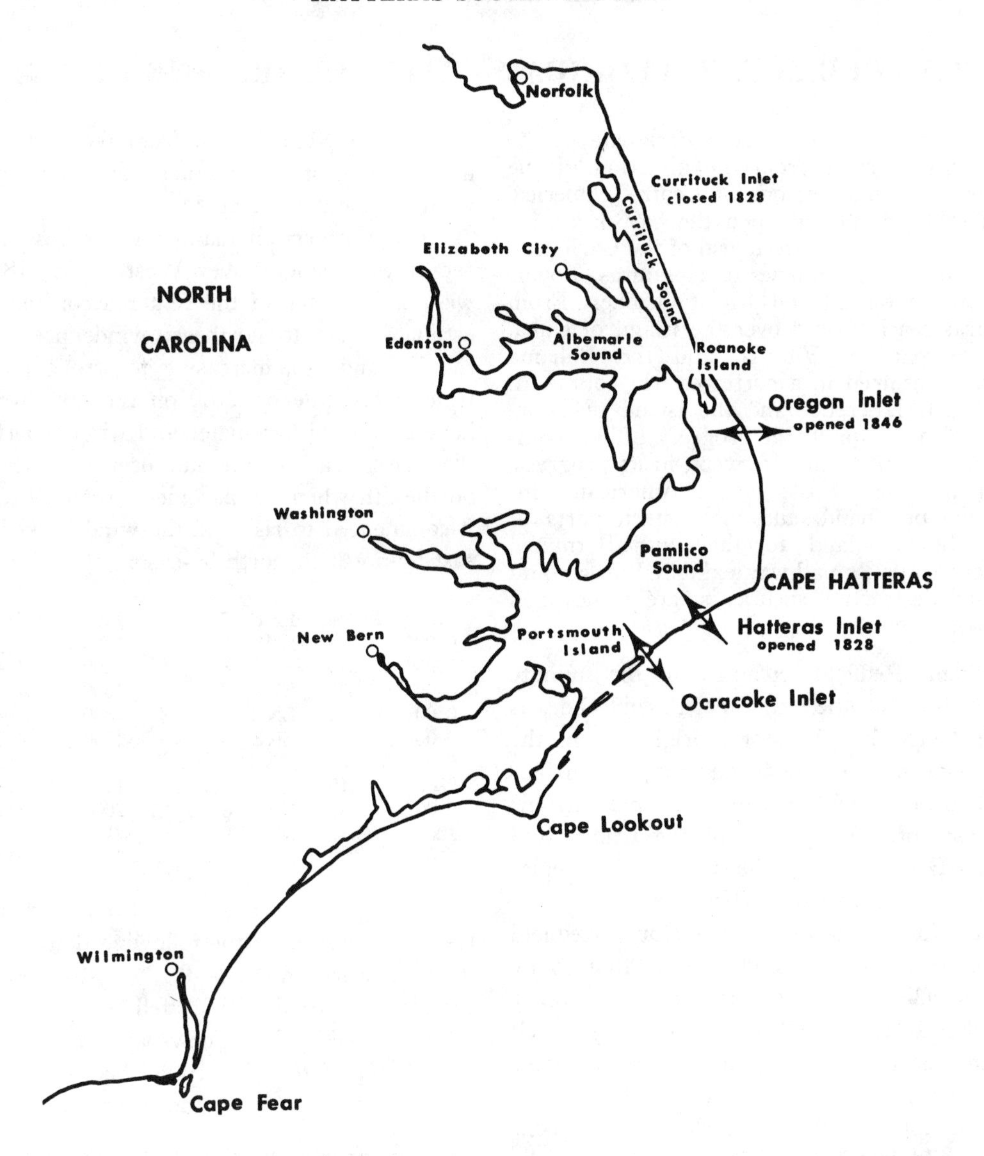

village were strewn with fallen trees and felled limbs.[6] Near Edenton on an arm of Albemarle Sound the float bridge at Hertford was carried away, the mail packet washed high and dry, and severe crop damage inflicted on farm areas. All vessels at the Edenton anchorage except one were driven from their moorings and piled up on shore.[7] Southward the Wilmington to Raleigh railroad was washed out, interrupting the flow of mail to Charleston for five days.[8]

Authentic wind information in the Cape Hatteras area is not available from a regular weather reporter. Two reports from mariners speak of northwest and north-northwest gales. Thus, the course of the center of the hurricane cannot be traced, though it is assumed that the center remained just offshore during the destructive period on the 12th. According to Lorin Blodget, writing a dozen years later in his *Climatology of the United States,* the storm moved inland near Norfolk and passed over Washington still pursuing a northwesterly course.[9] Heavy rains descended on both Virginia and Pennsylvania. The James River at Richmond rose six feet in the first hour of flood on the 14th, and eventually reached a stage higher than in 1836 and about equal to the great freshet of 1814. Next day the Schuylkill River near Philadelphia rose five feet over its banks.[10]

[1] *N. Y. Ship List,* 23 July 1842.
[2] David Stick. *Graveyard of the Atlantic.* Chapel Hill, 1952. 45.
[3] *Old North State* (Elizabeth City) in *N. Y. Trib.,* 25 July 1842.
[4] Stick, 45.
[5] *Ibid.,* 44.
[6] *Republican* (Washington, N. C.) in *N. Y. Trib.,* 25 July 1842.
[7] *Harp & Compiler* (Richmond) in *N. Y. Trib.,* 21 July 1842.
[8] *N. Y. Trib.,* 22 July 1842.
[9] Lorin Blodget. *Climatology of the United States.* Phila., 1857. 400.
[10] *N. Y. Trib.,* 18, 20 July 1842.

THE CUBAN AND FLORIDA STRAITS HURRICANE OF 1844

This remarkable storm, which I designate as the *Cuban Hurricane,* came from the direction of the Pacific Ocean and the regions of Central America. It appears to have entered upon the bay or sea of Honduras, which is the western arm of the Caribbean sea, directly from the countries of Honduras, Poyais and Yucatan, on the 3d and 4th of October. From the Honduras sea it passed over the island of Cuba, the southern part of Florida, and the Bahama Islands, and continued in a northwesterly course, to the Gulf of St. Lawrence and the island of Newfoundland, with a rapidity of progress hitherto unknown in American storms. It swept, in its progress, the salient portions of the North American continent on the one hand, and the eastern parts of Cuba and Newfoundland, together with Bermuda, on the other; while its pathway exhibited an amount of injury and destruction such as is rarely known in the annals of commerce.[1]

Thus William Redfield commenced his minute examination of this gigantic hurricane. Possibly he was wrong in his belief that the storm originated in the Pacific Ocean—nevertheless, he made a major contribution to our knowledge of hurricane movement by his brilliant analysis of the progress of the Cuban and Florida Straits Hurricane from its birth in the tropics to its disappearance in the sub-Arctic.

In searching for the origin of the storm Redfield received notice of a violent thunderstorm attended by heavy rains which swept over Barbados and other Windward Islands on 25 September. This might well have been an easterly wave disturbance which later triggered the surface cyclonic circulation in the western Caribbean as it approached the mainland of the continent. "Very wild weather" was experienced on the coast of Honduras from the 1st to 4th of October; Montego Bay in northwestern Jamaica was hit by an enormous storm tide on the 4–5th; and the southern Cuban coast was ravaged on the 5th when 158 vessels were wrecked and 2546 houses destroyed by wind and water.[2]

The eye of the hurricane made a transit across Cuba about 85 miles east of Havana where a total of 76 vessels were said to have been lost. Matanzas, the seaport on the north-central coast of Cuba, lay close to the path of the disturbance as it crossed the island. Following a northeast hurricane, early on the 5th the blow abated about 1030 as the center reached the northern coast. The squally winds slowly veered around to west, where the hurricane resumed. The barometer commenced to rise about 1400 and an hour later the storm had "entirely yielded."[3]

A ship near Cat Key passed through the eye when close to the Double Shots between Salt Key Bank and Florida Reef (24 15 N, 80 30 W) about 1700 on the 5th. The wind had been from the east-southeast, and after a lull of 20 minutes came out of the north-northwest and northwest.[4]

On the American mainland our best meteorological report was made at Key West about 118 miles to the west of the track of the center according to Redfield's estimate. East to northeast winds prevailed there on the 3rd and 4th, increasing to strong on the 4th and to a severe gale by 2100 on the 5th. Fortunately, we have an hourly barometer and wind report; this shows that the "crisis" of the gale occurred there about 1400 on the 5th when the barometer read 29.13"; thereafter it commenced to rise and the wind backed from north-east-by-north to north-by-east:[5]

0100	29.768	ENE	7	1200	29.217	NE/N	9
0200	.729	ENE	7	1300	.166	NE/N	9
0300	.516	E/N	8	1400	.134	NE/N	9
0400	.457	ENE	8	1500	.185	N	9
0500	.433	ENE	8	1600	.264	N/W	9
0600	.402	ENE	8	1700	.410	NNW	8
0700	.335	NE/E	7	1800	.455	NW	8
0800	.418	NE	7	1900	.520	NW/W	6
0900	.536	NE	8	2000	.551	WNW	8
1000	.331	NE/N	9	2100	.587	W	7
1100	.272	NE/N	9	2200	.642	W	6

The large disturbance was also a good rain producer. Key West measured the following daily amounts as the storm approached: 2nd—0.17", 3rd—1.96", 4th—1.75". On the 5th a total of 9.62" fell.[6]

The local Key West newspaper with the appropriate title, *Light of the Reef,* published an extra edition to describe the passage of the hurricane:

> On Friday evening last it commenced blowing quite fresh from the North East and continued increasing in violence until 3 P.M. on Saturday, the 5th inst. During the time of the gale, which lasted about 18 hours, damage was done to property to an almost incredible extent. The unparalleled fury of the gale, when at its height can scarsely be conceived, it swept everything before it—houses, fences, trees, vessels, and almost everything in its course was leveled to the earth or borne off with frightful velocity . . . The general aspect of the island since the gale is dreary in the extreme. The beautiful gardens, which were, before, a pride of their owners and an ornament to their buildings, have been entirely swept away or killed by the force of the wind.[7]

At Indian Key, much closer to the center, all houses on the islet were blown down and all wharves washed away by the fury of the wind and tide. Northward at Jupiter Inlet, between Miami Beach and Palm Beach, a severe gale raged all day on the 5th culminating in a great flood tide.[8] Along the north Florida coast the influence of the vast storm was felt. St. Augustine had a gale on the 5th with northeast winds, mounting to "violent" from 1400 onward and reaching force 9 at 2200. The peak gales came during the night. By dawn of the 6th the direction had backed to northwest and the glass was rising as the storm moved off to the northeast well offshore now.[9]

The track of the hurricane had moved up the Florida Straits. Redfield estimated that the center passed 90 miles west of Nassau in the Bahamas, or about 100 miles east of Miami Beach and very close to the Biminis. From Nassau: "We experienced a severe hurricane on the Banks on the night of Oct. 5th, and the loss of lives and property has been greater than any previous gale for some years. In this harbor, from the quarter we had the wind, S.E., all was protected, and no damage of consequence." Another report from the central Bahamas gave the wind behavior as veering from southeast to south and ending at southwest, putting the area in the eastern semicircle.[10]

Northward along the Atlantic coast neither the gale nor the rain reached as far westward as Savannah or Charleston although lowered barometers at each place pointed to the passage of a large disturbance to the eastward. At Cape Fear in southeastern North Carolina a northeast gale at force 6 without rain marked the presence of the storm some 300 miles eastward of that point.[11] In the lower Delaware Bay there was a northeast gale, and New York had fresh winds from the northeast, mounting to force 5, on the night of the 6–7th.[12]

At Vineyard Haven on Martha's Vineyard two ships were dismasted on the night of the 6–7th by a northeast gale, and on Nantucket Island, perhaps 260 miles from the center according to Redfield, a wind increasing from the northeast and a falling barometer presaged the approach of a storm from the south.[13] The low point of William Mitchell's barometer was reached at midnight or soon thereafter when he noticed a reading of 29.50″ with the wind at his top rating, force 6, "with rain, gale extremely heavy." He measured a rainfall of 1.50″.[14]

[1] William C. Redfield. On Three several Hurricanes of the American Seas. *Amer. Jour. Science,* 2d ser., 1–3, May 1846, 333.
[2] *Ibid.,* 338–39.
[3] *Ibid.,* 341.
[4] *Ibid.,* 342.
[5] *Idem.*
[6] *Idem.*
[7] *Light of the Reef* (Key West) in *Florida Herald* (St. Augustine), 22 Oct 1844.
[8] Redfield, 343.
[9] *Idem.*
[10] Redfield, 343–44.
[11] Redfield, 348.
[12] *Ibid.,* 350.
[13] *Ibid.,* 352.
[14] *Idem.*

THE HATTERAS INLETS HURRICANE OF 1846

Two new inlets of major commercial importance were opened up as a result of a severe hurricane which brushed the Outer Banks of North Carolina in early September 1846. To the south of the Cape, a new Hatteras Inlet between Ocracoke and Hatteras Islands provided a new entrance into Pamlico Sound, while to the north, Oregon Inlet, so named for the first ship to pass through, split Bodie Island below Nags Head for a more direct route to Albemarle Sound ports. These inlets were cut out more as an aftermath rather than as a direct result of the wind force at the peak of the hurricane. It was the rush seaward of the piled-up waters of the sounds and bays, driven by a westerly wind in the rear of the storm system, that swept over the beaches and cut the new passageways through the sandy outer barriers.

September 1846 produced a number of storms of tropical origin as preliminaries to the main hurricane of the season to follow in mid-October. The origin of the Hatteras Inlets Hurricane under present consideration is not known. There was a storm of limited extent in the northwestern Gulf of Mexico on the night of September 6–7th that caused the steamship *New York* on the Galveston-New Orleans run to founder with the loss of 19 persons.[1] But this could hardly have been the powerful hurricane raging on Cape Hatteras 24 hours later; no reports have been uncovered that would suggest any meteorological continuity between the two.

Our best account of the approach of the Hatteras Hurricane came from the U. S. Brig *Washington* when off Cape Hatteras on the morning of the 8th. The wind increased to a hurricane at 0700 that morning to reach its peak force about 1100. An hour later the ship, knocked on her beam-ends, wallowed in the now diminishing gale until the next day.[2] Even ships within the protection of the bars suffered great distress as the wind direction changed constantly when the hurricane swept by a short distance seaward. According to the marine reporter of the *Newberian,* the gale caught 20 ships at Ocracoke Inlet and drove all but two of them ashore or out to sea. The wind there backed from northeast to northwest at the height of the gale. The small community of Hatteras just south of the Cape, fully exposed to the full brunt of hurricane winds regardless of direction, had all but six houses flattened by the blasts.[3] The storm effects on the barrier beaches to the north were of the same dimensions. The *Edenton Sentinel* vividly described the two aspects of the combined force of wind and waves:

THE STORM

Much damage has been done by the late storm to the shipping on the coast. Our Bay presented quite a novel appearance; nearly all the water was blown out of it, except immediately in the channel. The water in Perquimans River, near Hertford, fell seven feet, which was, as a gentleman living in that vicinity informed us, four feet lower than he had ever known it before. At Nag's Head the tide rose about nine feet higher than common tide, and destroyed the warehouse of Mr. Russell, proprietor of the Hotel, together with nearly all the stores which it contained, carrying it down the Beach about half a mile; swept away the market house; the house belonging to Dr. Wright was blown from its blocks; and nearly all the trees on the Hill were destroyed. Several families were compelled, for safety, to leave their houses and seek shelter in the Hotel. All the boats belonging to this place were carried off, depriving them of the means of fishing for a time. Persons living some four miles below Nag's Head, on the sea beach, found it necessary to flee to the garrets of their houses; to save themselves from drowning. They lost all they had to survive on—their clothing was all destroyed, and also their cooking utensils.[4]

The actual cutting out of one of the new inlets was witnessed by C. O. Boutelle, then assistant superintendent of the U. S. Coast Survey, who was busy running a base line on Bodie Island at the time:

On the morning of the September gale the sound waters were all piled up to the southwest, from the effects of the heavy northeast blow of the previous days. The weather was clear, nearly calm, until about 11 a.m., when a sudden squall came from the southwest, and the waters came upon the beach with such fury that Mr. Midgett, within three quarters of a mile from his house when the storm began, was unable to reach it until four in the afternoon. He sat upon his horse, on a small sand knoll, for five hours, and witnessed the destruction of his property, and (as he then supposed) of his family also, without the power to move a foot to their rescue, and, for two hours, expecting to be swept to sea himself.

The force of the water coming in so suddenly, and having a head of two to three feet, broke through the small portion of sea beach which formed since the March gale, and created the inlets. They were insignificant at first—not more than 20 feet wide—and the northern one much the deepest and widest. In the westerly winds which prevailed in September, the current from the sound gradually widened them; and, in the October gale, they became about as wide as they are now. The northern one has since been gradually filling, and is now a mere hole at low water . . . [but the southern one] between high water marks, measured on the line, is 202 yards [wide, and] between low water marks, 107 yards.[5]

The storm center apparently moved northeastward offshore without making any landfall. It was still close enough to the Delaware Capes and the South Jersey coast to trouble shipping there, but we have no reports from land stations in the Middle Atlantic States or New England of any severe storm activity at this time.[6] It is interesting to note that the current pride of the Atlantic, the large steamer *Great Western,* when in a severe storm on the 19th off Nova Scotia, suffered unusual structural damage for so large a ship.[7] On the same day a gale on the Grand Banks of Newfoundland was described as "the most severe they have ever experienced" there.[8] Though this could possibly have been the Hatteras Hurricane which had been blocked by an anticyclone to the north and delayed in its normal passage, it is more likely that the *Great Western* hurricane encounter of the 19th was related to another, probably the storm that hit Barbados on the 12th and then moved directly northward over the open seas to the east of the course of the Hatteras Hurricane.[9]

[1] *Galveston News* in *U. S. Gaz.* (Phila.), 21 Sept 1846.
[2] *Baltimore Patriot* in *North Carolina Register* (Raleigh), 2 Oct 1846.
[3] *Newberian* in *N. C. Reg.,* 23 Sept 1846.
[4] *Edenton Sentinel* in *N. C. Reg.,* 30 Sept 1846.
[5] David Stick. *The Outer Banks of North Carolina.* Chapel Hill, 1958. 279.
[6] *Daily Advertiser* (Newark), 12 Sept 1846.
[7] *N. Y. Courier and Enquirer* in *N. C. Reg.,* 16 Oct 1846.
[8] *Yarmouth Register,* 22 Oct 1846.
[9] *Bermudian,* 17 Oct, in *N. Y. Trib.,* 3 Nov 1846.

THE CAROLINA HURRICANE OF 1854

Just fifty years after the great Gale of 1804, coastal Georgia and South Carolina experienced a close repetition of the disasters attending that landmark in early nineteenth century storm reference. During the intervening years there had been several substantial storms that had seriously affected some parts of the coastal area, but none of these had been of the great physical proportions of the 1804 storm, nor had any followed a path so filled with danger for all the seaports of Georgia and the Carolinas.

The course of the early September 1854 hurricane has been adequately documented by Prof. Lewis R. Gibbs of Charleston College, who in the 1850's became a collector of local hurricane data in emulation of William C. Redfield of New York. Prof. Gibbs' summary of this storm was contained in a letter to Redfield, dated 23 October 1854, and has been preserved in the Redfield Collection at the Yale University Library.[1]

The 1854 hurricane approached southeastern United States from the south-southeast after moving through the northern Bahamas. The brig *Reindeer,* when some 60 miles northwest of Abaco Island, encountered a five-day storm with the gales commencing in the northeast, but veering to southeast early on the morning of 7 September. The ship's barometer then read 27.70 inches—indicative of the extreme intensity of the storm. The five-day duration of the gales in the vicinity of the *Reindeer* pointed to a slow-moving storm center of rather large dimensions.[2]

The hurricane made a landfall on the Georgia coast somewhere between Brunswick and Savannah. At St. Augustine, Jacksonville, and St. Simon Island the gales on the afternoon of the 8th backed from northeast into northwest; while at Savannah and Whitemarsh Island in northeast Georgia and at Beaufort in South Carolina, as well as at Charleston and other points northward, the gales veered from northeast to southeast and finally to south as the disturbance drew abreast to the west.[3]

St. Augustine reported a gale from the 5th through the 9th as the storm passed offshore, but no real damage occurred there. Jacksonville also had strong northeast to northwest gales, but the tide on the river did not rise high enough to cause trouble.[4]

As usual during the 1850's, we can turn to R. J. Gibson of Whitemarsh Island near Savannah whose valuable Smithsonian records of meteorological conditions at his exposed location served as an excellent checkpoint for tropical storms:

> The storm of the 8th was by much the severest which has occurred in the last twenty years. But one tide (that of the Gale of 1804) is remembered which overtopped that of the 8th. The barometer fell steadily until 4 P.M. of the 8th, then began to rise. The wind hauled toward East from N.E. about 1 P.M. and blew its heaviest puffs about 2 P.M. from S.E. by E. By midnight it had got to south or near it and blew at times very hard from that quarter.[5]

The Smithsonian observer at Savannah logged the following conditions during the two-day storm:[6]

		Corrected Barometer	Wind estimated	
Sept. 7	0700	29.847″	N	15 mph
	1400	29.778	N	35
	2100	29.735	NE	35
Sept. 8	0700	29.455	ENE	60
	1400	29.037	ENE	75
	1600	28.737	ESE	90
	1800	28.975	SE	75
	2000	29.167	SE	75
	2100	29.250	SE	70

Prof. Gibbs' collective is informative as to the course of the center of the hurricane through the Southeastern States:

> St. Simon's Island, Georgia. Gale began morning 7th from N.N.E. & N. On morning of 8th blowing a hurricane from N.; at 8 a.m. N. by W.; noon N.N.W.; 2 p.m. N. W. by N.; barometer lowest, 4 p.m. W.S.W.; 6 p.m. S.W. by S.; 10 p.m. S.W. At midnight gale abating, and on morning of 9th at S.W. gradually diminishing. Crops injured.
>
> Savannah, Georgia. Gale set in Thursday morning from N.E. and continued from that quarter until 3½ p.m. on Friday the 8th, when it slowly shifted to the E. At 10 p.m. it had shifted to the S.E. and gradually wore around to the S.W. Houses, churches, wharves and shipping extensively injured.
>
> Laurel Hill, South Carolina, near Savannah River. Wind from N.E. on Wednesday, from same quarter on Thursday increasing to a gale at night. On Friday still increasing blowing a terrific gale until night, when wind shifted to E., and then to S.E. Plantations submerged and crops injured.
>
> Charleston, South Carolina. Gale began early on Thursday morning the 7th from N.E.; at noon violent rain for an hour, and showers at intervals throughout the day and night. At 10 p.m. gale increased. On morning of 8th wind from same quarter; at 2 p.m. barometer lowest, and wind shifted to E.; at 9 p.m. again shifted to S.E. from which quarter it blew all night and also next day during the subsidence of the gale. The roofs of Charleston Hotel and other large buildings much injured, and the wharves, shipping, East Battery, and Bathing House sustained great damage. On Sullivan's Island many houses injured, and some totally destroyed, by flood and wind.
>
> Columbia, South Carolina. High wind with rain on 8th from N.E., veering to E., then on 9th to S.E. and at night to S., and finally on 10th to W. Barometer lowest 28.96 (uncorrected) at 4 p.m. on 9th.
>
> Asheville, North Carolina. Wind was light from N.E. on 7th and 8th. Barometer 27.60 (uncorrected), with a little rain on the 8th on afternoon of the 9th barometer was lowest 27.05, wind light N.W.[7]

The 1854 hurricane rather closely duplicated the path followed by the famous Gale of 1804. Both crossed the Georgia coast at a sharp angle while in the process of recurving—thus they tended to move northward then northeastward moving somewhat parallel to the northeast-trending coastline. This put all the important seaports from Savannah northward in the dangerous right semicircle of the disturbance, and at the same time the center remained close enough to its energy source in the warm coastal waters to sustain its force.

The impact of the storm at Charleston has been well described in the edition of the *Courier* published on the final day of its unusual three-day duration:

> The first indication and warning appeared at a very early hour of Thursday morning, or soon after the midnight of Wednesday. The breeze was from the northeast and gradually increasing amounted to a gale about midnight on Thursday. From this period it continued with occasional intermissions and with violent accessions throughout the whole of yesterday. The direction was changed frequently, and there were brief periods of comparative quiet and as has been usual in all our severe gales, the wind, for the greater

part of its violent duration, settled down at, or nearly, in the South-east.

* * *

The ebb of the tide on this occasion was marked by a phenomenon which is noted as something remarkable.

At 12 M on Saturday, the 8th Sept., the fall of that tide was only to the extent of two feet. It was naturally apprehended under these circumstances that the next tide would be beyond the point previously reached, on the contrary, however, through the influence of the wind, the fall continued throughout the regular period of the flood, and at high water on Sunday, the water was not as high as it had been at the lowest stage on Saturday.

The previous gales recorded as the most violent and destructive occurred in 1752 and 1783. The subsequent gales that were specially memorable, were in 1811 and 1822. [1804 had been mentioned earlier in article.] [8]

The southeasterly blasts along the South Carolina coast drove the waters of the Atlantic Ocean into all the bays and inlets that abound there, over some of the low-lying islands, and into the tidal lowlands that fringe all the rivers and streams. Edisto Island near Charleston suffered severely, as did Port Royal and Beaufort to the south.[9] The massive extent of the disturbance is indicated by the vast inundations that took place in the Winyah Bay area a hundred miles north of Charleston. At Georgetown, according to the *Pee Dee Times,* though the tide was as high as in the disaster of 1822, the wind was thought to have been lower. The storm at that northeastern South Carolina location commenced on Thursday, the 7th, and did not end until Saturday night: "being the longest continuous blow in the remembrance of any inhabitant." [10] A graphic description of the inundations attending the hurricane has been preserved in the correspondence of Adele Petigru Allston:

Since I wrote you last we have had a great blow, Storm. It commenced on the 7th inst and lasted until the night of 9th. The tide was higher than has been known since the Storm of 1822. Harvest had just commenced generally and the damage to the crops is immense. From Waverly to Pee Dee on the 8th not one head of rice was to be seen above the water, not a bank or any appearance of the land was to be seen. It was one rolling dashing Sea, and the water was Salt as the Sea. You will see at once that the crops must have been terribly injured. Many persons had rice cut and stacked in the field, which was all swept away by the flood. Your papa had none exposed in that way for he apprehended high tides from the state of the moon, and prepared as far as possible for it. Mr. J J Middleton had 40 acres of very superior rice swept away, a total loss, and many others have suffered in the same way, tho' not to the same extent.[11]

The extremely long continuance of hurricane force winds at coastal South Carolina points can be attributed to the slow recurvature of the storm center after passing well inland through central Georgia and the western Carolinas. The meager wind data available from Camden, Columbia, and Aiken placed the center as moving to the west of those South Carolina locations.[12] Dr. Young, the Smithsonian observer at Camden, reported "the equinoctial gale" as commencing at 0300/8th and continuing throughout the daylight hours of the 9th. His wind, which had been at northeast at all observations on the 8th, veered to southeast on the 9th and finally into southwest at 1600/9th soon after he had registered his lowest barometer reading of 29.47".[13]

Farther north the hurricane retained much of its punch as it curved through North Carolina toward the coast again. At Wilmington a tornado roared through the eastern section of the town, cutting a swath about the width of a city street, in the midst of a southeasterly gale on the 9th.[14]

[1] Lewis R. Gibbs to W. C. Redfield. Redfield Ms. Letters Received. 4, 425–28. (Yale).
[2] *Idem.*
[3] *Idem.*
[4] *Jacksonville News* in *Daily Picayune,* 19 Sept 1854; Blodget, 400.
[5] SI.
[6] SI.
[7] Gibbs.
[8] *Courier* (Charleston), 9 Sept 1854.
[9] *Ibid.,* 15 Sept 1854.
[10] *Pee Dee Times* (Georgetown), 13 Sept, in *Courier,* 15 Sept 1854.
[11] Adele Petigru Allston to Benjamin Allston, Beach, 20th Sept 1854, in *The South Carolina Rice Plantation.* J. H. Easterby, ed. Chicago, 1945. 408.
[12] Gibbs.
[13] SI.
[14] Wilmington dispatch in *Daily Picayune,* 18 Sept 1854.

THE CENTRAL AMERICA DISASTER IN 1857

The *S.S. Central America* left the port of Havana on September 8th with a large passenger list containing several prominent Americans and with a valuable cargo list including much bullion. The ship was some thirty years old, but had been recently refitted for modern passenger service and given a prestige name. Despite her new trappings the veteran of many a storm was to meet a fatal rendezvous with a small, but vigorous hurricane off the North Carolina shore.

Soon after leaving port fresh westerly breezes sprang up and with the help of auxiliary sails pushed the ship along at a good clip toward New York. On the 9th the wind increased in strength to strong, and by the morning of Thursday, the 10th, a full hurricane was raging. The previous history of this storm is not known. It is thought to have been a relatively small disturbance that churned around the shipping lanes off the Carolina capes from the 10th to the 13th, before coming under the influence of a steering current which carried it northeastward away from the shore. Our only land reports of gale strength winds ashore came from Georgetown and Wilmington. Several ships near Frying Pan Shoals off those ports also encountered hurricane winds on these days.

The *Central America* sprang a serious leak at the height of the gale on Friday, the 11th. All the rest of that day and on Saturday the crew and passengers manned the pumps in an effort to keep the leaky hulk afloat. Slowly the water gained. The ship sank lower and lower in the high-running seas. Fortunately, the storm lulled somewhat about 1400 on Saturday afternoon, but the fate of the ship had already been sealed. At 1900 the luckless craft sank beneath the surface. Over 400 persons were reported in the press to have lost their lives, only a few being saved—the greatest single ship disaster of a commercial vessel attributable to a hurricane in our period.[1]

Other vessels were caught in the same storm off the Carolina capes. One near Frying Pan Shoals apparently passed through the center of the disturbance early on the 12th. She had been taking gales from the east-northeast, then east-southeast on the previous day. From 0200 to 0900 on Saturday, the vessel wallowed in heavy seas with "the wind down"—no doubt in the eye of a very slow moving hurricane. After 0900 the hurricane blasts again struck, now from the west to north, and in the afternoon the wind went around to south and by midnight had slacked off to a mere breeze as the small storm center moved away.

The steamship *Southerner* out of New York for Charleston also encountered a severe blow when about 20 miles south of Cape Lookout. A heavy southeast gale there "blew a perfect hurricane" on Friday evening, but shifted around to northwest by 0800 on Saturday morning to continue at gale strength until Sunday morning. A religious service was held aboard ship later in the day as a thanksgiving for deliverance from the clutches of the storm. Later in the week at Charleston the passengers again assembled at a local hotel to pay testimony and praise to the ship's captain whose handling of his vessel during the gales was thought to have saved her from the *Central America's* fate.

Ashore on the 11th the weather was described as "stormy" by the Charleston observer with winds mounting to force 5 from the northeast. The barometer continued to sink slowly until the afternoon of the 12th, reaching the lowest point of 29.88 inches; by then the winds had already backed into northeast.[4] At Georgetown "a gale" raged from before midnight on the 11th until the night of the 12th, with the wind slowly backing from north at 1100 to northwest at 1300 and finally to west at 2100. The barometer reached its lowest point of 29.61″ at 1530 and commenced to rise at 1900. Rain continued throughout the entire day.[5]

[1] *Courier* (Charleston), 23 Sept 1857.
[2] *N. Y. Ship List,* 23 Sept 1857.
[3] *Daily Advertiser* (Newark), 19 Sept 1857.
[4] Board of Health, Charleston. Ms. Met. Reg. (National Archives).
[5] SI.

THE GULF COAST: 1815–1870

THE LAFITTE HURRICANE AT GALVESTON IN 1818

Most of the books about the Lafitte brothers and their allegedly piratical activities mention the occurrence of a severe storm in the autumn of 1818 when the Lafittes and other freebooters were holding forth on the ill-fated sandy spit of land which separates Galveston Bay from the Gulf of Mexico. None of these authors ascribed any date for the storm, and none of our recent compilers of hurricane chronologies have been more definite.[1] But in a valuable article in the *Louisiana Historical Quarterly* of July 1940, Stanley Faye, an indefatigible historian of the Gulf Coast, named the day and described the event:

> On Sept 12 a storm of extraordinary violence swept up the Gulf. It struck the war ships from Vera Cruz and put them out of commission for months to come. Along the coast of Texas it flooded the barrier beaches. At Galveston the disaster differed only in degree from that of a more famous storm eight decades later. Salt water flowed four feet deep in the village. Only six buildings remained habitable. Of the six vessels and two coasting barges in the harbor, even the two not seriously injured were reduced to dismasted hulks. To one of the hulks Jean Lafitte removed himself in order that his Red House might serve as a hospital for the French colonists.[2]

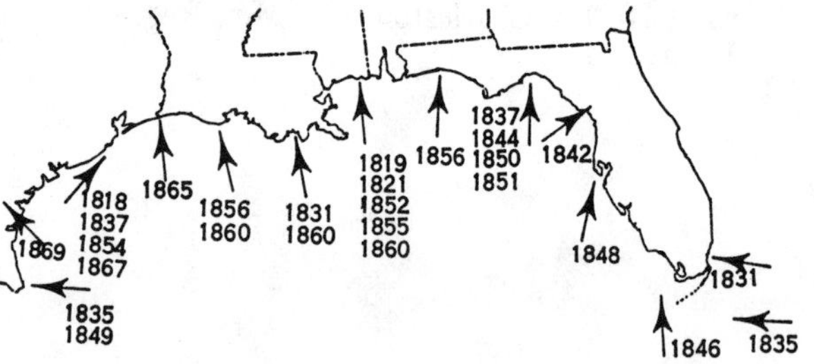

Though Faye gave no particular source of his document, he probably drew his dating of the storm from the manuscripts of Antonio Martinez, an early Spanish governor of Texas. In the latter's letters, first published in 1957, appeared the following meteorological complaint: "As I have previously stated, the greater part of the harvest has been lost as a result of the extremely dry weather and the hurricane of the 12th, 13th, and 14th of last September." [3]

No detailed information has been uncovered about the nature or course of this first known hurricane to afflict the island of Galveston. Two years later, Colonel W. C. D. Hall, when cruising along that stretch of the Mexican coastline, saw the wrecks of the four ships which once served Lafitte and his adventurers.[4] Poey listed a storm in the vicinity of the Cayman Islands south of Cuba and in the Gulf of Campeche on 10–12 September, and this very likely was the same disturbance which struck Galveston.

[1] W. Armstrong Price. *Hurricanes Affecting the Coast of Texas from Galveston to Rio Grande*. Beach Erosion Board, Corps of Engineers. Technical Memorandum No. 78. Washington, 1956. A-2.

[2] Stanley Faye. The Great Stroke of Pierre Lafitte. *Louisiana Historical Quarterly*. 23–3, July 1940. 795–96.

[3] Bexar, Oct 20, 1818. *The Letters of Antonio Martinez*. Trans. & ed. by Virginia H. Taylor. Austin, 1957. 189.

[4] S. W. Geiser. Racer's Storm (1837), with Notes on Other Texas Hurricanes in the Period 1819–1886. *Field and Laboratory* (Dallas). 12–2, June 1944. 64.

THE BAY ST. LOUIS HURRICANE OF 1819

J. C. Moret, writing in 1868 about his fifty years on the Mississippi Gulf Coast, thought the hurricane of 27–28 July 1819 was "by far the most severe and strongest that ever blew on this coast since I came to it." [1] His opinion was seconded by others who had passed through the horrors of 1837 and 1855.[2] Moret had taken up residence in 1817 at Shieldsborough, then the only sizable community at Bay St. Louis, the best-sheltered harbor from the mouth of the Mississippi to Mobile. As skipper of coastal vessels, he had weathered all the storms, big and little, that had visited the area for half a century. He described the meteorological conditions attending the late July hurricane:

> The storm began during the night of the twenty-seventh of July, blowing first from the east, and after day-light increasing continually; at about 8 o'clock A.M. veered to E.S.E. and later to N.E. until about 11 or 12 o'clock at night, when it suddenly died away to a perfect calm for about ten minutes, then as suddenly sprang up from the S.W. and for about an hour blew twice as hard as it had blown from any other points.[3]

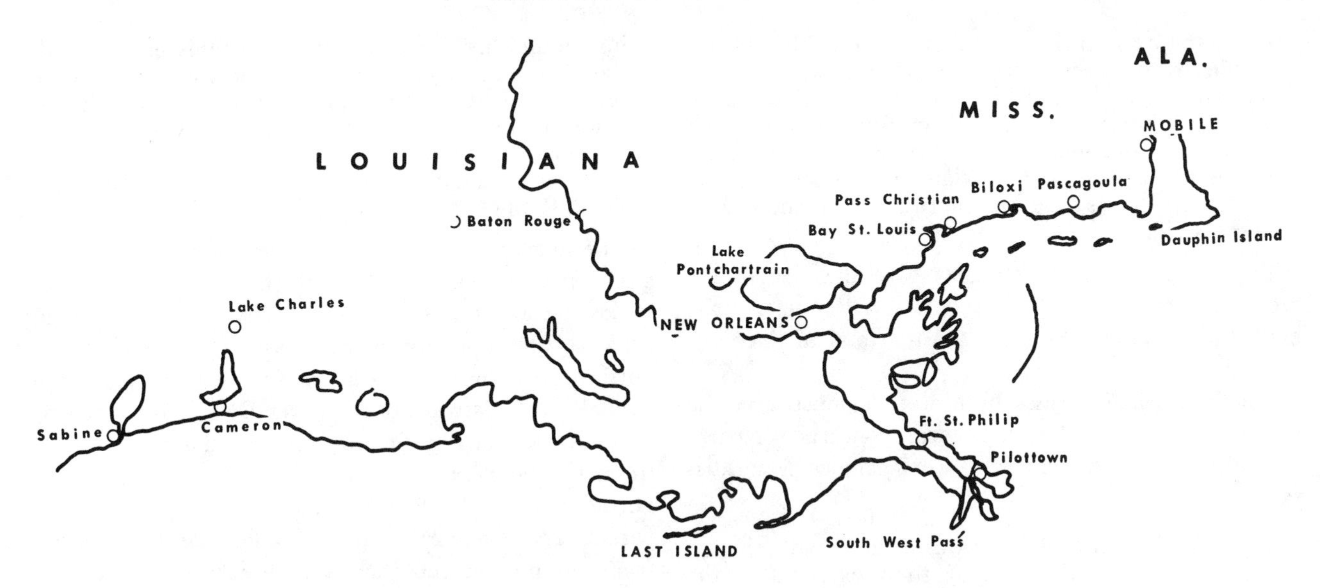

The *Louisiana Gazette* declared: "the hurricane was sensibly felt in New Orleans, but occasioned no serious damage in its vicinity or in the city itself." [4] Mention was made two days later that ships at the Belize, at the mouth of the Mississippi River, had suffered little injury: "The gale was very severe for 24 hours from NNE to SSE—no other damage done (except three ships ashore). The lighthouse building on Frank's Island received no injury whatsoever." [5]

The main impact of the storm lay eastward. In a dispatch entitled, "The Late Hurricane," the *Gazette* continued:

> We are told that the storm was not less overwhelming on shore. At the Pass of Christian and Bay St. Louis, most of the houses have greatly suffered, and many were blown down. The waters rose five or six feet and destroyed a great deal of stock along the coasts of Lake Borgne and Pontchartrain . . .
>
> All the houses at Bay St. Louis were seriously damaged; and most of them blown down; at the Pass also, only three houses were left standing: no lives, however, were lost at either place. The whole coast from Rigoulets to Mobile, to which later place only our intelligence reaches, is a scene of desolation, covered with fragments of vessels and houses, the bodies of human beings, and the carcasses of cattle.[6]

The Mobile Bay area suffered more seriously than the Mississippi River section since the center of the storm apparently passed over Bay St. Louis and this would put all of the Alabama and West Florida coast in the dangerous eastern semicircle of the storm. Turtles and alligators were washed up into the streets of Mobile at the height of the blow and a large brig stranded on Dauphin Street, just east of Water Street. The tide was thought to have been nearly as high as in the great storm and high water of August 1852.[7] We have no report from Pensacola, still in Spanish hands, as no newspaper was published there until mid-summer 1821.

Skipper Moret described his experiences on the night and day of the storm:

> On the night of the twenty-seventh of July, 1819, I was on board of a schooner named the *Peacock of Pearl River,* lying in company with another schooner, *The Odd of Pearl River,* among the Malheureaux Islands, near the south-shore of Lake Borgne, to the eastward of the place where the terminus of the Mexican Gulf Railroad from New Orleans, has since been established. About 8 o'clock A.M. finding that the wind was still increasing and had wore to E.S.E., we got under weigh, by hoisting a piece of one of our sails, and ran into Pearl River, and there rode the balance of the gale in safety. During the beginning of the gale several coasting vessels made harbor in Bay St. Louis and escaped uninjured. But the schooner *Favorite,* captain Michel Eldridge, was driven on the East point now known as Henderson's Point, where she remained for a considerable length of time. Several other vessels were entirely lost at different places along the coast.[8]

The principal marine incident which made the hurricane live long in memory was recalled by Moret:

> In that storm the U. S. Man-of-War Schooner *Firebrand* was lost; she was a vessel of about 150 tons, carrying twelve guns, drawing about eleven or twelve feet of water, and having a crew of about seventy-five or eighty men, commanded by Lieutenant Cunningham, who had set out two or three days before the storm, (leaving her lying at anchor between Ship Island and the northend of the Chandeleur Islands) in a small boat, with a crew of six men and coxswain for New Orleans, where he was during the gale, and he and his men composing the boat's crew were the only ones who escaped out of her whole equipage. No one can tell what took place on board of her after he left, every man in her perished.
>
> She had been seen the morning of the twenty-ninth of July, after the gale had subsided, capsized, bottom upwards, lying on "the square Handkerchief," a shoal of sand between the Mississippi and Louisiana shores, off the west end of Cat Island; and how and

when she came there no one ever knew; the greater number of her crew were supposd to have been confined in her hull, as she lay with her bottom upward and for a considerable length of time emitted great stench, and but few of the bodies were found; some on the western end of Cat Island, others on the shore of Bay St. Louis, and one or two on the shore west of Pass Christian.[9]

Contemporary press estimates give the size of the crew and the number lost as only thirty-nine. Apparently the passing years increased the total of lives lost by two.

Another incident related by Moret, demonstrating the magnitude and extent of the storm, concerned a party of United States soldiers who were camped some distance inland:

> The early part of the day was cloudy, and although there was little wind felt by this command in the valley where it was encamped and surrounded by the pine forest, it was evident from the rapid motion of the clouds that strong currents of air were passing from the southeast to the northwest during the day. At night there was a strong breeze through the camp, increasing gradually in force and attended by pretty copious rain. About 8 o'clock it amounted to a gale, and the tents of both officers and men were swept away. This gale continued through the night, prostrating the forest trees, driving the soldiers before it, some of them to a great distance, killing one and wounding about 20 others, some of them severely.
>
> * * *
>
> Much damage was done beyond this point by the tempest, but between this place and the gulf shore, evidences of stronger and stronger winds were met with every mile. The roads were obstructed by fallen timber so that it was with extreme difficulty that a horseman found his way; and the heavy rains uniting with the waters of the sea, driven by the winds into every branch and bayou and submerging all the low lands, made it necessary occasionally to swim. All the streams leading into were in this way charged with salt water, and the sea spray was noticed for more than 50 miles from the coast.[10]

The hurricane of 1819 was probably of rather small size, but its intensity rated with the severest storms to strike the Middle Gulf coast in our period. The wind behavior at the mouth of the Mississippi, veering from northeast to south-southeast, would seem to place the landfall just westward of the lower Delta. Moret's mention of a calm at Bay St. Louis indicated a passage over the mainland near that point. A track trending north-northeast over the lower Delta, crossing St. Bernard parish, and over Mississippi Sound can be deduced from our meager reports. New Orleans remained well to the west and suffered little damage, but Mobile and probably Pensacola felt the full lash of the easterly and southeasterly blasts in the eastern sector of the storm.

[1] J. C. Moret. Storms on the Sea Coast of the Mississippi. *De Bow's Review* (New Orleans), 38–9, ns. 5–9, Sept 1868, 791.
[2] *Daily Picayune* (New Orleans), 19 Sept 1855; *Daily Crescent* (New Orleans), 16 Aug 1856.
[3] Moret, 792.
[4] *Louisiana Gazette,* 4 Aug 1819.
[5] *Ibid.,* 6 Aug 1819.
[6] *Courier for the Country* (New Orleans), 4 Aug 1819.
[7] W. H. Gardner. *Record of the Weather from 1701 to 1885.* Auburn, Ala., 1886?. 7.
[8] Moret, 793.
[9] Moret, 792–93.
[10] A. P. Merrill. The Hurricane of 1819. *De Bow's Review.* Sept 1868, 790.

THE SEPTEMBER HURRICANE OF 1821

A second major hurricane in two years visited the Middle Gulf coast in September 1821. Its course followed closely the famous July 1819 storm in that the center, as evidenced by a calm period, also passed over Bay St. Louis, according to our contemporary witness J. C. Moret.[1] The backing of the winds at Ft. St. Philip at the English Turn on the Mississippi from easterly to northerly and the southeasterly flow at Mobile confirm that the center passed between those points and made a landfall close to the Alabama-Mississippi boundary. The 1821 storm, though less intense locally than its predecessor, seemed to be of great extent as we hear of its effect all the way from Louisiana to Apalachee Bay in Middle Florida. It raised a tide at Mobile one foot higher than in 1819, but the damage reports from wind force onshore were much less than reported in the first storm.

By September 1821 the transfer of West Florida to U. S. Government forces had been completed and American soldiers and civilians were established at the principal ports along the Middle Gulf. There was considerable more shipping activity in the area than previously to report storm behavior, and the setting up of local printing presses provided a source of local news. Thus, we are able to follow the September 1821 hurricane at a number of points along the shore.

NEW ORLEANS

> In our immediate vicinity, little damage was done by the gale of wind which commenced in the night of the 15th and continued to the 17th inst. Some 12 or 15 large willows on Maligny's Canal were torn up by the roots, and a few pirogues were filled with water by the market house. Other, but trivial injury was felt in the gardens in the city and suburbs.
>
> It appears, however, that the storm was more vio-

lent to the N.E. We are told that four soldiers were drowned at Petite Coquille and the buildings erected to shelter the troops entirely demolished. Both lakes Borgne and Pontchartrain are now very high; the swamps are filled with salt water, and even the high roads in some places partly covered.[2]

PETITE COQUILLE ISLAND

The late gale was very severe at Petite Coquille from seven to eleven P.M. The water rose from six to eight feet on the whole island and a tremendous surf broke over the fort which swept away nearly half the bank on which it was situated, totally demolishing one half of the parapet and the whole of the officers quarters, and had the storm continued two to three hours longer, the remaining buildings would have been carried away, as their foundations were beginning to undermine, and the whole garrison probably perished.[3]

FORT ST. PHILIP

Sept. 15—On this evening a hurricane commenced from the East and continued the 16th veering to the North and Northwest; it ceased about midnight on the evening of the 16th having blown violently from the south about three hours. Water flowed in both gates of the fort.[4]

BAY ST. LOUIS

The second storm, or hurricane, took place sometime in September, 1821—I think on the 21st [actually from the 15th to 17th]—it began before day, the wind blowing from the East, then wore to N.E., and after 12 o'clock chopped to S.E., and continued changing from one point to another until about 9 o'clock P.M., when it fell to a dead calm for about 30 minutes, and subsided gradually, so that by 12 o'clock at night there was no wind from any quarter.

Several coasting schooners ran in the Bay of St. Louis in the morning of the gale, and sheltered themselves in the "Portage," a deep cove, in the eastern shore of the Bay St. Louis, and there rode out the storm at anchor without accident. The schooner *Washington,* a vessel of about 80 or 100 tons, had anchored during the night the gale began, outside of Bay St. Louis towards Pass Christian. She remained at her anchorage until about 9 o'clock A.M. where she began to drag her anchors, with the wind at East, and had nearly reached the western shore of the Bay St. Louis when the wind suddenly wearing to N.E., she was driven, still dragging her anchors, along the shore within about 150 yards of the land, until she came to a bend in the coast, where General E. W. Ripley had a military encampment; there her anchors held her until night, at which time she was last seen on water.

The next morning she was seen bottom upwards, with her bows open from her deck to her keel, lying about 50 or 60 yards from the water, on the beach, over which she had been driven until stopped by the trees, at a place near where the house of James Philips now stands. Every person on board perished; some of the dead bodies were found lying from four to five hundred yards in the woods where they had been carried by the waves; no one escaped, and from the number of bodies found, it was supposed that there must have been about twenty persons on board.

During that storm several other vessels were scattered along the coast, from the Rigolets to Mobile Bay; not less than four or five were driven on shore at and near Pass Christian, among them were the sloop *James,* Captain Pigeon, a vessel of about 60 tons, and the schooner *Hookey* of Sandwich, a vessel of about 80 tons, and two or three others, names not recollected. During that gale and that of 1819 a greater extent of bank along the whole coast was washed away than has been by all the storms that have taken place since.[5]

CANTONMENT BAY ST. LOUIS

We have just experienced one of the severest gales of wind, accompanied with rain, ever known by the inhabitants of this part of the country. It commenced on Saturday the 15th inst. from the N.E. and continued to blow without intermission until 4 A.M. on the 17th. I fear much damage has been done to the vessels navigating the lakes between New Orleans, Mobile, Pensacola, &c.

* * *

Our barracks are in a complete state of ruins, not a house in camp but was either unroofed or inundated; timber two feet over has been left on the highest part of our parade. Fortunately, no lives were lost; and, although every person was exposed to the wind and rain during the whole night of the 16th, I am in hopes no other bad effects will result from it, with the exception of the loss of part of the public property and a portion of the officers' baggage.[6]

(Col.) Z. [Zachary] Taylor

MOBILE

On Saturday evening last, a gale commenced from S.E. and has continued until this morning, with dreadful violence, carrying almost everything before it, and injuring the shipping, the wharves, &c. very much. The water was at one time one foot higher than it rose in the gale of 1819. The floors of the stores bordering on Water Street, were covered with water. The ship *Hope* lies high and dry, within 30 feet of Water Street, and will not, it is thought, be got off. Many small vessels are destroyed and are on shore. The U. S. Cutter *Alabama* went past the city in the gale last night, dismasted and firing distress guns; likewise many other vessels went up, none of which are yet heard from; but, being up the river, no serious fears are entertained for them. Fortunately, no lives have been lost that we have heard of yet.[7]

PENSACOLA

About ten o'clock on Saturday night, 15th, it commenced blowing quite fresh, and continued with increased and increasing violence until in a few hours it raged a most destructive storm. The gale continued with unabated fury until about three o'clock on Monday morning, when it moderated to something more than our usual breeze. We have heard of no lives lost, and little or no damage was done to the houses here or in the neighborhood; but the shipping suffered very severely [of thirteen brigs, schooners, and sloops in the harbor: 6 were driven high ashore, 5 were able to get off after beaching, and the others rode it out].[8]

FORT ST. MARKS

On the morning of the 16th Sept. the tide was uncommonly high which completely inundated the whole of the adjacent country and left but a few yards of dry ground within the walls of the fort. The water flowed through all the quarters excepting two, both of the officers and men, in some of them it was two feet deep. [An east wind prevailed on the 16th with rain.] [9]

[1] J. C. Moret, 793.
[2] *Louisiana Gazette,* 20 Sept 1821.
[3] *Louisiana Courier,* 24 Sept 1821.
[4] SG.
[5] Moret, 792.
[6] *Nat. Int.* (Wash., D. C.), 23 Oct 1821.
[7] *Ibid.,* 17 Oct 1821.
[8] *Floridian* (Pensacola), 22 Sept 1821.
[9] SG.

THE EARLY TROPICAL STORM OF 1822

An early season tropical storm of probably less than full hurricane intensity drove shoreward between Mobile and New Orleans on the 7th and 8th of July 1822. The principal incident concerned the sloop *Lady Washington* which was caught off Pensacola on the 7th by the rising gale. After weathering the blow for two days, the sloop finally ran ashore on Ship Island. Though no lives were ultimately lost, rumors spread that there had been a serious disaster until later reports clarified the situation.[1]

At Dauphin Island astride the mouth of Mobile Bay the surgeon weather observer at Fort Gaines noted in his weather diary: "8th—east blowing a gale; 9th—southeast gale continues; 10th—variable but more moderate." [2] Back in the harbor of Mobile the force of the gale, probably from the southeast, pushed a flood tide four feet over the wharves which served the small port. Two brigs were driven high and dry.[3]

To the eastward at Pensacola, both the 8th and 9th were described by the local observer as stormy during the day with rain and wind from the southeast.[4] There were no damage reports other than that to the *Lady Washington.*

To the westward of the storm track at Fort St. Philip in the Mississippi Delta area, the weather observer at the post noted the following:

> July 7th—High wind which increased to great violence on the 8th at 6 P.M. the water began to rise which continued to increase with heavy rains until 8 A.M. on the 9th. the wind NW. at 12 the wind W at which time the hurricane began to abate, and at night the water was nearly off the grounds. At 6 the wind SW. It continued to blow and rain throughout the night of the 9th, on the 10th rain in the forenoon at 12. The water during the hurricane was one foot higher than it has been since the Aug. storm of 1812.[5]

The New Orleans press carried no reports on any local damage or effects of the storm raging to the eastward.

[1] *Pensacola Gazette,* 20 July 1822.
[2] SG.
[3] Capt. Webster's dispatch to New York, *Nat. Int.* (Wash., D. C.), 10 Aug 1822.
[4] *Pensacola Gazette,* 3 Aug 1822.
[5] SG.

THE BARBADOS TO LOUISIANA HURRICANE OF 1831

During the mid-August days of 1831 a tremendous hurricane, outstanding for physical size and length of destructive path, spread ruin from east of the Leeward Islands, through most of the Greater Antilles, across the Gulf of Mexico, and to the American mainland in the Mississippi Delta area. It was one of the *great* hurricanes of the century, or any century.

The storm's first victim was the island of Barbados where the aggregate destruction caused by all previous hurricanes, including the Great Hurricane of 1780, was thought not to have equaled the havoc wrought by this single visitation. More than 1500 persons succumbed in the disaster there on the night of 10–11 August 1831, and property damage estimated by survey engineers ran in excess of $7 million.[1]

To direct the task of reconstruction, King William IV dispatched Lt. Colonel William Reid of the Royal Engineers to the Leeward Islands. During his two-and-a-half-year stay Reid became absorbed in trying to understand the nature of the great physical forces that had caused such a calamity, and this led to a lifelong study of tropical storms. In his major work, *An Attempt to Develop the Law of Storms,* first published in 1838, Reid made use of a vivid eyewitness account of the complete devastation attending the passage of the 1831 hurricane over Barbados; this had appeared in the Bridgetown press following the storm:

> On reaching the summit of the cathedral tower, to whichever point of the compass the eye was directed, a grand but distressing ruin presented itself. The whole face of the country was laid waste; no sign of vegetation was apparent, except here and there small patches of a sickly green. The surface of the ground appeared as if fire had run through the land, scorching and burning up the production of the earth. The few remaining trees, stripped of the bows and foliage, wore a cold and wintry aspect; and numerous seats in the environs of Bridgetown, formerly concealed amid thick groves, were now exposed and in ruins.[2]

The course of this hurricane through the West Indies and into the Gulf of Mexico was traced by William C. Redfield of New York City, then just commencing his study of hurricane tracks. He placed the center as approaching Haiti on the 12th, at the eastern tip of Cuba on the 13th, at Matanzas on the 14th, off Dry Tortugas on the 15th, and well into the Gulf of Mexico on the 16th.[3] The storm proved particularly severe in its passage across the length of Cuba, and numerous ships were lost along the northern shore of that isle. Key West experienced a stormy day on the 14th with variable winds, but lay far enough to the north of the storm track to escape major damage and injury.[4]

The storm's approach to the American Gulf Coast was announced by rising tides and increasing northeast winds. Fort Pike, at the entrance to Lake Pontchartrain about 25 miles northeast of New Orleans, had "storm, rain, hail, high winds, tide overflowing" on the 16th, with the wind out of the northeast. On the 17th the rising gales had veered into the southeast and heavy rain fell, as the medical officer noted: "storm continues with great violence." By the 18th, though the air flow out of the southeast continued and rain fell, the "storm moderates, tide falling slowly." [5] At nearby Fort Wood on Lake Borgne the storm tide on the 16th and 17th had inundated the hospital area with four feet of water. The only wind direction mentioned in the Fort Wood report was southeast.[6]

J. C. Moret in his storm memoirs recalled that the 1831 Hurricane differed from all others in that the wind blew from only one direction and continued there for three full days. Moret commented on its effect at Bay St. Louis: "Though this gale did not equal, in violence, those of 1819 and 1821, yet owing to its protracted duration the waters rose considerably higher, than it ever done either before, or since." [7]

New Orleans, Lake Pontchartrain and the mailboat ports eastward to Mobile lay in the advancing hurricane's dangerous eastern sector where a tremendous storm tide was added to the fury of the winds. The long sweep of the storm whirl across the open sea piled up the Gulf waters to unusual heights. As this surge reached the coastline, harbor and shore installations and ships in port took a severe beating. At Pascagoula on the Mississippi shore the tidal sweep washed away the main steamboat wharf, leaving the place without any port facilities.[8] Mobile, well to the east of the storm center's track, came off with relatively light losses—*The Register* declared: "not much damage has been sustained." Since there was no major shipping in port, injuries in the harbor were confined to small craft. At flood tide, however, the southeasterly air flow had pushed the waters of the harbor over Water Street from its intersection with St. Francis Street north and from Conti Street south.[9]

At the city of New Orleans the strength of the gale was considered the greatest since that of August 1812. Small shipping along both shores of the river suffered severely, and a breach was made in the levee in the lower part of the city.[10] The major damage in the area occurred in the northeastern section of the city when a tidal overflow from Lake Pontchartrain engulfed the lowlying parts bordering that body of water. The three-day blow from the southeast had forced Gulf waters through Lake Borgne into Lake Pontchartrain where the confined waters rose to an unprecedented height. All flimsy landing docks and bath houses recently constructed along the lake shore were swept away by the rampaging waters. In the city, itself, the lake waters inundated the back parts with the flood standing three feet deep in Treme Street and Esplanade Street at its peak.[11] The *Louisiana Courier* described the scene on the day of the storm:

> A storm of wind and rain of unexampled violence for many years, commenced yesterday morning and has continued without much abatement, up to the present moment. The custom house has been partially unroofed, and various damage has been sustained by buildings in the city. Scarcely a single vessel in port has escaped serious damage, and they have nearly all been drifted ashore, having broken

from their moorings and thrown down the stagings by which they were laden or unladened.[12]

The lightly-populated area in the bayou country to the south and southwest of New Orleans also suffered severely from the storm tide. The Island of Barrataria, famous in the early days of the century for pirate activities, was completely inundated when the waters of the nearby Gulf rising six feet invaded the bayous and rolled over the lowlying land. The inhabitants sought safety in trees and small boats. Early reports indicated that 150 persons were drowned, but that proved false and apparently no one was lost despite the immensity of the tide. Plantations in the Terre aux Boeuf area were submerged under three to four feet of water, but the sugar cane, the principal product of the region, suffered only light losses due to the early season.[13]

The force of the hurricane was felt to the north at Baton Rouge and Natchez, as well as a short distance west of New Orleans.[14] From the meager wind data available, the storm center probably made a landfall just to the west of Last Island and drove northward through the Atchafalaya area.

Two weeks later a second tropical storm swept in from the Gulf. It raised high tides along the coast from Lake Borgne westward, but the main impact of the winds struck much farther westward than the mid-August blow. New Orleans had a gale on Sunday night through Monday, the 28–29th.[15] But in the Opelousas and Attakapes region in central Louisiana, as well as to the westward toward the Sabine, the gale raged severely and did considerable damage. Cotton in the Alexandria and Baton Rouge parishes was ruined by the heavy rains and high winds; as much as one-third of the crop was feared lost in the Clinton sector.[16] At Fort Jessup, southwest of Nachitoches, heavy rains fell during Sunday night and on Monday accompanied by high winds from the northeast.[17]

[1] I. R. Tannehill. *Hurricanes.* 9th ed. Princeton, 1956. 152.
[2] William Reid. *An Attempt to Develop the Laws of Storms.* London, 1838.
[3] William C. Redfield in *N. Y. Journal of Commerce,* 27 Sept 1831; also *Amer. Jour. Science,* 21, Jan 1832, 192–93.
[4] SG.
[5] SG.
[6] SG.
[7] Moret, 794.
[8] *Mobile Register,* 20 Aug 1831, in *Courier* (New Orleans), 24 Aug 1831.
[9] *Idem.*
[10] *La. Adv.,* 22 Aug 1831.
[11] *La. Courier,* 19 Aug 1831.
[12] *Ibid.,* 17 Aug 1831.
[13] *La Adv.,* 22 Aug 1831.
[14] *Baton Rouge Gazette* in *Courier,* 29 Aug 1831.
[15] *La. Adv.,* 31 Aug 1831.
[16] *Daily Picayune,* 16 July 1865.
[17] SG.

THE ANTIGUA-GULF OF MEXICO-RIO GRANDE HURRICANE OF 1835

The major tropical storm of the 1835 season has been named "The Antigua Hurricane" by its historian, Lt. Colonel Reid. The center appears to have passed directly over that Leeward Island on 12 August when a 20-minute lull occurred between full hurricane strength winds, first coming from the north as the storm approached from the east, and then from the south as it departed westward. The barometer fell 1.40 inches. Trees were blown down in lanes, while nearby stretches were seemingly untouched—an effect noticed in other descriptions of tropical storms in the area. At the commencement the wind was described as coming in great gusts, rather than with steady force.[1]

The center of the disturbance moved northwest-by-west across Puerto Rico, Hispaniola, and Cuba. Ships were driven out of Turks Island in the southeastern Bahamas on the 15th, and others piled up on the shore of southern Cuba near Trinidad the same day.[2] The forefront of the storm brushed Key West and Dry Tortugas as the center, still pursuing a west-northwest course, cut through the northern part of Cuba to pass just south of Havana.

Three days later, on the 18th, after crossing the width of the Gulf, the still violent storm struck a mighty blow at the Mexican coast near the mouth of the Rio Grande River. A press report thought the storm "more severe than any within the recollection of the oldest inhabitant." In its 28-hour duration, many houses were blown down at Matamoros, the principal settlement of the area. A vast storm tide engulfed all the lowlying sand islands along the immediate shore, and water covered the lowlands along the river. Every vessel in the nearby harbor of Brazos Santiago was either driven out to sea or beached high and dry by the shifting gales. One ship was carried out to sea without a soul aboard, the crew having jumped overboard into the teeth of the storm when it was obvious that the gale and tide were driving the ship out.[3]

Another witness at Matamoros described the "dreadful hurricane" as commencing the night of the 18th: "Many houses fell, three hundred were damaged. The violence of the storm was tremendous; nothing could resist it; trees were twisted and torn out of the ground, and carried away. The rain was heavy; the river rose to a fearful height. Four lives only were lost; but more dreadful was the destruction of both lives and property in the Brassos de San Jago, and in the Boca del Rio. Many vessels were stranded and dismasted. There was not a house left standing in the Bonita or the Boca Chica." [4]

The hurricane appears to have continued westward, without recurving, into the mountains of northern Mexico.

[1] William Reid. *The Law of Storms.* 43.
[2] *N. Y. Ship List,* 2, 9, and 19 Sept 1835.
[3] *Nat. Int.* (Wash., D. C.), 3 Oct 1835.
[4] *The Nautical Magazine* (London) for 1848, 528, quoted by S. W. Geiser. Racer's Storm (1837), with Notes on Other Texas Hurricanes in the Period 1818–1886. *Field and Laboratory* (Dallas). 12:2. June 1944. 59–67. Geiser quotes the month incorrectly as September rather than August.

1837 NO. 6—THE APALACHEE BAY STORM

When the historian of early territorial days in Florida, John Lee Williams, published his volume in 1837, he observed: "Severe storms are usually experienced about the equinoxes, though several successive years often intervene without a gale. They rarely penetrate far inland. Although a few vestiges of severe hurricanes are seen, which must have prostrated all the timber on extensive tracks of country, yet none have been experienced, since the Americans have taken possession of Florida." [1] The author's opinion, presumably applicable only to Middle and West Florida, was to be dramatically controverted by the hurricane visitations in late summer and early autumn of 1837. Early in October the editor of the *Pensacola Gazette* would comment: "Between the 7th of August and 7th of October, the northern coast of the Gulf has been swept by three distinct gales, either of which was more severe than any gale that has been known here since 1821." [2]

Storm No. 6 of the 1837 season originated far to the west of the previous disturbances of that year and pursued a very different course. It was one of a family of late August storms which were to plague the Florida and Middle Gulf coasts for the next fifteen years. Spawned somewhere in the Gulf of Mexico or western Caribbean, Storm No. 6 was felt on 30 August at Cape St. George where West Florida bulges southward into the Gulf. The rival commercial ports of St. Joseph, Apalachicola, and St. Marks lay close to the path of the advancing whirl.[3] The storm, itself, appeared to have been of small diameter as no wind worthy of mention occurred at this time at Pensacola in extreme West Florida, only 150 miles west of Apalachee Bay.

APALACHICOLA

"I write from the midst of ruins. A hurricane yesterday swept our town and half destroyed it. Nearly every house is unroofed; a number of upper stories are blown down, and many houses leveled.

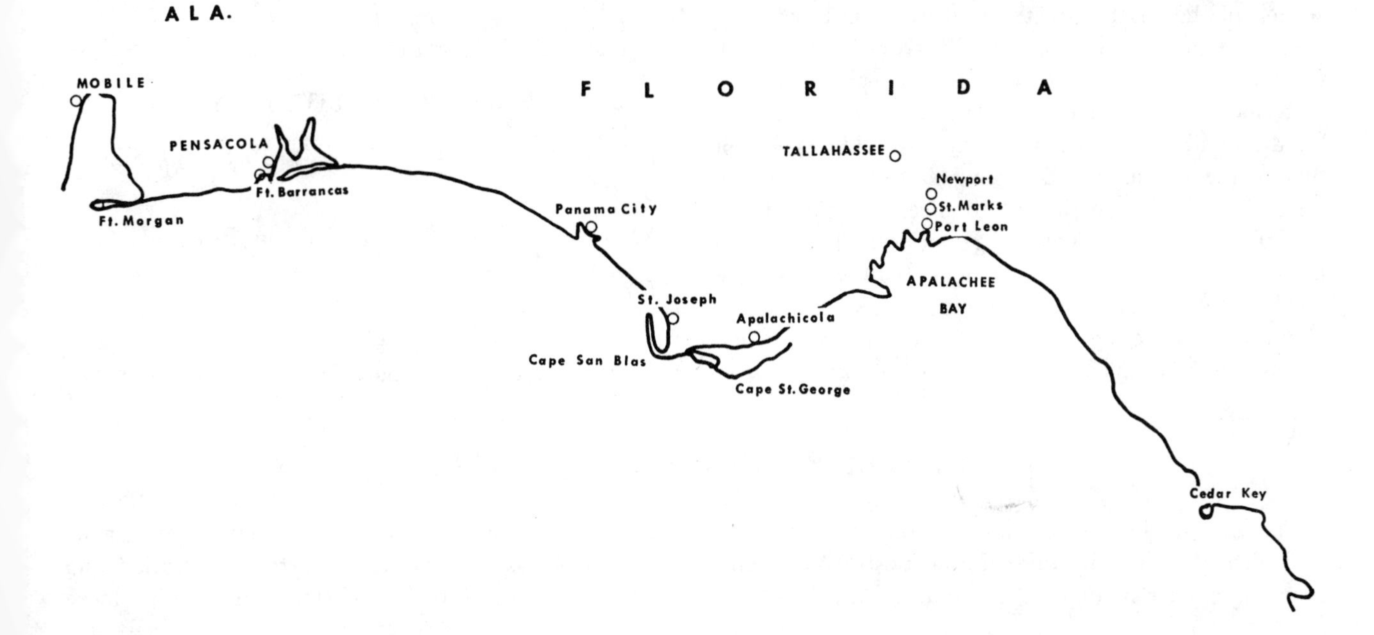

The storm commenced on the afternoon of the 30th August, but was not severe until 4 A.M. on the morning of the 31st when it became very violent until 7 P.M. The wind was from the southeast to the north."

So did the editor of the *Apalachicola Gazette* inform the world in an early dispatch of the disaster which had befallen the small port on the Gulf.[4] In a second account he filled in more details. A squall had arisen from the southeast on Wednesday, the 30th, and continued for two hours. A shift to east followed, and by 2300 it was blowing "a pretty hard gale" which gradually increased to hurricane strength by sunrise. Early in the morning the tide rose six feet and surged over the wharves. At its peak the water was estimated to stand from ten to fifteen feet above normal low water. About noon of the 31st the wind backed to the north, with the gale continuing at full blast, until a final move to northwest took place near midnight as the storm center passed to the northeast. Twenty to thirty buildings in Apalachicola were unroofed, and many stores and homes were defiled by the inundation from the sea. General damage estimates in the press ran as high as $200,000.[5]

ST. MARKS

"There has been a severe storm at St. Marks, which commenced about sunrise on the morning of the 31st August, 1837, the wind being from northeast. At 8 A.M. the wind was north, and it had increased in violence: only one wharf has been left standing. At the lighthouse the sea rose eight feet higher than usual." [6]

The local correspondent of the Tallahassee *Floridian* also sent further details. The water at St. Marks rose three feet higher than in the storm of August 7th and was considered the severest gale ever felt on that section of the coast since the settlement of Florida by the Americans. The schooner *Washington* was torn from its moorings and driven into the pine woods a half-mile from the river. A storm wave ten feet high hit the lighthouse, sweeping away all buildings except that of the keeper and drowning eight persons there. Several houses were dashed to pieces as the village was inundated to a depth of seven feet. There were three to four feet in the warehouses. In view of the northeast winds prevailing at high water, the editor found it hard to account for the depth of the tide unless it had been caused by a strong southeast flow in the Gulf. All the inhabitants dashed for the nearby fort as the waters rose. Total damage was estimated at $30,000.[7]

ST. JOSEPH

On Thursday last the 31st ultimo we were visited by the severest gale felt on this coast by the earliest settlers in Florida. In fact, many of our oldest inhabitants pronounced it the most violent storm they have ever witnessed.[8]

According to the *St. Joseph Times,* the storm commenced from the southeast about daybreak. During the day it backed around to east, north, and finally to northwest, blowing at full strength all day until 1800 when it abated somewhat. A three-story building was demolished at St. Joseph along with several smaller houses. Damage to the important commercial wharf there, fortunately, proved slight.[9]

The exact landfall of the storm center cannot be determined as there were few inhabitants southeast of Apalachee Bay along the northern reaches of the Florida West Coast. Over on the eastern side of the peninsula, St. Augustine had a gale with rain on the 31st and 1st, but the blow had abated on the 2d.[10]

Both Charleston and Savannah lay to the north and west of the storm track where high winds were experienced but little damage resulted. At the latter the wind rose early on the 31st and continued to increase in strength until noon when it backed around to west. At 1400, however, it was still blowing a severe gale.[11] Shipping southeast of Cape Hatteras encountered the disturbance on the 2d in the vicinity of 33N 74W. It was the commencement of this gale that the *Calypso* encountered as she neared the safe haven of Wilmington on the 1st.[12]

[1] John Lee Williams. *The Territory of Florida.* New York, 1837. 17.
[2] *Pensacola Gaz.,* 14 Oct 1837.
[3] William Reid. *The Law of Storms.* London, 1838. 119.
[4] *Apalachicola Gaz.,* 2 Sept 1837, in *Pensacola Gaz.,* 9 Sept 1837.
[5] *Ibid.* in *Floridian* (Tallahassee), 16 Sept 1837.
[6] *Floridian,* 2 Sept 1837.
[7] *Idem.*
[8] *St. Joseph Times,* 6 Sept 1837, in *Pensacola Gaz.,* 9 Sept 1837.
[9] *Idem.*
[10] SG.
[11] *Courier* (Charleston), 1 Sept 1837.
[12] *N. Y. Ship List,* 13 Sept 1837.

RACER'S HURRICANE OF 1837

The eventful hurricane season of 1837 was brought to a fitting climax by one of the most famous and destructive hurricanes of the century. Racer's Storm has lived long in memory, partly from its apt name, but more so as a result of its extreme duration and the immensity of its path of destruction covering more than two thousand miles. The disturbance was first encountered, according to Lt. Col. Reid, its able historian, in the central Caribbean by the sloop-of-war *H. M. S. Racer* on the 28th of September.[1] The island of Jamaica had undergone high winds and very heavy rains flooding downtown streets on the previous two days, perhaps indicative of the presence of an upper-air pressure trough that could have been the agency spawning the famous hurricane.[2]

H. M. S. Racer ran with the storm under an east-northeast wind and bare poles until early on 1 October when she had reached a position off the east coast of the Yucatan peninsula. After a slice across the northern part of that promontory, the storm center continued on a northwesterly track across the Gulf of Campeche toward the mouth of the Rio Grande River. In approaching the coast early on the 3rd, the dynamics of recurvature slowed the storm's forward progress and turned it gradually into the north and then northeast. Matamoros, then the principal settlement of the area and the Mexican port of entry for the lower Rio Grande country, endured a three-day lashing from the now slow-moving atmospheric whirl, hesitantly searching for a steering course of westerlies which would carry the storm system off to the northeast. All vessels lying in the harbor of Brazos Santiago were either driven ashore or sunk. The often-battered Mexican customhouse on the sand spit lying across the river entrance was swept away; after this disaster the entry post was rebuilt at a safer location in the low hills inland.[3]

Late on the 4th the storm center was near Matagorda Bay, halfway from Corpus Christi to Galveston, where the settlements all suffered severely as a storm tide surged over the flimsy wharfs and buildings. A Galveston-bound ship, which had just discharged some New Orleans passengers on Matagorda peninsula, was blown off shore by the gale and never heard from again.[4]

Galveston Island lay directly in the path of the large and vigorous hurricane, now constantly curving more and more to the northeast following the trend of the Texas shoreline. This island, long before the haunt of pirates and filibusters, had been selected without much exercise of weather wisdom by Americans now pouring into the newly-proclaimed Republic of Texas as a site to build a major settlement. Work on many new projects was under way in early October 1837 when the storm struck. There were several eyewitnesses of the resulting disaster whose records have been preserved. Perhaps it will be best to present first the basic details which appeared in the *Houston Telegraph and Texas Register,* the nearest press then operating:

> The late accounts from the seaboard are of the most distressing character. A tremendous gale appears to have swept the whole line of the coast and destroyed an immense amount of property. It commenced on the 1st and increased in violence until the 6th. At Velasco four houses were blown down; the whole country for miles inundated and all the vessels in the harbor, consisting of the brig *Sam Houston,* and the schooners *De Kalb, Fannin, Texas,* and *Caldwell,* were driven ashore, the last named has since been got off and cleared on Sunday last for New Orleans.
>
> At Galveston the waters were driven in with such violence that they rose 6 or 7 feet higher than ordinary spring tide. They inundated a large portion of the east end of the island, and compelled the soldiers of the garrison to desert their barracks and seek shelter on the elevated ground near the intended site of Galveston City. The large new warehouse of Mr. Thomas McKinney and the new customhouse were completely destroyed and the goods scattered over the island.
>
> The brigs *Perseverance, Jane,* and *Elbe* were driven ashore, and are complete wrecks; the *Phoenix* is also ashore, but slightly injured, and may easily be set afloat again. The schooners *Select, Henry, Star, Lady of the Lake,* and the prize schooner *Correo* are ashore, some of them high and dry. The *Tom Toby* (privateer) is a wreck, and the *Brutus* (Texan naval schooner) is considerably damaged. The schooner *Helen* is the only vessel which has received no damage. So far as we have been able to learn only two individuals have perished. . . .[5]

> There were about 30 vessels in Galveston Harbor when the great storm commenced on October 1, 1837. It began with wind from the southeast and held to that quarter mostly for three days; then it veered a little to the east and so continued until the sixth day, filling the bay very full and making a 4-foot rise at Houston. On the evening of the 6th [5th?] the wind veered to the northeast and blew very strong. The schooner, *Tom Toby,* a privateer, parted her cable and went ashore on Virginia Point. About sunset the wind, veering all the time to north, and, if possible, increasing, brought the large volume of water from the bay onto the island with such force and violence as to sweep everything in its course. On land every house, camp, sod house, and inhabited structure was swept away, except the old Mexican customhouse. Only one of the vessels held to its moorings.[6]

> The scene upon the land was equally terrible. The brig *Jane* of Saybrook, (Conn.) was dashed against a large three story ware-house (of F. T. McKinney and S. M. Williams) which had just been enclosed, and the whole fell with an awful crash into a heap of promiscuous ruin. Not a stick of its timbers after the gale subsided could be seen . . . The new (Texan) custom house was swept from its foundations, and but two houses in the whole island survived from the wreck. Human suffering in the meantime was immense. Men, women, and children were seen floating upon boards, logs and small boats, for days and nights, in every part of the island. But one life . . . was lost, which must be regarded as provi-

dential, when we consider the great destruction of property, and the imminent perils which were encountered everywhere. The scene upon the island after the storm was over, was one of utter desolation. Provisions, furniture, and goods of all kinds, had either been swept off, or were found in a ruined condition, scattered over the island: and the homeless inhabitants were seen wandering about in despair, gathering something from the wreck to hide their nakedness, or save them from starvation. . . .[7]

The next check point along the path of the storm came at the mouth of the Sabine River which forms the boundary between Texas and Louisiana.[8] Here the wind was reported at east and northeast with the height of the gale on the night of the 5–6th. Eastward at New Orleans the wind set in first from the southeast, but on Friday morning the 6th backed into the northeast as the storm center approached on a path taking it slightly south of the Crescent City, probably very close to the mouth of the Missisippi River. Maximum wind strength there was attained late in the day on the 6th and raged until the morning of the 7th. The *True American* of New Orleans described the scene:

The Gale—On Thursday as we stated in our last, the rain poured down in torrents, accompanied by a very high wind from the South East; the streets were flooded and remained so several hours. On Friday morning the wind chopped around to North East, and continued in increasing strength, until it blew a perfect hurricane. This forced the waters of the Lakes through the swamp into the back part of the city as high up as Phillippa and Burgundy streets, overflowing all the yards, many houses having one to two feet of water in them.

The water at Lake Pontchartrain rose eight feet above ordinary high water mark and covered the Rail Road for nearly two miles.

The damage done was very great, the steamboats *Merchant, Columbia, Pontchartrain,* and *Mobile* are total wrecks, the break water and pier at Port Pontchartrain are damaged to the extent of fifty thousand dollars, all the houses but two are washed away, and one or two lives were lost.

At the end of the Bayou the damage is also very great, the Light House is blown down, the pier ruined, the bath houses washed off and many houses injured. The inhabitants of the Bayou were obliged to leave their beds and we understand several of them are missing.[10]

During the heaviest part of the storm late on the 6th another press report stated that water backed up into the city from Lake Pontchartrain as far as Rampart Street. The northeast gales on the Lake took a heavy toll. Seventeen buildings at Port Pontchartrain at the eastern end of the Lake were swept into the swamp by the surging tide, and damage at Milneburg was heavy. The railroad serving Lake resorts suffered damage estimated at $100,000, as much of its right of way was undermined.[11]

Eastward along the Gulf Coast the steamboat wharves at Bay St. Louis, Pass Christian, Biloxi, and Pascagoula were destroyed as the center of the vigorous hurricane moved east-northeastward probably very close to the shoreline.[12] At Mobile the tide rose and inundated streets around the harbor and water entered the cellars of stores and warehouses, but there was little structural damage from wind, and shipping in the harbor and bay suffered very little.[13]

The center of the storm must have passed between Mobile and Pensacola. At the latter point the blow commenced on the 6th with a southeasterly wind component, but shortly before noon on the 7th shifted to south, and later moved into the west. Farther eastward at St. Joseph some damage was sustained from the winds on the 7th. The Company wharf and warehouses were "in part destroyed" and all ships but one were aground.[14] One must be cautious in relying on damage reports from the Florida Middle Coast where there existed great commercial rivalry and jealousy among the small aspiring ports. Editors tended to play down or remain silent about damage to their own ports, while playing up the difficulties of their rivals.

Inland evidence pointed to the presence of much higher wind speeds than were general along the coasts. At Baton Rouge the air flow came out of the east and northeast before backing to north. Very heavy rains fell over the area north of the storm track. At Clinton, just north of Baton Rouge, the heavy rains commenced at 1600 on the 5th and continued until the morning of the 7th. The bottom lands were deluged to such an extent that some farmers expected to gather only one-third of the anticipated crop. The forests in the area were mangled by the winds with fallen trees blocking roads for many days afterwards.[15]

As the eye of the hurricane moved across Alabama, central Georgia, and central South Carolina and headed for the Carolina coast near Wilmington, great damage struck the cotton crop, especially to the north of the storm track. It was characteristic of this storm that the winds in the northern sector were the strongest. Reports from Baton Rouge in Louisiana and from eastern Tennessee and upper Carolina spoke of damaging winds. The presence of a large high pressure area over the Ohio Valley and Northeast would account for the high winds through the Upper South and would also provide a steering current that carried Racer's Storm east-northeastward instead of northward. Redfield informed us that his barometric pressure at New York City read 30.73″ on the 9th when the storm center was almost directly south over North Carolina.[16] This would have created a very steep barometric gradient with attendant heavy gales between the two pressure features. Norfolk experienced a northeast gale on the 8th and 9th which prevented the steamboats from leaving dock for the usual runs up Chesapeake Bay.[17]

Another dramatic marine incident occurred when the hurricane was leaving the mainland of North Carolina to pass into the Atlantic shipping lanes. The *S. S. Home,* a newly-built steamboat costing $115,000, left New York City on Saturday afternoon bound for Southern ports.[18] As she sailed southward on Sunday, a steadily mounting northeast gale drove the ship along and it looked as though the *Home* would set another record for the Charleston run as she had done on the previous trip. By Monday morning, however, it was necessary to run ashore some 22 miles north of Cape Hatteras as the water in the hull was gaining on the pumps. Later in the day the *Home* got off and stood out to sea, though the badly leaking hull required all hands, including passengers, to bail. The immediate objective was to clear the Cape southward, then run ashore in its lee where the surf would be much less dangerous. Unfortunately, the now rapidly-filling ship, sinking lower and lower with its engine rooms flooded, grounded well short of the beaches just south of the Cape. She struck about 2200 on the night of the 9–10th and soon commenced to go to pieces. Of approximately 130 persons aboard, only forty made it safely to shore. Among these was Captain Carleton White who survived and published a bestseller on the subject a few weeks later.[19]

[1] William Reid. *The Law of Storms*. 3rd ed. London, 1850, 133.

[2] *Jamaica Dispatch* in *ibid.*, 133.

[3] *New York Ship List*, 8 Nov 1837.

[4] *Matagorda Bulletin*, 11 Oct 1837; *N. Y. Ship List*, 15 Nov 1837.

[5] *Telegraph and Texas Register* (Houston), 11 Oct 1837.

[6] Ben C. Stuart. Early Texas Coast Storms. *Monthly Weather Review* (Washington). 47–9, Sept 1919, 641.

[7] "R". Original Papers and Notes on Texas, by a citizen of Ohio. *The Hesperian, or Western Monthly Magazine* (Cincinnati). 1, 1838. 353, quoted by S. W. Geiser. Racer's Storm (1837), with notes on other Texas Hurricanes in the period 1818–1886. *Field and Laboratory* (Dallas) 12–2, June 1944. 62.

[8] William C. Redfield. On Three Several Hurricanes of the American Seas. *Amer. Jour. Sci.* ns. 1–2, March 1846, 167.

[9] *Bee* (New Orleans) in *Nat. Int.* (Wash., D. C.), 16 Oct 1837.

[10] *True American* (New Orleans) in *Pensacola Gaz.*, 14 Oct 1837.

[11] *Daily Picayune* (New Orleans), 11 Oct 1837.

[12] *Louisiana Courier* (New Orleans), 11 Oct 1837.

[13] *Idem.*

[14] *Pensacola Gaz.*, 7 & 14 Oct 1837.

[15] *Louisiana Courier*, 18 Oct 1837.

[16] Redfield. *Op. cit.*, 168.

[17] *Nat. Int.*, 19 Oct 1837.

[18] *N. Y. Express* in *Nat. Int.*, 26 Oct 1837.

[19] Capt. Carleton White. *Narrative of the Loss of the Steam-Packet Home*. New York, 1837.

THE LATE GALE AT ST. JOSEPH—SEPTEMBER 1841

Our equinoctial storm came off on the night of the 14th inst. [September]. The tide was higher here than ever before known, and a considerable portion of the old wharf (the pens of which had been cut apart for the piles of the new wharf) was carried away. No other material damage was done.

At Apalachicola the storm was of equal violence, unroofing several of the slate covered brick stores, and blowing down the market house and swamping all the small boats in the harbor.

All of the wharves were more or less injured—some entirely destroyed. The two steamboats lying at Sand Island were scuttled and thus saved from total destruction. The steamer *Chamois,* on the ways about seven miles above the town, had her cabin blown off.

At St. Andrews, the water was four feet higher than in any former gale, and the only craft in the harbor, the schooner *Clementina,* of Apalachicola, was blown high and dry opposite Porter's landing.[1]

[1] *St. Joseph Times* in *Pensacola Gazette,* 9 Oct 1841.

THE KEY WEST HURRICANE OF 1841

A late October hurricane swept northward out of the Caribbean Sea to batter Havana and Key West on the 18th and 19th. At the American outpost a south wind, blowing steadily for two days, raised a tide in the harbor to the highest point remembered by the oldest inhabitants, according to the local press. Many ships piled up on the shores of the Keys, especially at Mango Key, nine miles east of Key West.[1] Later reports spoke of numerous vessels dismasted in the Bahamas, indicating that the storm pursued a normal October track to the northeast. Cape Lookout in North Carolina had shifting gales from the 19th through the 23d as the disturbance passed seaward.[2]

[1] Key West press, 21 Oct, in *N. Y. Ship List,* 10 & 17 Nov 1841.

[2] *N. Y. Ship List,* 30 Oct 1841.

THE SEASON OF 1842

ANTJE'S HURRICANE

The Gulf of Mexico in 1842 provided a scene of great hurricane activity from early September to late October. The season commenced, as far as records indicate, with Antje's Hurricane, so named by Redfield, since it was first reported by H.M.S. *Antje* on 30 August at 26°N, 63°W, or approximately halfway between Bermuda and Puerto Rico. The storm picked up an easterly steering current and never deviated from a westward course. It passed between Key West and Havana on 4 September. At the latter it was considered to be more severe than that of 1821—the barometer read 28.93".

At Key West the gale of the 4th blew directly out of the harbor, a direction which greatly lessened damage to shipping there. Nevertheless, great losses were experienced on the neighboring islands. Half of Sand Key blew away. The light-tender's house was demolished. The storm tide also washed away the Dove Key Beacon nearby.

On the 5th the gale dismasted a ship about 100 miles west of the Tortugas. Finally, after pursuing almost a straight course westward across the Gulf close to latitude 25°N, the hurricane struck a hard blow at the Mexican coast between Matamoros and Tampico. At Victoria, the calm center arrived about 1300/8th—after a five minute lull, a southerly gale commenced. Very heavy rains fell along the lower Texas coast as the storm finally dissipated in the hills south of the Rio Grande River.[1]

A GALVESTON STORM

Another tropical storm moved about the northwestern Gulf in mid-September. It was noticed at Fort Morgan off Mobile harbor on the 17–18th where an east wind pushed water into the fosse to a depth of 18 inches.[2] Farther west at Galveston "a tremendous gale" drove Gulf water over the island into the bay behind. The walls and steeple of the Episcopal Church were rent and the interior much damaged, and the Catholic Church also suffered structural damage during the night of 17–18 September. The combined forces of wind and waves also demolished several smaller buildings and houses. People were forced to evacuate to safety in the dead of night. Before dawn the wind lulled and the waters quickly subsided. Damage was estimated at $10,000. No lives were reported lost. Some small boats ended up high and dry, but there was no major damage to shipping.[3]

A PENSACOLA STORM

Another disturbance of less than hurricane strength swept the northeastern Gulf later in the week. It could have been related to the Galveston storm, but more likely was a new cyclone born out of the same conditions. At Pensacola the gales of the 22d and 23d were described as "severe," though no lives were lost nor did any serious damage occur. It was not felt on these days to the westward of Pensacola.[4]

AN ATLANTIC COAST STORM

A final tropical storm raged along the Atlantic Coast of Florida and the Carolinas late in October. The gale of the 26th at St. Augustine was "severe," but "not as violent as that of the 5th." Very little damage occurred at this time on land. Charleston also noticed a gale on the 29–30th, presumably from the same offshore disturbance, but little harm befel that port.[5]

[1] William C. Redfield. On Three Several Hurricanes of the American Seas and their Relation to Northers. *Amer. Jour. Sci.* 2nd ser. 1. Mar 1846. 2–15.
[2] SG.
[3] *Galveston Citizen,* 21 Sept, in *N. Y. Ship List,* 8 Oct 1842; *Niles National Register* (Balto.), 8 Oct 1842.
[4] *Pensacola Gazette* in *N. Y. Ship List,* 15 Oct 1842.
[5] *Florida Herald* (St. Aug.), 31 Oct 1842; *N. Y. Ship List,* 9 Nov 1842.

THE GULF TO BERMUDA HURRICANE OF 1842

This storm originated on the last days of September 1842 in the Gulf of Campeche west of the Yucatan peninsula in the southwestern extremity of the Gulf of Mexico. Strong north and northwest winds at Vera Cruz on 30 September to 2 October signaled the formation of the tropical low pressure area. The storm system moved northeastward across the Gulf buffeting many ships along the way. The effects of the gale were felt in the middle Gulf from 100 miles east of the mouth of the Rio Grande to about 40 miles south of the Mississippi Delta, as well as eastward to within 100 miles of the Tortugas. Thus, in its sweep diagonally across the Gulf, the circling winds reached all but the land areas. Many dead birds were later found in the waters of the Gulf as the converging wind flow prevented them from reaching a haven of safety.

William Redfield made a most minute study of ship and land reports of this storm.[1] He pinpointed the landfall of the hurricane eye in the extreme northeastern Gulf, in Apalachee Bay of Middle Florida close to the seaport of St. Marks. The skipper of the Brig *Sampson*, anchored off St. Marks, contributed a valuable meteorological report:

> Brig Sampson came to anchor off St. Marks on the 4th, weather squally from the eastward with much thunder and lightning. Every squall became harder till at 4 a.m. on the 5th, it was blowing a severe hurricane from E. by S. which drove the brig from her anchors, and continued, varying one or two points eastward, till 4 p.m. when it suddenly died away calm. This lasted about fifteen minutes, when the wind came from N.N.W. and blew with increasing fury for about three hours, when it began to abate to a common gale of wind, which after twelve hours moderated. During the hurricane no canvas could stand a minute.[2]

Damage to land installations along the northern sector of the hurricane was heavy as extreme wind speeds mounted offshore. At Apalachicola the wind blew "a perfect hurricane" and was "thought to be one of the severest gales on record." [3] The lighthouse at East Pass was ruined when 20 to 30 feet of the tower blew down. As the keeper's house was swept off the island, his wife was drowned along with several other residents of low-lying areas nearby. The press reported the wind peak at 1600 on the 5th when many houses in Apalachicola were unroofed by the force of the gale.[4] At Tallahassee first damage estimates ran as high as $500,000 as the hurricane tore off roofs, smashed walls, and shattered windows. Roads in all directions from the Florida capital were blocked with thousands of fallen trees.[5] The damaging winds did not reach as far west as Pensacola where there were northerly winds on the 4th and an easterly flow on the 5th as the storm passed almost 200 miles to the east.[6]

To the south of the storm track enormous tides took their usual toll. At Cedar Keys "the hurricane commenced on the 4th from E. and S.E. and continued on the 5th, from S.E. and S. with heavy rain until late at night. On the 6th cloudy with high winds from the northeast. The water is stated to have risen twenty feet above low water mark, and within six feet of covering the island." [7]

The storm's path through the northern part of the peninsula took the center across the Suwanee River watershed, probably over the Okefinokee Swamp, to emerge on the south Georgia coast just north of the St. Marys River.

St. Augustine experienced "the greatest gale ever remembered, at least for 15 years," according to the editor of the *Florida Herald*. On the 5th "there was a great gale from the E. with heavy rain. During the evening the wind varied from S.E. to S. and continued with violence during the night. Early on the 6th, the wind hauled around to the west, and somewhat abated." The statement added that "but about ten at night it shifted to due N. and blew much severer than had been experienced for years." Most probably this referred to the night of the 5–6th as the center passed due north.[8] The reference to excessive wind speeds from the north confirmed the prevalence of higher winds in the northern sector, as evidenced by the great destruction at Tallahassee and Apalachicola, a similar condition to that which we met with Racer's Storm in 1837.

On the northern side of the storm as it neared the Atlantic seaboard, a weather observer at Savannah on the afternoon of the 5th noted: "the wind commenced blowing hard from the S.E. and E., with rain, until about 4 or 5 P.M. when it shifted to N.E. and increased in violence during the night to a hurricane, and continued with torrents of rain, to 5 or 6 P.M. of the 6th."[9]

Charleston ran a high barometer at 30.20" all day the 5th. Toward evening it commenced to blow from the N.N.E. and continued through the night. In the course of the morning of the 6th the wind changed to N.E. and N.N.E. raging at gale strength throughout the entire day with very high tides overflowing several streets along the bay. The barometer remained close to 30.00".[10]

Northward along the coast we have no land reports of destruction. At Wilmington, North Carolina, the sea was deemed too rough to permit the departure of the packet boat for Charleston, and off Cape Lookout and Cape Hatteras several ship disasters occurred as the storm veered gradually away from the shoreline.[11] The ultimate path of the storm moved close to and north of Bermuda where on the 8th and 9th southeast gales mounted as high as force 9 soon after midnight. The lowest barometer did not come until 1600 with the wind out of the south; thereafter, it veered around to southwest.[12]

[1] William C. Redfield. On Three several Hurricanes on the American Seas and their relation to the Northers, so called, of the Gulf of Mexico and the Bay of Honduras, with charts illustrating the same. *Amer. Jour. Science* (New Haven), 2d ser., 1–2, March 1846, 1.

[2] *Ibid.*, 22.

[3] *Apalachicola Journal* in *N. Y. Ship List,* 19 Oct 1842.

[4] *Apala. Jour.* in *Fla. Herald* (St. Augustine), 14 Nov 1842.

[5] *Tallahassee Sentinel,* 7 Oct, in *N. Y. Trib.*, 18 Oct 1842.

[6] Redfield, 157.

[7] *Idem.*
[8] *Fla. Herald,* 10 Oct 1842; also SG.
[9] Redfield Ms. Storm Book (Yale).
[10] *Idem.*
[11] *Courier* (Charleston) in *N. Y. Trib.,* 18 Oct 1842.
[12] Redfield, 159.

GREAT STORM—PORT LEON—1843

Our city is in ruins! We have been visited by one of the most horrible storms that it ever before devolved upon us to chronicle.—On Wednesday [13 September] about 11 o'clock A.M. the wind commenced blowing fresh from the south-east, bringing up a high tide, but nothing alarming—at 5 P.M. the wind lulled and the tide fell, the weather still continued lowering. At 11 at night, the wind freshened, and the tide commenced flowing, and by 12 o'clock it blew a perfect hurricane, and the whole town was inundated. The gale continued with unabated violence until 2 o'clock, the water making a perfect breach ten feet deep over our town. The wind suddenly lulled for a few minutes, and then came from southwest with redoubled violence and blew until day light.

Every warehouse in the town was laid flat with the ground except one, Messers Hamlin & Shell's, and a part of that also fell. Nearly every dwelling was thrown from its foundations, and many of them crushed to atoms. The loss of property is immense. Every inhabitant participated in the loss more or less. None have escaped, many with only the clothes they stand in. St. Marks suffered in like proportion with ourselves. But our losses are nothing compared with those at the lighthouse. Every building but the lighthouse gone—and dreadful to relate fourteen lives lost! and among them some of our most valued citizens. We cannot attempt to estimate the loss of each individual at this time, but shall reserve it until our feelings will better enable us to investigate it.[1]

[1] *Commercial Gazette* (Port Leon), 15 Sept 1843, in *Niles National Register,* 30 Sept 1843, 70.

MATAMOROS—1844

It appears that on the night of 4th August, about 10 o'clock a violent hurricane commenced, which continued until 10 o'clock the next morning, and that throughout the whole town, with the exception of a new church and two dwellings occupied by Dom Pedro Jose de la Gaza, Donna Juana Perea, there is not a building which is not damaged, or lies a heap of ruins. Several persons were killed, and some dreadfully injured.[1]

W. Armstrong Price in his history of Texas hurricanes relates that this storm which struck the Rio Grande coastal area did not leave a house standing at the mouth of the river or at Brazos Santiago on the north end of the barrier island. The waters cut a pass clear through the former site of the settlement. Lt. J. Edward Blake of the Topographic Engineers, U. S. Army, submitted a report the following year telling of the disaster. He reported the loss of life at 70 persons. As a result of the frequent inundations, the Mexican customs office was removed from the island to the mainland.[2]

[1] *Tropic* (New Orleans) in *Commercial Advertiser* (Apalachicola), 7 Oct 1844.
[2] W. Armstrong Price. *Hurricanes Affecting the Coast of Texas from Galveston to Rio Grande.* Tech. Mem. No. 78. Beach Erosion Board, Corps of Engineers. Washington?, 1956. A-3.

APALACHICOLA—1844

TREMENDOUS GALE—GREAT LOSS OF PROPERTY!

As was anticipated yesterday morning [8 September] we were visited last evening by a tremendous Gale, which has done extensive damage throughout the city. It commenced blowing from the N.E. about three o'clock, and gradually increased to a perfect hurricane, till about half past five, when it suddenly lulled, and became a dead calm, which continued about an hour—the wind then hauled around to the N.W. and blew with extreme violence till 8 o'clock when it ceased altogether. The tin roofs of the brick stores, torn in pieces, were flying in the air like scraps of paper—boards, bricks, and everything which the wind could reach, were sent flying in every direction. Fortunately, no lives were lost, and very little bodily injury sustained, though several had very narrow escapes.

(A long list of damaged buildings and property follows.)

* * * * *

The wharves are considerably injured, but not as much as they would have been, had the wind not been accompanied by a heavy rain, which beat down the sea; the tide did not rise high enough to cover them—a few planks but no timbers were displaced.[1]

(Damage was estimated at $18,000 to $20,000.)

TALLAHASSEE

Last night our annual September gale came in all its fury. The wind blew in fitful and furious gusts, driving before it a deluge of rain, from about half past seven P.M. to about one o'clock this morning, veering in its direction from N.N.E. to S.S.E. We dread to hear of its effect on the cotton crop, fearing they must be disastrous.[2]

[1] *Commercial Advertiser* (Apalachicola), Tuesday Morn 9 A.M., 9 Sept 1844.

[2] *Tallahassee Sentinel,* 9 Sept 1844, in *Commercial Advertiser* (Apalachicola), 16 Sept 1844.

THE GREAT HURRICANE OF 1846—I

Colonel Walter C. Maloney in his history of early times at Key West considered the hurricane of 11–12 October 1846 as "the most destructive of any that has ever visited these latitudes in the memory of man." [1] This was a truly *great* hurricane whose trail of devastation, stretching from Cuba northward along the West Coast of Florida and then northeastward along the entire Atlantic sea plain to Canada, stood unrivaled in disastrous extent during our period. Col. Maloney's opinion was well substantiated by the contemporary testimony to follow.

The Isle of Cuba was the first to report the presence of this already fully mature hurricane which had probably originated in the Caribbean area, a favorite haunt of severe October tropical disturbances. In the vicinity of Havana the 1846 hurricane proved the worst since 1821. Its impact has been described in two publications appearing soon afterwards: *Huracan de 1846. Resena de Sus Estragos en las Isla de Cuba* and *Relation de los Estragos Causados por El Temporal del Once de Octobre del Corriente Ano.*[2]

The high winds at the Cuban capital commenced shortly before midnight of the 10th and reached their peak force about 0900 the next morning. The gale began in the northeast, but backed gradually to northwest and west, placing the track of the eye close but to the east of the city and harbor. The lowest barometric reading was listed at 27.06", but the exact time was not given—we may assume that it occurred in mid-morning, probably near 0900, as the glass had recovered to 28.35" by noontime. Of the 104 vessels in the harbor at the time, all but 12 were either sunk, wrecked, dismasted, or otherwise seriously injured. In addition, some 40 to 50 small coasting vessels were destroyed. Damage ashore to structures was comparable.[3]

Winds began to pick up at Key West during the early morning of the 11th as the hurricane eye raged northward along the 82° meridian with no diminution of energy. Its impact on the military and marine community at the end of the Florida Keys was detailed in several contemporary communications to the press. These were widely copied in American newspapers. Unfortunately, *The Light of the Reef,* which had chronicled the passage of the hurricane two years before, had now been extinguished, and the original meteorological records at Fort Taylor have not been preserved in the National Archives. Perhaps they were lost during the gale and flood. Our best eyewitness account came from the pen of a Lt. Pease who was aboard a small naval vessel in the Key West harbor when the hurricane struck:

> The gale commenced about 10 o'clock A.M. on the 11th instant. and about 2 P.M. it blew a perfect hurricane. I was on board the revenue cutter *Morris* about one mile from Key West at anchor with 150 fathoms of chain out, yards down on deck, and every preparation made for the storm. Our riding bitts were working, and it became necessary to back them with deck tackles, the current was now moving by us at the rate of 12 miles per hour: the *Morris* laying broadside to it as well as the wind, made her labor very heavy, and in danger of parting her chains, when we were compelled to cut away the mainmast for the safety of our lives as well as the vessel. When the mast went over the side it hung by the triatic-stay, and in danger of falling on us every moment: a man could not get aloft, and we were anxious to hold on the foremast as the last resort in case the schooner should founder at her anchors. After a few moments a man made out to get aloft and cut the stay, when the mast fortunately fell clear of us.
>
> It was a narrow escape. Thirty men tossing to and fro on the deck of a small vessel, with a mast suspended over their heads as it were by a thread, made the situation anything but enviable. We now battened down the hatches, and all hands passed through the wardroom. The vessel continued to labor very heavily, and the sea made a complete breach over us. It was with difficulty we could keep her free with both pumps going and bailing from wardroom and birth-deck. At 4 P.M. the air was full of water, and no man could look windward for a second. Houses, lumber, and vessels drifting by us—some large sticks of lumber turned end over end by the force of the current, and the sea running so high and breaking over us brought lumber, casks, &c. on board of us and across our decks. At quarter past four the water was up to our lowest half-ports inboard, and gaining on us when our starboard chain parted: and we commenced dragging we know not which way, as

our compasses flew around in such a manner that they became useless for that object. Now our fears were that we should go out over the Reef and into the Gulf, and before we got into the Gulf the vessel must strike and bilge: but that would not save her. At this time we cut away the foremast, when a sea struck us, knocking the schooner on her beam-ends, carrying away our bulwarks, cranes, larboard boat, quarter house, swinging boom, and everything movable off deck: and to right the vessel we hove the lee guns overboard and knocked out the ports—all expecting momentarily to go to the bottom. We were in this suspense for about one hour, when we struck on some reef unknown, when our larboard chain parted, and we made preparations to scuttle the vessel. The hurricane gradually subsided, although at midnight we were sticking heavy, and blowing a gale from S.E.

On the morning of the 12th the scene was anything but agreeable: we had drifted about three miles, and about one-half of that distance over a shoal with only about two feet of water on it at ordinary tides—this is the depth of the water around the *Morris* when I left her. Around her, large wrecks of all descriptions: one ship on her beam-ends, three brigs dismasted, also three schooners; three vessels sunk in a small channel, and four vessels bottom up. How many persons attached to these vessels had been drowned I am unable to say. We have picked up only two—one of them a young man who I knew intimately. The light-ship at the N.W. Pass had gone from or sunk at her mooring. The light-house at Key West and Sand Key washed away, and Key West in ruins. A white sand beach covers the spot where Key West light-house stood, and waves roll over the spot where Sand Key was.

Fourteen persons were either killed or drowned at Key West Light-House, and not a soul escaped to tell the tale. The only vestige of the Light-House to be seen is a portion of the iron posts of the lantern, and some pieces of soap stone which have washed one hundred yards from the spot where they fell.

At Sand Key, six persons were killed or drowned—most likely the former, as the general impression is that they fled to the stone Light-House for refuge, the Key being very low. Poor old Capt. Appleby—I knew him very well: he told me the first hurricane would sweep all to destruction, and alas! his prediction is verified.

At Key West the tide was five feet high, and running six miles an hour through the center of the town. The citizens fled to the back part of the town, which is rather higher than the rest, into the bushes—laid down and held on, expecting every moment the waves would reach them. Parents were separated from their children, husbands from their wives, and all was confusion, terror, and dismay. The island trembled to its very center.—A few hours more and a white sand beach would have covered the now desolate remains of Key West. The occupants of the marine hospital were expecting every moment to go into eternity. A large stone building, surrounded with five feet of water running by six miles an hour, cutting the sand from under the foundation, made the situation awful. Thirty feet of the stone washed away from one corner, fifteen from the other, and the roof blown off.

All the wharves are washed away or injured—not one warehouse escaped the fury of the storm—wood and stone seemed all alike going to destruction. There are not more than 6 out of 600 houses but are unroofed or blown down!

The public buildings at the Fort, as well as the wharf, are all gone, and the Fort itself is a mass of ruins. It is estimated that the Government alone will loose at least $200,000 by the hurricane.

The Custom-House is much injured, but the U. S. Barracks at the East end of the town sustained no injury and are occupied by the crew of the brig *Perry* and the revenue cutter *Morris,* and those whose houses have been blown down.

The streets and roads are impassable, being filled up with lumber and the ruins of the fallen houses.[4]

Some meteorological details were included in a letter of George O. McMullin to the editor of the Tallahassee *Floridian*:

We had a delightful passage to Key West, only 7 days, arriving on Saturday last about 11 o'clock. The wind was not very high, but was increasing and continued to increase, the barometer was falling, short time after dinner rose, then fell again. 12 o'clock Saturday night, every indication of a gale, barometer still falling; about day-light Sunday morning Capt. Minor made moor fast to the wharf; before breakfast all passengers left the vessel and went into town, it blowing a perfect gale. At 12 o'clock wind still increasing; after dinner the whole island was in a commotion, the sea was all up in the town; houses in every direction were falling; women, and children running in every direction; it is impossible for language to express the condition of the whole island. There is not a dwelling, store, warehouse, or wharf uninjured, nor a vessel of any size in the harbor or on the coast that is not a wreck.[5]

A summary of the devastation was provided by an officer on the Brig *Perry* in the Key West harbor:

But of all the scenes of desolation Key West presents the most ruinous. Not a house is left uninjured. Half of them are unroofed—blown down or gone to sea. The lighthouse is gone. Sand Key light is gone, and with it every vestige of the Islet. Many families have not a shelter to cover their heads. The streets are piled with house tops, etc. rendering them literally impassable. *Every* vessel in the harbor is either sunk or driven ashore, and most melancholy of all, more than forty lives have been lost . . . As yet there have been more than 20 wrecks heard from, and the loss of life truely lamentable. But no vessel could have lived in that storm without foundering or running ashore.[6]

Our next checkpoint is Fort Brooke on the shores of Tampa Bay, some two hundred miles to the north. The surgeon weather observer there noted: "severe storm commenced at 4 P.M. on 11th and increased until 6 A.M. on 12th. A large number of live oaks in the cantonment are prostrate. Some have been righted up but little prospect of living. A large number of forest trees have been broken off at the ground & uprooted. A number of houses blown down here and all the fencing." Wind flow on the 11th was constantly northeast or east. At the last observation, at 2100, it was northeast, force 4, and the barometer read 29.84"—rain was falling. Next sunrise, the glass had plummeted to 28.93" and a northeast whole gale at force 8 was raging. A shift to southeast, also at force 8, had occurred by 0900—the barometer read the same as before, though most likely it had dipped much lower sometime in the intervening three hours. Rainfall ended at 0600 after 2.55 inches had been registered.[7] Thus, the center of the hurricane passed northward in the Gulf only a short distance west of the Tampa Bay area.

The exact course of the hurricane northward over Middle Florida cannot be accurately ascertained now. We have no reports between Tampa and St. Augustine. We do know that the waters of Ochlockonee Bay on the west side of Apalachee Bay near St. Marks rushed seaward as the hurricane passed to the east, leaving the flats bare of water but covered with dying fish. The water was at least three feet lower than known for years.[8] A storm track over Cedar Keys is very probable, and then northeastward between Tallahassee and Jacksonville. A press report from the latter definitely placed the path to the westward:

> On Monday last, this place was visited with a gale, which, in severity, was beyond anything in the recollection of that ancient individual, the oldest inhabitant. It commenced to blow Sunday evening. The gale increased until Monday, the wind coming from the eastward. The river rose six feet above high water mark, at two o'clock on Monday afternoon had flooded all the wharves, and had entered the lower floor of nearly all the stores on Bay Street which runs parallel to the river.
>
> On Monday night the wind chopped around to the South East, and the gale was at its height. The violence of the storm and the height of the water united in producing a fearful scene of devastation. All the wharves in town were carried away, and several buildings contiguous to the river, were destroyed. The saw mills in the vicinity of the town have lost all their lumber and logs.[9]

All of the principal ports of the South Atlantic seaboard lay to the east of the storm's course, but the wind flow from an easterly quadrant did not continue long enough to raise tides that would spell disaster for shipping. At the Savannah the blow, though judged the worst since 1824, caused only one marine casualty; on land fences and trees were blown down and some damage of a not too serious nature was inflicted on roofs of structures.[10] Another press report from Charleston described the nature of the gale there:

> FURIOUS GALE
>
> The dry gale which commenced blowing on Saturday evening last, continued until Monday evening, when it was accompanied by a fall of rain, continuing throughout the day, the wind increasing considerably. Towards nightfall on Monday a regular northern gale set in, and continued to rage with great violence until about three o'clock yesterday morning, the wind changing early in the night from N.E. to S.E. It was the most severe gale we have experienced for some dozen years past, but fortunately mainly to be attributed to the quarter from which the wind blew, and the short neap tides, there has been less damage sustained by the shipping in port than was anticipated.
>
> The wharves have suffered severely in some cases, a number of roofs have been injured, trees & fences blown down.[11]

In addition, we have the Board of Health meteorological report for these days. With northeast winds prevailing on the 11th and the morning of the 12th, the barometer commenced to fall steadily from a height of 29.86 inches as the storm approached from the south. By 1600 the wind was in the southeast at force 6. Next morning at 0700 the barometer read 29.09 inches, and this very probably was not the overnight low. By the morning of the 13th the wind had moved around to the southwest while the glass pursued a steady rise. Total rainfall at Charleston was measured at 2.03 inches.[12]

The center of the storm passed between Augusta and Savannah, between Columbia and Charleston. The weather report in the press of the South Carolina capital indicated a heavy northeast windstorm with torrential rains (4.50 inches reported), but no serious damage other than to crops.[13]

The hurricane was also felt at coastal points farther north. At Georgetown the wind backed from northeast to southeast at 0400 on the 13th as the storm center sped by to the westward of that place. A tide two feet above normal did no material damage.[14] It was the same story at Wilmington in North Carolina.

The hurricane apparently pursued an accelerated pace northward through east-central North Carolina and burst into the Chesapeake Bay area early in the morning of the 13th. It passed to the west of Norfolk where the wind came around from northeast to southeast and heavy rain fell. No major damage resulted to shipping there. From here northward the track of the 1846 Hurricane closely paralleled those of late September 1896 and mid-October 1954, and damage was equally extensive over the Middle Atlantic States and western New England.

[1] W. C. Maloney. *A Sketch of the History of Key West*. Newark, N. J., 1876. 41.

[2] Jose Carlos Millas. *Huracanes que han afectado a Cuba desde el 1494 al 1856*. Habana, 1926.

[3] *Dairio de la Marino* in *Niles Weekly Register* (Balto.), 14 Nov 1846.

[4] Particulars of the Late Gale. Lieut. Pease, U.S.N. furnishes the *Delta* (New Orleans), 23 Oct, with the following account of the late terrific gale in the Gulf. *N. Y. Daily Tribune,* 2 Nov 1846.

[5] George O. McMullin of Tallahassee in *The Floridian* (Tallahassee), 31 Oct 1846.

[6] Officer of Brig. *Perry,* Key West, 16 Oct, in *Weekly Raleigh Register,* 6 Nov 1846.

[7] SG.

[8] *Sentinel* (Tallahassee), 20 Oct, in *Daily Picayune,* 31 Oct 1846.

[9] *News* (Jacksonville) in *Courier* (Charleston), 24 Oct 1846.

[10] Savannah, 14 Oct 1846, in *Daily Picayune,* 20 Oct 1846.

[11] *Courier* (Charleston), 14 Oct 1846.

[12] Ms. Met. Reg. Board of Health, Charleston. (National Archives.)

[13] Columbia dispatch in *Courier* (Charleston), 22 Oct 1846.

[14] Georgetown dispatch in *Courier,* 16 Oct 1846.

THE TAMPA BAY HURRICANE OF 1848

Hdq. Ft. Brooke, Florida Sept. 26, 1848.

I have to report that yesterday a very severe equinoctial storm, from the Southeast, destroyed all the wharves and most of the public buildings at this post. The Commissary and quartermaster store-house with all their contents were swept away, and a few damaged provisions &c., only can be recovered.

The officers quarters except Hdq. are destroyed or very much damaged and the barracks are beyond repair.

The storm began about 8 A.M. from the Southeast and raged with great violence until past 4 P.M. after which it veered to the south & southwest and lulled very much toward 8 P.M. Its greatest force was from 1 to 3 P.M.

The waters rose to an unprecedented height, and the waves swept away the wharves and all the buildings that were near the Bay or river.

The command was turned out early in the storm, but such was the violence of the wind and irresistible force of the waters, that no property could be saved.

I am happy to report that no lives were lost although some of the people were rescued with difficulty.

Maj. R. D. S. Wade [1]

Major Wade's letter to his Commanding General in Washington contained most of our basic information of the damage caused by what must rate as the severest hurricane of this period to strike directly at the central part of the West Coast of Florida. Nothing about the origin or previous course of the storm has been uncovered. To the south it was reported light at Key West, but at Charlotte Harbor near Ft. Myers "considerable damage" resulted.[2] To the north at Cedar Keys a brig was dismasted when off that point and within 80 miles of St. Marks in Apalachee Bay.[3] Nearby Tampa the sugar works on the Manatee River were completely destroyed by the onslaught of wind and waves.[4]

We have two estimates of the magnitude of the tidal influx: "water out of the bay rose 10 to 12 feet higher than ever known," [5] and "the tide rose 15 feet above low water." [6] The latter estimate, submitted by the post surgeon, also stated that "the water commenced rising very fast at 10 A.M. and continued to rise until 2 P.M." All accounts seem to agree that the peak of the storm came between 1400 and 1500. The surgeon's report quoted above noted the fall of the barometer from a prestorm reading of 30.12″ at 0900/24th and 29.92″ at 2100 to a low of 28.18″ sometime prior to 1500/25th, indicative of a storm of the severest type.[7] By 1500 the glass had recovered to 28.55″ and the wind was coming out of the south. Observations were then discontinued due "to the exposed condition of the barometer," i.e. the building was unroofed, the doors blown in, and the windows demolished.

The veering of the gale from northeast, to southeast and south during the heaviest part, and then to southwest would place the center of the disturbance as approaching from the south or southeast and passing to the west of Tampa Bay. The center probably curved inland from the Gulf to the north of Tampa Bay and crossed the peninsula diagonally moving northeastward. At Jacksonville two houses were blown down, and the streets of St. Augustine were flooded at this time, pointing to an easterly flow there.[8]

Some additional information was contained in an excerpt from a letter written at Tampa only one day after the storm and published in *The Daily Picayune* at New Orleans three weeks later:

> Yesterday we had the equinoctial, and here it far exceeded that of 1846. This port and the neighboring towns are *utterly wrecked.* The public storehouses and their contents were carried off by the breakers, and but little subsistence has been recovered. Most of the poor people near here have only the clothes on their backs, and but for the mild climate there would be dreadful suffering. As it is, I do not see how food is to be procured, except beef. Nearly all the buildings are beyond repair.
>
> The two preceding days, the 23d and 24th, were overclouded, sultry, peculiarly oppressive—so as to be noticed, and were considered as ominous. The nights were notable for an unusual phosphorescence on the waters, like a fire on the horizon. In this bay such an appearance is unusual. The wind was N.E. until 8 or 9 A.M. yesterday, when, after a heavy, rainy night, it veered to S.E. and increased rapidly till 1 P.M. (the barometer fell from 30° to 28°) and roared till 3 P.M. Then it began slowly to veer to S. and S.W. and was less steady, and towards sunset there was only a strong W. wind left, and the water

ran out. The barometer returned to 30°, and at 8 P.M. to 30½° (sic). The storehouse, wharves, &c. were all swept clean, leaving not a trace behind. We have not heard from the neighborhood of the damages done.

The *W. H. Gatzmer,* Capt. McKay's vessel, daily expected, is not yet in and apprehensions for her safety are of course felt. The *John T. Sprague* schooner arrived from New Orleans the day before the storm luckily, and she had some little flour, &c., on board.[9]

THE GULF OF MEXICO—1848

There were at least two other tropical storms in the Gulf during the season of 1848. New Orleans and the towns farther up the river had "a heavy blow" on Friday night and Saturday morning, 18–19 August. Torrential rains damaged cotton crops generally.[10] Baton Rouge had 5.50 inches overnight.[11] At Natchez it was described as one of the "most violent rain and wind storms we have experienced for many years. The rain fell in torrents from dark until daylight, and the wind blew a gale from the southeast the whole time."[12] The local press carried no accounts of damage to shipping along the coast.

Following the major storm of the 25th of September, another disturbance moved through the northeastern Gulf on 10–12 October, the second anniversary of the Great Key West Hurricane. Fort Brooke at Tampa Bay again experienced a very high tide, but this failed by five feet to reach the peak of the storm of the 25th. In the northeastern Gulf off Cape San Blas, the brig *Peconic* met an "awful hurricane" about 1000/11th and fought gigantic seas and hurricane force winds until 1600/12th when the "hurricane broke" and the barometer commenced to rise. It had been at 28.50 inches and the wind at northeast earlier in the day. The captain thought the storm surpassed that of 1846 at Key West which he had experienced.[13] Another ship met a northeast hurricane when only 20 miles southeast of Cape St. George and had her lee rail under water for eight hours. Another brig was knocked on her beam-ends for 36 hours on the 11–12th.[14] Tallahassee had "a regular gale from the northeast" on Thursday, the 12th, and the storm was noticed at Jacksonville where it comprised the last of the remarkable series of October gales that distinguished the 1840's in Florida.[15]

[1] Ms. Major R. D. S. Wade. Early War Records Section (National Archives).
[2] *Floridian* (Tallahassee), 14 Oct 1848.
[3] *Idem.*
[4] *Idem.*
[5] *Jacksonville News,* 7 Oct, in *Floridian,* 21 Oct 1848.
[6] Wade Ms.
[7] *Idem.*
[8] *Sentinel* (Tallahassee), 10 Oct, in *Daily Picayune,* 20 Oct 1848.
[9] *Daily Picayune,* 17 Oct 1848.
[10] *New Orleans Bulletin,* 22 Aug 1848, clipping in Redfield Ms. (Yale).
[11] *Baton Rouge Advocate,* 23 Aug, in *Daily Picayune,* 26 Aug 1848.
[12] *Natchez Free Trader,* 22 Aug, in *Daily Picayune,* 26 Aug 1848.
[13] *Daily Picayune,* 25 Oct 1848.
[14] *Floridian,* 21 Oct 1848.
[15] *Ibid.,* 14 Oct 1848; and SI.

GALE AT BRAZOS SANTIAGO IN 1849

One of the severest gales and the longest in its duration, that has occurred on the Rio Grande for several years took place on the 13th and 14th inst. The weather bore a threatening aspect for several days previously. On Thursday the 13th it commenced with a brisk gale from the north, which kept gradually increasing through the day, so much so that the steamship *Globe*—it being her day of sailing—did not put to sea. The lagoon being full of water from the previous easterly winds, the water now began to flow back on Brazos Island, covering it to the sand hills, to the depth of three to four feet. The wind, though at north, moderated somewhat through the night, but on Friday morning it recommenced in good earnest, accompanied with heavy squalls of rain: there being an unusual number of vessels in Brazos, apprehensions began to be felt for their safety. Meanwhile, the islanders, in still greater fear of being swept away by the water, commenced moving their property back to the sand hills and such other places as appeared to be beyond the reach of the water. The wind, however, created such a powerful ebb tide the water gained but very little on them. The gale continued to increase, and some of the vessels began to drag their anchors.

* * *

The wind now veered to northwest, then west and southwest; at 2 o'clock the gale broke, and continued from that time to moderate. All the vessels not mentioned among the disasters rode the gale safely.[1]

There were sixteen vessels in the harbor of Brazos when the September gale came on. Several ships dragged anchors or broke loose and were driven over the bar and out to sea. The crew of one jumped to a nearby ship when it became apparent that it was headed seaward, and the ghost ship disappeared from sight. Several smaller boats capsized; and others were severely chaffed at their docks so as to render them unseaworthy. The ship which excited the most interest was the *Neptune,* a beautiful little pilot boat running between Brazos and Galveston. The captain had gone ashore in the afternoon of the 12th, leaving the ship in charge of a 15-year-old boy. The captain tried to reach his ship, but had to seek safety on the *Globe*. From there he saw his own ship part cables and disappear into the gloom, headed for the bar and the open sea, crewed only by the boy.[2]

[1] The Gale at Brazos Santiago. (Furnished by an eyewitness.) *Daily Picayune*, 19 Sept 1849.
[2] *Idem.*

SEVERE STORM AT APALACHICOLA—AUGUST 1850

On Friday, the 23d of August, our city was visited by an extraordinary storm. It was not remarkable for the force of the wind as for the great height to which the tide rose. It was preceded by meteoric phenomena, which, though unobserved by instruments, were obvious to the senses. The general heat of the Summer, and more particularly of the present month, has led many to apprehend an equinoctial hurricane, but no one supposed it would take place before September. During the continuance of the gale the temperature of the atmosphere was oppressive; and both the maximum and minimum heat, indicated by the thermometer, seemed to be greater than on any previous occasion. There were frequent showers of rain throughout its continuance, but the sky was not wholly overcast with clouds till towards its termination. It is to be regretted that no barometrical observations were made.

We are told that an extraordinary influx of the tide was observed by persons living near the water, early in the morning, and long before the violence of the wind attracted their attention. Indeed, the latter never attained what we have been accustomed to regard as the characteristics of a hurricane; and appeared for several hours to consist of sudden gusts, followed by comparative lulls which usually accompany a subsiding gale, and which we suppose, marked that we were on the verge of the "annulus." The wind commenced at the eastward and pursuing its more normal course, terminated at the westward—manifesting in the latter direction, and throughout the night, its greatest violence.

We are unable as yet to form an idea of the direction of the tornado, the magnitude of its "annulus," or the position of its axis, though we were manifestly near one of its terminations. The temperature of the atmosphere would indicate its advance from the tropics; and the rise of the tide, before the wind was felt, showed that its force must have been directed against the current of the Gulf Stream.

The rise of water caused the whole of Water st., and a considerable part of Commerce st. to be overflowed. It even extended as far back as to the upper end of Market st. As a consequence of the high tide and the direction of the wind, Water st. was covered with, and rendered impassable by trunks of trees and logs floated from the bay and adjoining marshes.

The damage the storm occasioned to the town has not been general or in any instance great, however injurious it may have been to some individuals. The gardens on the margin of the bay, below the town, have been covered with drift wood and sand, and many fruit trees, grape vines, and other vegetable productions, that have been assidiously nursed for years, have been destroyed. Fences were generally prostrated or injured. We are sorry to observe that several shanties, at the lower end of Water st., occupied by industrious and poor fishermen, were washed away; and we learn that some cotton and other mechandise were damaged in the upper part of town. During the latter part of the storm two buildings on Water st.—Mr. Hamilton's and one belonging to the estate of Gen. McDougald—were partly unroofed.[1]

This storm probably originated in the Gulf of Mexico west of the 85th meridian. We have two ship reports close to that line at 26°N and 27°N indicating the existence of a southeast "gale" of 12 hours duration on the 22d.[2] Wind at Apalachicola and Newport in Middle Florida came from an easterly quarter, putting both places to the east of the storm center's track.[3] It is likely that the landfall was between Pensacola and Panama City with the line of advance north-northeastward.

The storm at St. Marks commenced on Friday with the wind direction given at southeast. By 1800 the water in the bay rose considerably higher than in 1846, but then receded despite the continuance of heavy rain. There was no damage of consequence along the shores at either St. Marks or Newport, but the bridge spanning the Wakulla River was carried away by the floods created by the heavy rains.[4] At Tallahassee the storm raged through Friday night and into Saturday, first from the east and later from the west. The editor of the *Southern Recorder* thought he had never experienced a more protracted gale. No lives were lost at the Florida capital.[5]

Montgomery, Alabama, reported a severe wind and rain storm from Friday night, the 23rd, until Saturday morning. The combined onslaught of the elements damaged the cotton, but the bolls were not yet mature so the losses were minimized. The editor of the *Montgomery Journal* thought the storm quite similar to that of 6 October 1837 (Racer's Storm).[6]

In Georgia reports of heavy wind damage came from

Columbus, Griffin, Waynesboro, and Augusta. Small houses were blown down, trees uprooted, and fences prostrated. Corn and cotton were laid flat in some places by the force of the wind blasts. Savannah felt the storm also from 1900/23rd until 1200/24th. In Burke County (northeast Georgia close to Augusta) the highest wind came from the southeast.[7]

In the Carolinas, very heavy rains and high winds swept the western Piedmont, while gales and light rains visited the coastal plain. The storm cut northeastward through central portions of both South and North Carolina. At Spartanburg the downpours carried away mills and dams throughout the area, and immense forest trees were uprooted.[8] Lower down at Camden on the Wateree the lowlands were flooded and crops damaged. It was the same story in central North Carolina—dams and bridges gone, trees uprooted, and crops laid flat.[9] The Dan River in North Carolina and Virginia rose 20 feet in many places and in narrow channels 40 feet above normal. Damage estimates ran as high as $7 million.

The only meteorological report located came from Charleston where a "heavy blow all day from the southwest" prevailed. The barometer dropped to 29.49″ late on the 24th.[10]

[1] *Commercial Advertiser* (Apalachicola) in *N. Y. Daily Trib.*, 12 Sept 1850.
[2] *N. Y. Ship List,* 14 Sept 1850.
[3] *Sentinel* (Tallahassee), 3 Sept 1850.
[4] *Wakulla Times* (Newport) in *Sentinel,* 3 Sept 1850.
[5] *Southern Recorder* (Tallahassee) in *Sentinel,* 3 Sept 1850.
[6] *Montgomery Journal,* 26 Aug, in *Daily Picayune* (New Orleans), 30 Aug 1850.
[7] *Savannah News,* 27 Aug, in *Daily Picayune,* 3 Sept 1850.
[8] *Daily Picayune,* 8 Sept 1850.
[9] U. S. Patent Office, *Annual Report*—1850, 325.
[10] Ms. Met. Reg. Board of Health, Charleston, S. C. (National Archives).

THE GREAT MIDDLE FLORIDA HURRICANE OF AUGUST 1851

TALLAHASSEE

On Friday evening last a lurid and threatening horizon attracted universal attention in town, and the weatherwise prognosticated a storm. About 6 o'clock there was a violent squall from the southeast and a tremendous shower. In a little while the squall ceased, but it continued to rain copiously all night till about 5 o'clock Saturday morning. Commencing again at half past 8 o'clock, the wind also began to rise from the eastward, the horizon to thicken up, scuds to fly, and every sign to betoken a blow. The rain poured down literally in torrents, with only occasional slight intermissions until about sunrise on Sunday morning. By 12 o'clock noon on Saturday, the wind had veered to southeast and was blowing a gale. At 1 o'clock trees began to give way, at 2 o'clock still veering southward, the blasts were rapidly increasing in violence, and worse evidently coming. From 3 to half past 6, the cry was stand from under. Tall forest oaks were uprooted or rudely snapped asunder; China trees stood no chance, fences were prostrated, tin roofing peeled up like paper, roofs torn up, brick bats flying; and altogether such a general scatteration taking place as is not often seen. From 6 to 10 o'clock there was no increase in the force of the gale, but if anything, an abatement. From 10 to 2 o'clock it piped up again and blew great guns. The night was too dark to note the mischief in progress, except as it happened just around one, but the howling, hissing, whistling, moaning and groaning of the blasts were very well calculated to excite lively apprehensions of general misfortune.

The gale had pretty much abated at 4, and daylight disclosed a novel spectacle. The streets were all obstructed with prostrate trees, and the ground covered with twigs, boughs, leaves, and "whiggings," from everything that grows.[1]

APALACHICOLA

This city was visited on the night of the 23d ult., by the most destructive storm it has ever witnessed. The wind blew for about twenty hours with a violence that nothing could resist, the town was flooded by water from the Bay, houses of all materials and all sizes leveled with the earth, stores washed away with their contents, leaving the inhabitants without shelter and almost without food. Not one building on Water Street remains uninjured, the Exchange buildings, the offices over the Hydraulic press, the building lately occupied by the Charleston Bank Agency, all in ruins. The store occupied as a Custom-House almost down; from that position of the town every house on Front or Commerce Street is in ruins. At the upper end of the town the wharves and small buildings are all gone. The Episcopal Church is almost a wreck and the Presbyterian Church totally demolished. At the Bluff the houses are all washed down. The steamer *Falcon* was washed off the ways; driven up the river by the tide and storm, and capsized on an island. All three of the lighthouses are blown down or washed away. At Dog Island five lives were lost; at Cape San Blas a Spanish brig-of-war was stranded on the beach and several lives lost.[2]

ST. MARKS AND NEWPORT

We learn from St. Marks that the water rose above all previous water-marks on the old fort. Portions of this old fortification, which had withstood the flood of ages, were swept away, and the buildings within it all destroyed. The custom-house books and papers are all gone, nothing being left of the establishment but the iron safe. The frame house occupied by Mr. Birchett, the deputy collector, on the point of junction of the St. Marks and Wakulla, was raised from its foundations, broke in the center, and dashed to fragments against the old fort. Mr. B. lost everything in the shape of clothing and household effects. All the houses at St. Marks (generally small frame houses of small value) were carried away. The water is stated to have risen at least twelve feet above ordinary tide levels.

At the terminus the railroad was washed up as far as the hotel. The water was two feet deep in the warehouses, parts of the floor were washed up, and the contents (principally bagging and salt) a good deal injured—a cotton shed was torn up, and 200 bales went adrift—the wharf gone, and seventy barrels of rosin with it—the cotton press injured—Holt's Hotel not much injured—the other buildings gone.

As the water rose a train of baggage cars was prepared, and when St. Marks and the terminus were no longer habitable the people repaired to the cars on planks and in boats, and pushed up into the higher pine woods.

At Newport there appears to have been an average of five or six feet of water. The wharves are all carried away, but no buildings destroyed as far as we can learn. The goods in storehouses have sustained much damage. Mr. Ladd's loss is estimated at five to six thousand, Messers. McNaught & Ormond three or four, Mr. Denham's two to three, and all the merchants have sustained serious losses.

An extra of the *Wakulla Times,* received this morning, estimates the loss at $12,000, which, from the reports we have received, we would infer was a very moderate estimate. The streets of the town are said to have been in good boating condition for the steamer *Spray,* which draws four feet.[3]

ST. MARKS LIGHT-HOUSE

We are glad to hear that all are safe at the light-house, and that the house and premises have suffered no material damage. The breakwater is carried away in two places, but the damage is supposed not to exceed $1,500.

There were, besides the keeper's family, some six or eight visitors at the light-house. They had, of course, a most anxious and alarming time, till about 10 o'clock Saturday night, when the veering of the wind a little to the westward permitted the water to recede, and they returned to the keeper's house. During the height of the gale and sea the light-house shook to its lowest stone, and the terror and discomfort of the fugitives was greatly increased by the forcing in of the iron door and the spray dashing up into the structure.[4]

The August hurricane of 1851 gave the Florida West Coast a good sideswipe before roaring inland near Apalachicola. Key West had its "August gale" on the 20th with the wind at northeast blowing heavy squalls. No damage was sustained there. The Sand Key Light, mindful of the October hurricane in 1846, slipped her chains and came into port.[5] At Fort Brooke on Tampa Bay a strong wind varying from southeast to south-southwest continued for 24 hours on the 22–23d and raised a very high tide.[6]

To the west of the landfall, Pensacola had very high tides, especially on the morning of the 23d when the wharf was overflowed and some damage done. Winds during the 23d were from the northeast but backed late in the day to northwest as the storm center passed inland well to the east. There was no report of wind damage.[7]

The storm center moved inland close, but to the west of Apalachicola, and then headed northeastward through Georgia. It passed southeast of Griffin County near Atlanta where the wind shifted on late Saturday from easterly to north and northwest.[8] Near the coast at Savannah the height of the gale came from daylight to noon on Sunday with the direction southeast and south. Seventy-five trees were blown down in one street in the Georgia seaport as the storm passed to the west.[9] At nearby Whitemarsh Island the peak of the blow came at about 1100 from the south, after which it abated a little and moved gradually to south-southwest and west-southwest.[10]

Charleston experienced a "severe storm from SE entire day" on the 24th. Wind was at force 6 at both Board of Health observation times. Rainfall for the 24 hours measured 2.10″.[11] The storm passed between the South Carolina port and Greenville where a northeaster prevailed all Sunday. Considerable damage resulted from the wind and rain to corn and cotton in the Edgefield and Newfield areas of the Palmetto State.[12]

[1] *Sentinel* (Tallahassee), 26 Aug 1851.
[2] *Commercial Advertiser* (Apalachicola), in *Sentinel,* 2 Sept 1851.
[3] *Sentinel,* 26 Aug 1851.
[4] *Idem.*
[5] Key West dispatch to *Savannah Republican,* 9 Sept 1851.
[6] SG.
[7] *Pensacola Gazette,* 23 Aug 1851.
[8] *Union* (Griffin?), in *Sentinel,* 9 Sept 1851.
[9] *Savannah Republican,* 26 Aug 1851.
[10] SI.
[11] Ms. Met. Reg. Board of Health, Charleston (National Archives).
[12] *Greenville Mountaineer* in *Sentinel,* 9 Sept 1851.

THE GREAT MOBILE HURRICANE OF 1852

About midnight of Tuesday, the 24th of August 1852, the wind commenced blowing strongly from the northeast at Mobile and a steady, driving rain set in. Around 0900 on Wednesday morning the wind shifted to southeast and continued blowing strong throughout the daylight hours.[1] This was the preliminary to "the highest and most disastrous flood ever known to Mobile." [2] We can follow the story as told by the editor of the *Daily Advertiser* on the day of the storm:

Thursday, 10 o'clock, A.M. When we concluded penning our account on yesterday, of the great storm, we were under the impression that the worst was over, and that the coming day would witness the abatement of the waters, and allow a return to ordinary business. In this, however, we were greatly mistaken. The wind which had comparatively lulled at 11 o'clock last evening, soon after blew with renewed violence, but yet, little serious apprehension of a further rise of water was manifested. Merchants on Water street—which, as yet, the flood had not reached—visited their stores about this time and returned with a feeling of security.

At 1 o'clock this morning the gale raged with terrible fury. Immense trees were torn up by the roots, falling into the streets or into yards, crushing fences or whatever impeded their descent; houses were unroofed and their helpless inmates were exposed to the pitiless pelting of the storm; and, added to this came the unwelcome assurance that, with the rising tide, the flood was progressing with alarming rapidity. By 2 o'clock it became apparent that Water street could not escape submersion, and some gentlemen whose anxiety would not permit them to rest, barely had time to reach their stores and secure their books before they were entirely ruined.

At sunrise, and later this morning, the aspect of the city was gloomy indeed. The flood having submerged Water street, forced its way into the stores, offices, and dwellings of St. Francis, Dauphin, Conti, and Government streets, nearly up to their several intersections with Royal.

The lower portion of the market house was filled to the depth of several feet. Awnings and awning-frames were scattered indiscriminately about, the zinc and tin covering of several buildings was torn up and deposited in rolls in various places, and, not least, the streets most devoted to residences, were every few yards obstructed by the prostrated forms of venerable and beautiful shade trees.

* * *

Twelve o'clock, M.—The flood is thought to be at a stand, though the storm still prevails.

Six P.M. The water is now rapidly subsiding. It has retreated below Water street, and shopmen on Dauphin and St. Francis streets are busy in clearing their stores and examining into their losses. Logs, wood, timber, &c. are left lying in the streets, and the foot of Dauphin street is pretty effectually blocked by an accumulation of barrels, hogsheads, &c. wedged in with the drift wood.

* * *

LATER

The wind has now shifted to the W.N.W. and the rain continues to fall moderately. Intelligence from the country, as far as received, indicates great rains and high water all about us.

STILL LATER

Eleven o'clock, P.M.—Mournful accounts of the loss of life and property still reach us. We have just learned of the destruction of the dwelling attached to Choctaw Point Light-House and the loss of five lives. About 10 o'clock A.M. the building gave way . . .[3]

The editor of *The Tribune* made a survey of the damage late on Thursday:

On St. Francis street, in the north part of the city, the waters came up to the office of St. John, Powers & Co., where we saw a large steamboat barge moored. Farther up, it reached St. Joseph street to a depth sufficient to float a boat. The railroad depot was completely surrounded by water, and the railroad for a distance of two miles, the only part examined, was covered to a depth of one to two feet.

On Dauphin street the water came up as far as this printing office. On Conti street it reached about the same height. On Government street near the lower part of the market-house, a small steamboat might have plied. In the southern part of the city the flood was still greater. At Spanish Alley the houses almost floated, and men were called there to rescue the inmates from drowning.

Along the whole length of Front, Commerce and Water streets, the water lay to a great depth—in the first street, ten or twelve feet in places. The lower floors of the store-houses were all covered to a considerable depth.

The walls of the new store, in course of erection by Messers. Dorrance & Sons, partly fell in, and, we imagine, the foundations of nearly all the new buildings in that quarter are very much impaired.

Innumerable quantities of trees were blown down, the China trees particularly suffered. Outside of the city, too, large pines were forced up from their roots, and some of the roads were made impassable by them. All along the bay this is the case. The fences in every direction also suffered greatly.

In the city the awnings and the awning posts were generally destroyed. Those iron fastenings, which have lately come into fashion, suffered considerably. Some of them were bent up as made of wire. Nearly every tin roof within the city was ripped off. Some of them were rolled up like scrolls, and cast into the streets. Among these is the roof of the Government Street Church, and the roof of Hall's Restaurant on Conti street. His dining room was flooded.

In the western parts of the city all the low places

were turned into lakes. For a quarter of a mile on the lower part of the Shell road and the road below it, leading to Dauphin street, the water was from two to three feet deep.

The scaffolding of the Battle House was blown, and the rooms of the building fronting south and east were considerably wetted, but without any apparent damage.

The losses in the stores on the east of the city must be immense. Very few supposed that there was danger, or took precautions against it, and the consequences is a vast destruction of all perishable articles in groceries, hardware, clothing, &c. This loss is estimated at from a half to a million dollars.[4]

The storm was not felt on Lake Pontchartrain or westward. New Orleans had a "strong breeze from the north" on the 25th—the editor of the *Picayune* remarked on its unusual direction but not its speed.[5] But from Bay St. Louis eastward the hurricane raged with great fury. At the latter point "it blew very hard, but entirely without any variation from the N.W., which being off the land and blowing directly out of the Bay, drove the waters away and made no sea; as the greatest damage from these storms are caused by waves, this gale did no harm at Bay St. Louis, though the wind was very high and quite a number of coasting vessels, both loaded and empty, were lying there at anchor, yet none suffered damage."[6] At Pass Christian, Missisippi City, and Biloxi winds were also from a northerly quarter and the water was driven out of the harbors.[7] But offshore at the Chandeleur Islands the north wind created a huge sea and the waters overran the island and piled up around the lighthouse. On Thursday morning, the 26th, the foundation was washed away and the keeper and family narrowly escaped a catastrophe as they evacuated the building only a few minutes before. They retreated to a point three-quarters of a mile away, the only place on the island above water, and anxiously awaited forty-eight hours there until rescue came.[8]

The center of the hurricane must have come onshore close to Pascagoula; from there eastward damage to harbor installations, shipping, buildings, and forests was immense. At West Pascagoula the main wharf was carried away, and many thousands of trees were uprooted, many of them solid oaks which had withstood previous tempests.[9]

On the eastern side of the hurricane Pensacola shared the fate of Mobile with hurricane winds raging for almost forty-eight hours. The 24th was described by the local observer: "rain with short intervals and hurricane through day and night"—25th: "rain and hurricane continue"—26th: "rain and hurricane continued." The winds were out of the northeast on the 24th, mounting to force 10 in the evening; from southeast and south on the 25th, varying southeast and south; from south veering to southwest late in the day on the 26th. His rain gage caught 13.20″ from late on the 23rd through the 26th, but "there was much more rainfall than indicated by the gage. The driving wind prevented the gage from holding the water, and at night the water was blown out when less than partly full, *at least 18 inches fell during the storm.*"[10]

The *Pensacola Gazette* briefly summarized the storm the following day:

Severe Gale—On Tuesday last there commenced one of the severest storms that we have witnessed for many years. The wind at first easterly gradually shifted around first southerly then westwardly. Wednesday and Thursday the wind blew steadily with tremendous force, and the rain fell in tremendous quantities. From two-thirds to three-quarters of the main wharf was carried away—leaving only the outer end remaining. All the bathing houses have been swept away. We have not heard of much damage done to shipping, but fear that we may yet hear of many wrecks upon our coast.[11]

Northward through Alabama and Georgia the storm raged with great fury. Columbus, Georgia, had high winds, heavy rains, and floods that carried out many bridges on the 27th.[12] Along the coast both Savannah and Charleston felt the force of the gale though little damage occurred.[13] Inland the rain caused a memorable flood. Augusta saw two main bridges over the Savannah River carried away. Losses of $50,000 were estimated in the city in the greatest freshet since 1840. At Hamburg water ran six feet deep in the streets, and the railroad depot was submerged under twelve feet.[14] The heavy rains extended northward into Carolina and Virginia, but the wind force subsided as there are no reports of high winds or damage north of Georgia.

[1] *Mobile Daily Adv.* in *N. Y. Weekly Times,* 11 Sept 1852.
[2] *The Mobile Tribune,* 27 Aug, in *The Evening Picayune,* 28 Aug 1852.
[3] *Daily Adv.,* 27 Aug, in *idem.*
[4] *Tribune,* 27 Aug, in *idem.*
[5] *Daily Picayune,* 26 Aug 1852.
[6] *DeBow's Review,* 38–9, Sept 1868, 791.
[7] *Daily Picayune,* 31 Aug 1852.
[8] *Idem.*
[9] *Tribune* in *Daily Picayune,* 3 Sept 1852.
[10] SI.
[11] *Pensacola Gazette,* 28 Aug 1852.
[12] *Daily Picayune,* 3 Sept 1852.
[13] Ms. Met. Reg., Board of Health (National Archives).
[14] *N. Y. Weekly Times,* 11 Sept 1852.

THE MIDDLE FLORIDA STORM OF OCTOBER 1852

TERRIBLE GALE AT NEWPORT

On Saturday morning last about sunrise a few drops of rain began to fall, accompanied with a light wind, varying from the E., S.E., and S.S.E. which gradually increased until about half past two P.M. when it became a violent hurricane.

About half past 2 o'clock the western end of the Court House was blown out; at about a quarter past 3 o'clock, the whole structure was blown down. The Judge of Probate, Mr. Wm. J. Councill, narrowly escaped being crushed under the fallen timber. The gable end of the dwelling of Mr. Haywood was also blown down. For some 29 minutes the wind blew, we think, harder than we ever before experienced. About 6 o'clock the wind began to lull, and shift to S.W. and the tide commenced falling.

The water at Newport rose about 7 feet above ordinary spring tide doing considerable damage to the goods in some of the warehouses, and some slight injury to the wharves, carrying off large quantities of wood, shingles &c.

The Plank Road track was washed up as far as the toll-gate. Hands are engaged in relaying it.

Some damage was done to the bridge across the St. Marks River which has been partially repaired.

At St. Marks the damage done to shipping lying at the wharves was most disastrous. All the wharves at St. Marks are gone. About forty feet of the Western Railroad warehouse, in which was the office of the Collector of Customs, was entirely unroofed . . . The railroad track was washed up for the distance of half mile from the river. From the lighthouse we learn that the keeper's house and breakwater were washed down . . . The Wakulla bridge is gone.[1]

Information about this October hurricane in the Gulf is scanty. Fort Brooke at Tampa had "a strong wind from SE with rain" on the 8th. The ship *Astoria,* when 250 miles southeast of the Mississippi Passes, was struck by a hurricane from the east and knocked on her beam-ends for 12 hours.[2] Another ship apparently went through the eye of the storm when at 25 30N, 86 30W, also well southeast of the Passes and south of Pensacola. An east-northeast hurricane first buffeted the *Hebe,* the calm came at 1900, and was followed by a south-southwest hurricane. The wind finally commenced to lull at midnight.[3]

The Pensacola area on the 8th had a northerly flow gradually shifting to northeast and finally to southeast at dark. Next morning the wind commenced to back to east, north, and finally to northwest as the hurricane moved inland in the Apalachicola and St. Marks area.[4] High winds were felt in the vicinity of Tallahassee. The main structural damage occurred on the State Capitol building when a large chimney crashed through the roof and into the Senate chamber. The storm lasted only six hours.[5] In the neighboring counties and up into southern Georgia great losses to the cotton crop resulted. Jacksonville, well to the east of the storm track, had high winds from the southeast on the 9th—the local weather observer listed the blow among his tabulation of severe October storms.[6]

The Savannah area had a violent storm from the southeast on Saturday night, the 9th, with rain commencing at 2300. Damage, however, proved less than in the much larger storm of August 1851.[7]

[1] *Wakulla Times* (Newport, Fla.), 13 Oct, in *N. Y. Weekly Times,* 30 Oct 1852.
[2] *Daily Picayune* (New Orleans), 16 Oct 1852.
[3] *Idem.*
[4] SG.
[5] *Daily Adv.* (Newark), 20 Oct 1852.
[6] SI.
[7] *Savannah Republican* in *Evening Bulletin* (Phila.), 22 Oct 1852.

THE MATAGORDA HURRICANE OF 1854

Lorin Blodget, the able author of the pathfinding *Climatology of the United States* (1857), made a special study of the small, but vigorous hurricane which struck the central Texas coast on 17–18 September 1854. Blodget collected a number of observations made by the U. S. Medical Corps at military posts in Texas and throughout the Southwest, inland as well as along the Gulf Coast, to delineate the path and extent of the storm.[1] This disturbance appeared to have originated in the Gulf of Mexico as there were no reports of its presence in the West Indies or surrounding waters. After striking the coastline near Matagorda, the hurricane moved inland northwestward over Columbus, west-southwest of Houston, and in its dissipating stage became a widespread rainstorm over the entire west and central Gulf area.

We can follow the initial impact of the storm, which raised a storm tide higher than any since 1842, through the columns of the *Galveston News*:

At Galveston the wind commenced blowing pretty strong from the northeast, on the 15th inst., and continued from the same quarter till the 17th, when it increased to a gale and blew with great violence from the east and southeast. Torrents of rain fell during the night. The gale continued on the 18th and 19th. On the 18th the tide took a clear sweep across the island. The merchants on the Strand and up as far as Market Street, suffered much loss from the damaging of their goods by the water.[2]

* * * * *

The water from the bay had come up, when at its highest, to the Tremont House, so that light skiffs were paddled freely from that point to the Strand. The water continued to fall until 10 or 11 o'clock on Monday morning, when the fall was about twelve or fifteen inches, but then it again commenced rising, and continued to rise until 1 or 2 P.M., and then fell again about 12 inches, as at first. At this point it has continued with slight variations up to this time (Tuesday 9 A.M.), covering the Strand and Mechanic Street, and partly covering Market Street. The wind has continued to blow constantly from the same quarter, namely, about southeast, but its violence has at no time been as great as on Sunday night, though it has been very heavy.

The water that came in from the Gulf on Sunday night ran off very rapidly through the streets into the bay, and nearly all of it was gone by 12 M., on Monday, and none has come in since from the Gulf side. It has been fortunate for us that the wind has not yet changed round to the north or northwest, as in that case a most disastrous overflow would have followed. But fears are entertained that such a change of the wind may yet take place. The shipping thus far has escaped any material damage, and so have the buildings of the city. Though the wind is not now sufficiently heavy to do much injury, yet the storm still continues, and heavy showers of rain continue to fall at intervals.[3]

* * * * *

The wind continued to increase in violence from about 1 P.M. till 12 o'clock at night, accompanied with torrents of rain till night. The wind very fortunately changed very nearly to the south in the afternoon, which had the effect to drive the water from the bay, faster than it had come in, so that the streets of the city were entirely drained before night. After night, the wind changed still farther to the west, and finally came around to northwest, or nearly north, from which quarter it blows this morning. This wind has again caused the overflow of the Strand, but as the water is very low in the bay it will soon again recede. No further danger is now apprehended.[4]

Galveston and the villages and towns southward along the coast to Matagorda lay in the right-hand semicircle of the dangerous storm. At Brazoria, just inland from present-day Freeport, on the Brazos River about 50 miles southwest of Galveston, Blodget secured the following report: "17th violent N. wind at evening; 18th still more violent through entire day with heavy rain, and changing to northeast; 19th violent wind and rain from S.E. and S." [5]

The main impact of the hurricane fell in Matagorda and Lavaca Bays. Evidence pointed to the storm's having crossed over upper Matagorda Bay and Peninsula and followed a course northwestward roughly parallel to the course of the Colorado River of Texas. At the southern end of the Bay, at Port Lavaca and Powderhorn, damage was relatively light, as might be expected to the south of the storm path where winds would have been offshore. At the exposed points along the northern end of the Bay, open to the sweep of northerly and easterly gales from the Gulf, great damage resulted to the ocean-going vessels which were tied up awaiting transfer of their cargoes to river boats. At Saluria the figure was fixed at $20,000 as the main wharf and attendant buildings were washed away. At Dekro's Point many houses were carried away by wind and tide, some unroofed, and others knocked from their blocks. Not a single structure escaped major damage. Quite a few were flattened to the ground. Near Dekro's Point two schooners dragged their anchors, then capsized with the loss of their entire crews. On Matagorda Peninsula water was reported to be two feet deep with only the higher ridges remaining above water.[6] A boundary survey team on their way to the Rio Grande put into the Bay after the storm. They found that Matagorda was "leveled"—the main damage occurred when the waters of the Bay, backed up by the onshore easterly flow, were released and driven seaward over the lowlying land by the following westerly winds. The channel was both straightened out and deepened to 9 to 11 feet over the bar.[7]

An account of the scene at Matagorda was brought to New Orleans by a Mr. Poole, purser on the steamship *Louisiana*:

The gale visited Matagorda with almost unparalleled fury, destroying nearly all the buildings in the place, those of Col. Williams, Mrs. Sartwell, and one or two small ones being all that escaped prostration or unroofing. The stocks of goods of the merchants have been mostly ruined, with one exception, Messers. Sheppard & Burkhart, whose store not only was unroofed, but goods materially damaged. Four lives were lost in the town: Mrs. Duffey, Mr. Merriam, and a negro woman and child.

The steamboat *Kate Ward* was entirely wrecked near the town, Capt. Ward, his brother, and nine of the crew perishing. But three only escaped by clinging to one of the wheels and were taken off on the 22nd.

Schooner *Tom Paine,* Capt. F. Hulsemann, owned in Matagorda, was totally lost with captain and crew.

A vessel from Sabine, with lumber, lost on the peninsula; crew saved.

The new steamer *Colorado,* built at Matagorda and nearly finished, lies high and dry in the prairie, 300 yards from the bayou.

Crops of cane and cotton are blown down and nearly ruined. In fact, it is said not a bale of cotton is left in the county.

Quite a number of small craft are reported lost with all their crews.

Trespallacious, and the houses on the peninsula opposite, were all swept away, except Col. Lewis's and two others not recollected. Several lives are reported to have been lost, among which were two children of Capt. Jn. Rugely, an old and much esteemed planter, who were killed by the house being blown down upon them.

An eyewitness to the devastation of Matagorda, says that he never could have conceived of such a sight as he witnessed between four and five o'clock on Monday morning; houses crashing and breaking up, their materials flying through the air, women and children screaming and running wither they knew not, seeking protection, and when found, only to be driven forth again after a short lapse of time to find a new one, and in many instances in nearly a denuded state.[8]

The relatively small areal extent of destructive winds to the southward was confirmed by the Smithsonian observer at Corpus Christi. He reported north winds at force 6 all day on the 18th, but made no special mention of rain or storm.[9] At Fort Brown near Brownsville the highest winds were merely force 2, although a hurricane was raging some 200 miles to the north-northeast.[10]

Inland at Columbus, on the Colorado River about 70 miles west of Houston, Blodget obtained the following: "17th violent rain with east wind at evening; 18th continued and increasing; 19th gale from E. in morning; then calm; then south a gale, then west; and finally to N. from whence it blew severely as it had in the morning from the east." [11] The now slow-moving storm center must have passed just to the east of Columbus. As the storm commenced to dissolve and spread out, very heavy rains fell over the Gulf Coast area. To the north, Baton Rouge reported five days of storm "with heavy rain 17th to 21st (5.55 inches) wind east but not violent." [12] Moderately heavy rains descended as far east as Pensacola where 1.85 inches fell on the 19th–20th. Mobile had 3.68 inches on the same days.

[1] Lorin Blodget. *Climatology of the United States.* Philadelphia, 1857. 392–93.
[2] *Galveston News* in *Daily Picayune,* 25 Sept 1854.
[3] *Idem.*
[4] W. Armstrong Price. *Hurricanes Affecting the Coast of Texas from Galveston to Rio Grande.* Washington, 1956. A-3.
[5] Blodget, 393.
[6] Matagorda, 28 Sept, in *Victoria Advocate* in *The Galveston Weekly News,* 31 Oct 1854.
[7] *Idem.*
[8] Account of Mr. Poole, purser of the steamship *Louisiana, Daily Picayune,* 27 Sept 1854.
[9] SI.
[10] SG.
[11] Blodget, 393.
[12] *Daily Picayune,* 26 Sept 1854.

THE MIDDLE GULF SHORE HURRICANE OF 1855

THE MISSISSIPPI-LOUISIANA SHORE

On Saturday evening (15 September) the wind being from ENE began to freshen and increase until 0800, Sunday morning, when it temporarily lulled. It came out again, fresher than ever, between 1000 and 1100, from SE, and then from SSW and continued up to 1500 to 1600, when it in a measure abated and hauled more to the south. At sundown, on Sunday, the wind was from SSW.[1]

THE STORM ON THE LAKE

From Proctorville to Biloxi, all along the Lake coast as far as heard from, the storm of last Saturday night has proved the most destructive and disastrous since that of 1819. Even that of 1837 is said not to have been equal to it, although how severe that was few who recollect it will ever forget.

From Capt. Frost, of the Mobile mail steamboat *California,* we have a very concise general account of the effects of the storm. He left the Lake end of the Pontchartrain Railroad at noon on Saturday last. During the evening it commenced blowing very freely, and by midnight the wind had increased to such a gale that Capt. Frost judged it prudent to make Round Island, which he reached at 3 o'clock on Sunday morning, and came to anchor. The storm continued to rage so furiously that he had to let out a second anchor and to lay there for sixteen hours, and consequently did not get to Mobile until 9 o'clock on Monday morning.

At Pass Christian, Bay St. Louis, Mississippi City and Biloxi, he says, everything accessible to injury by storm is gone, with the exception of two wharves at Biloxi. At Point Clear, however, there appears to have been little or no damage done; and at Ocean Springs, Montrose, Hollywood, &c., nothing beyond the washing away of a wharf or two and some bathhouses. Of all the ordinary stopping places of the Mobile mail boats there is no chance left of landing anything at present, except at Pascagoula and Biloxi. The *California* landed her passengers for the former point with ease and safety, as no damage of consequence had been experienced there.

A number of schooners, Capt. Frost reports, have been driven ashore everywhere along the whole portion of the coast visited by the storm. He states that a woman was said to have been lost on the Bay St. Louis wharf. It was believed that she was a peddler woman who had come thither to wait for the boat.[2]

PASS CHRISTIAN

The Pass Christians retired to their slumbers on Saturday night little dreaming of the scene that the morrow's dawn was to reveal to their bewildered senses. The *California* landed her passengers here between 9 and 10 o'clock, amid a brisk shower, and what the salts call "a spanking breeze"—lucky souls to have reached shore and shelter in "the nick of time," even with soaked garments. The breeze increased to a blow and warned those whose lot it was to be out upon the waters to prepare for a lively time. By 3 o'clock A.M. on Sunday, the blow had increased to a gale, and from gale to hurricane. The waters lashed and foamed as the power of the dread winds lashed them into angry motion and swept off everything before them. The bath houses and wharves which lined the water edges of the shore, proved but puny obstacles to the combined elements, and were scattered in detail upon the beach from one extremity to the other, not a single one escaping. The wind and rain continued from the southeast until 11 A.M. yesterday, when they reached a height and fury beyond the power of description.[3]

BAY ST. LOUIS

The *Creole* landed her passengers about a quarter to one o'clock on Sunday morning at Bay St. Louis. The hurricane was not then at its height, but was truly terrible. The sea was perfectly illuminated and ran mountains high. . . .

The storm increased to a tornado by half-past two o'clock, sweeping away in its fury every wharf, bathing and summer house along the shore, uprooting a greater portion of the splendid live oak, magnolia, Pride of China, and drooping willow trees, twisting down fences, blowing out windows, casting off slates and shingles, and making sad havoc with all the fine gardens, choice exotics and fruit trees, paying no respect to highways, bridges, blowing down and upsetting three dwellings, besides demolishing all the small stores and oyster houses fronting the bay . . . At 11 o'clock on Sunday the tornado was at its height, the wind suddenly shifting around to southwest. The scene at this time was truly awful.

The wind continued to blow in its utmost violence for three hours, and it was during this time that the greatest danger to life existed. The waves broke over the high bluffs, and it was feared the waves would rise above the road and embankments and wash away foundations, &c. &c. . .[4]

* * * * *

Mr. Toulmin, one of the oldest inhabitants on the Lake shores, states that the hurricane has been equaled in severity only twice in forty-three years: first, in 1812, when several American gunboats went on shore, while watching the British fleet, and again in 1819, when nearly every dwelling was blown down.

BAY ST. LOUIS

The storm of 1856 [1855], as almost every other which preceded it, began with wind at the east during the day of September 15th, increasing toward night and blew from that quarter the whole night still increasing till daylight on the 16th when it began to moderate and by eight o'clock A.M. had nearly subsided, and had fallen to a dead calm by nine o'clock; when the rain began to fall in torrents, and as quick as though the wind had sprung up from southwest, and from the quarter blew for about three-quarters of an hour fully twice as hard as it had blown during the whole storm. The water on the Bay St. Louis was raised by the violence of the southwest wind, during the short time it lasted, more than four feet and a half. The wharves and bathing houses had all been swept away long before daylight, yet no other damage had been done on shore until the wind sprang up from the southwest, then it tore up, broke up, and blew down trees and fences, and destroyed nearly all the orchards around Shieldsborough.[5]

LAKE BORGNE

At Lake Borgne on the afternoon of Saturday, the water in the Lake began to rise, showing that it must be blowing outside. About sunset a smart breeze sprang up and continued to increase till midnight, by which time it blew a perfect hurricane, destroying the following property: First the wharf and bath-houses adjoining Pajol's hotel, and then Orfelia & Brother's large grocery store. The following houses were much injured, viz.: those belonging to Messers. Ritchie, Turner, Williams and Joyce, besides those of others whose names are not recollected.

All the oyster and other small boats found safe refuge in Proctor's Bayou, and large vessels rode in safety outside where there is excellent anchorage.

The train that left New Orleans at a quarter past five P.M. on Saturday, was obliged to stop half a mile on this side of the terminus, the water putting the fires out; and it was obliged to remain there till Monday about noon. Mr. Harper and one or two others who had gone down on the train, left it, determined to reach their families, which they did, though not without having sometimes to swim for it.

The conductor of the cars states that he could distinctly hear the screams of the people at Proctorville, although half a mile distant.

At Proctor's landing the water was four feet deep. On many parts of Mr. Proctor's sugar plantation it was two feet deep.[6]

THE GULF SHORE

The tide at places between Lake Pontchartrain and Bay St. Louis was said to have risen ten to fifteen feet above normal high tide. The railroad was covered with water as far as four miles back from the coast, and plantations as far as six to eight miles back were under two to three feet of water. The front road along Bay St. Louis, some fifteen feet high on an embankment, was cut up and washed away so as to be impassable; and spars, trees, and miscellaneous debris covered the back road halting regular traffic there. Most of the improvements recently made to properties along the ocean front were completely destroyed.[7]

The storm was reported to have reached full strength there about 0430 on Sunday morning and did not lull completely until well after noon. One unweatherwise plantation owner sent his Negroes out on Sunday forenoon to clear up debris along the front road as the storm appeared over, but they "were surprised by the sudden renewal of the storm, and found it necessary to use ropes they had with them to lash themselves to trees to prevent themselves from being blown off into the lake." [8] This would suggest that the center passed close to the eastern end of Lake Pontchartrain and directly over Bay St. Louis near mid-morning on Sunday. Since neither New Orleans nor Mobile suffered any destructive winds, the disturbance must have been of only moderate dimensions.

Eastward along the coast Mobile experienced rain and a rising wind on Saturday and fears were expressed that the city might be in for a repetition of August 1852. The wind, however, when at its peak on Sunday morning, blew from south-by-west. This raised a high tide in Mobile Bay, but did little damage in the city.[9] Along the eastern shore of the Bay the usual damages to bath houses and wharfs were reported. High seas kicked up as far as Apalachee Bay. The *S. S. Florida* ran into Pensacola Harbor on the 15th to escape 30-foot waves and tremendous gusts from the southeast. The gale was not severely felt at Pensacola.[10]

[1] *Daily Crescent* (New Orleans), 19 Sept 1855.
[2] *The Daily Picayune* (New Orleans), 19 Sept 1855.
[3] *Idem.*
[4] *Idem.*
[5] Moret, 796.
[6] *Picayune,* 19 Sept 1855.
[7] *Idem.*
[8] *Idem.*
[9] *Mobile Adv.* in *Picayune,* 19 Sept 1855.
[10] *Picayune,* 20 Sept 1855.

THE LAST ISLAND DISASTER

The city of New Orleans and surrounding river country experienced an increasing tropical storm with heavy rains and rising gales on Sunday, 10 August 1856. Surgeon J. B. Porter at the U. S. Army Hospital had noted an easterly flow at moderate speeds on Saturday, varying from force 3 to force 5, but by 0400 on Sunday morning an increase to gale force took place with a heavy rain commencing a half hour later. The wind mounted to force 8 by early afternoon as the heavy downpour continued.[1] Signs were blown down and awnings demolished, as often happens in such storms. The *Daily Crescent,* however, was pleased to report next morning: "We have as yet heard of no serious accident or damage." [2]

The tropical tempest continued to rage around New Orleans through Sunday night and into the next afternoon, with the rain ceasing about 1600 when Dr. Porter measured 4.82 inches in his rain gage. The wind, having shifted from east to southeast Sunday evening, held to the latter direction throughout Monday—on Tuesday morning it veered farther into the south and the rain recommenced. Wednesday brought a subsequent shift into the southwest, but no cessation of the tropical deluge. When the skies finally cleared on Thursday at 1800, an additional 6.32 inches were measured, making a storm total of 13.14 inches.[3]

The local weather observer logged the following during the storm period:

Aug 10—High wind last night and during the day. Lightning during the evening.

Aug 11—High wind and rain last night. Heavy rain, thunder, lightning A.M. Lightning in the evening.

Aug 12—Light rain during day and heavy rain in evening. Lightning in the evening at 2100.

Aug 13—High wind and heavy rain last night. Heavy rain A.M. and P.M. Sharp lightning during evening.

Aug 14—Thunder and lightning & heavy rain 3 P.M. to 4.

The wind log at the three daily observations showed the following:[4]

	0700	1400	2100
Aug 9	NE 3	E 5	E 3
10	E 5	E 8	SE 6
11	SE 6	SE 3	SE 4
12	S 5	S 4	SE 5
13	SW 6	SW 5	SW 2
14	SW 3	NW 4	S 2

This tropical cyclone at New Orleans might have been dismissed as a worthy representative of a type not unusual at this season. It produced gale force winds in the city and environs, but nothing approaching a destructive hurricane. Though very heavy downpours accompanied the disturbance, such amounts were not unprecedented. The four-day storm might have been quickly forgotten. But soon ominous rumors began to filter in from areas to the south and southwest of the city. These indicated that New Orleans might have been only on the fringe of a violent hurricane—there were "melancholy intimations" of possible disaster for the residents of the lowlying exposures along the bayous and islands making up the Gulf Coast of central Louisiana.

Let us follow the story as it unfolded to the residents of New Orleans in the columns of the daily press, commencing on the morning of Wednesday, the 13th, the third day following the initial impact of the storm:

We learn from a passenger who came down on the Opelousas Railroad yesterday, that the storm had raged with great violence on the line of the road, doing much serious damage. Chimneys and houses were blown down. The crops are completely ruined on the line of the road. It is reported that several lives were lost. New Orleans *Daily Crescent,* 13 Aug 1856.

Brashear City Hotel
Wednesday, August 13—4 A.M.

Eds. Pic.—John Davis has just got here from Last Island in a small sail boat, and reports Last Island entirely swept of all the houses by the storm of Sunday night, and that 137 lives were lost by the disaster. This is the amount hurriedly ascertained at present.

Berwick's Bay, Aug. 13, 1856

In great haste. We have just sent the *Major Aubrey* to the assistance of the sufferers, who are now clinging to the hull of the steamboat *Star.* She starts hence in one hour, only waiting to wood at this place. Respectfully yours,

Eugene Daly
Evening Picayune, 13 Aug 1856

Mr. Starr S. Jones, of the Opelousas Railroad Express, yesterday brought us the melancholy news of the complete overflow of Last Island and the destruction of every house and probable drowning of every person on the Island. He says the steamer *Star,* which was lying there at the time of the gale, went ashore on the Island and bilged near where Muggah's Hotel had stood. As there was at the time of the gale, upwards of one hundred and forty persons on the Island, we await further intelligence from them with the greatest anxiety. New Orleans *Daily Crescent,* 14 Aug 1856.

LAST ISLAND INUNDATED

Shocking Loss of Life

The rumor which prevailed yesterday on the destruction of Last Island in the late storm is probably too true. We have only some general reports of the greatness of the disaster, and a few vague particulars of the loss of individuals and families. The accounts brought from Thibodaux and Berwick's Bay, by the Opelousas Railroad last evening, are confirmatory of the inundation of the island, the destruction of the buildings, and the probable loss of a great many lives, reaching, perhaps, six or seven score. In the meantime, the anxiety to learn the particulars is very great; and the means of communication between the city and the scene of the suffering are very slight . . . By the arrival of the Opelousas cars this afternoon we hope to have further details, and pray that the accounts heretofore received may have been exaggerated.

In the meantime we subjoin such items as we have been able to gather. The following letters will show the excitement caused by the reception of the intelligence of the disaster at Brashear City, and the promptness with which steps were taken to send relief to the survivors. *Daily Picayune,* 14 Aug 1856.

Bayou Boeuf, August 14, 1856

Dear Pic.—You may have heard ere this reaches you of the dreadful catastrophe which happened on Last Island on Sunday the 10th inst. As one of the sufferers it becomes my duty to chronicle one of the most melancholy events which have ever occurred. On Saturday night, the 9th inst., a heavy northeast wind prevailed, which excited the fears of a storm in the minds of many; the wind increased gradually until about ten o'clock Sunday morning, when there existed no longer any doubt that we were threatened with imminent danger. From that time the wind blew a perfect hurricane; every house upon the island giving way, one after another, until nothing remained. At this moment everyone sought the most elevated point on the island, exerting themselves at the same time to avoid the fragments of buildings, which were scattered in every direction by the wind. Many persons were wounded; some mortally. The water at this time (about 2 o'clock P.M.) commenced rising so rapidly from the bay side, that there could no longer be any doubt that the island would be submerged. The scene at this moment forbids description. Men, women, and children were seen running in every direction, in search of some means of salvation. The violence of the wind, together with the rain, which fell like hail, and the sand blinded their eyes, prevented many from reaching the objects they had aimed at.

At about 4 o'clock, the Bay and Gulf currents met and the sea washed over the whole island. Those who were so fortunate as to find some object to cling to, were seen floating in all directions. Many of them, however, were separated from the straw to which they clung for life, and launched into eternity;

others were washed away by the rapid current and drowned before they could reach their point of destination. Many were drowned from being stunned by scattered fragments of the buildings, which had been blown asunder by the storm; many others were crushed by floating timbers and logs, which were removed from the beach, and met them on their journey. To attempt a description of this sad event would be useless. No words could depict the awful scene which occurred on the night between the 10th and 11th inst. It was not until the next morning the 11th, that we could ascertain the extent of the disaster. Upon my return, after having drifted for about twenty hours, I found the steamer *Star,* which had arrived the day before, and was lying at anchor, a perfect wreck, nothing but her hull and boilers, and a portion of her machinery remaining. Upon this wreck the lives of a large number were saved. Toward her each one directed his path as he was recovered from the deep, and was welcomed with tears by his fellow-sufferers, who had been so fortunate as to escape. The scene was heart-rending; the good fortune of many an individual in being saved, was blighted by the news of the loss of a father, brother, sister, wife or some near relative.

* * * * *

As I stated before, not a single building withstood the storm. The loss of property is immense, amounting to at least $100,000; the principal sufferers being John Muggah & Co., Thomas Maskel, P. C. Bithel, Gov. Hebert, Thos. Mille, L. Desobry, Lynch, Nash, A. Comeau, and others. The loss of baggage belonging to visitors on the island at the time, which is complete, amounts to at least $5,000, besides about $10,000 in money on those who were drowned, which was nearly all recovered by a set of pirates who inhabit the island. The bodies of those who were recovered had been invariably robbed by these men. It was an awful scene to see the avidity of these heartless beings to pillage the dead. I hope that the hand of justice will take hold of them and dispose of them as they deserve. *Daily Picayune,* 17 Aug 1856.

THE STAR

The steamboat *Star,* on Saturday, started on her regular semi-weekly trip from Bayou Boeuf to Last Island, with a fair complement of passengers aboard. She arrived outside the bar at an early hour on Sunday morning. Finding the water low, she was compelled to remain outside to wait for the tide. The weather at this time was moderate, though showing some signs of an approaching storm. The storm soon came on in all its fury, as already stated, and water enough came to carry her first over the bar and then close up to the hotel, which was still standing. Capt. Smith, with equal presence of mind and energy, set to work to combat the danger which he saw threatening all, had both anchors thrown overboard, and seeing that she dragged somewhat, the cabin began to yield, he ordered all the passengers down to the main deck, and set the crew to work to cut away all the upper woodwork of the vessel. Had not this been done she would soon have been torn to pieces or driven out to sea in such a condition that all on board must have perished.

The residences on the island immediately after this commenced yielding, and the first was borne completely away by the wind and wave, with all its inhabitants. The hotel, it is said, stood the force of the elements remarkably well, being among the last to be swept off, and going only by piecemeal. On the storm's approaching its height the inmates betook themselves to a large room upstairs, considered very secure; but the violence of the elements increasing, the bar-room was recommended as preferable for safety, and thither they all therefore went. The upper part of the hotel was soon blown away, and the water making its way into the lower part, they were all driven from that place of retreat. The steamboat *Star,* now a wreck, as already described, lay close to the door, and all endeavored to get on board her. The distance was short but the traverse perilous, and it was in making it that the infant child of the Hon. W. W. Pugh was rested from its mother's arms and borne to the realms of eternity by the ruthless waters. There was, of course, a desperate effort to reach the boat by the majority of the apparently doomed ones; and were it not for the strenuous exertions and the noble courage of a few the result even on this account must have formed a sad addition to the features of the disaster.

* * * * *

[Accounts of some rescues follow.]

Fortunately there was sufficient food on the *Star* for immediate necessities, and the people on her lived through the terrifying storm till Monday morning. . . . During Monday a cow, some sheep and one horse were found alive on the island. The cow was killed, and furnished sufficient meat for all the survivors. On Tuesday, after suffering intense anxiety, they were rescued. *Daily Picayune,* 17 Aug 1856.

The depth of the water on the island, referring as we understand it, to the highest portion, is reported at five feet. The wreck of the steamboat *Star* is stated to be lying on the island, near where the hotel of John Muggah formerly stood.

The rise of the inundation is said to have been of unparalleled rapidity, the height of five feet being reported to have been attained in two minutes! Although we were to read "hours" instead of "minutes," it will not be difficult to understand what a wild excitement and fearful havoc it must have caused, with no succor at hand.

All the houses on the island were swept away, and it is particularly reported that most of those staying at the hotel were drowned.

* * * * *

It is stated that there were about 400 persons on the island at the time of the disaster; and the number surviving on the wreck of the *Star* is estimated at from 250 to 275.

* * * * *

They inform us that the storm commenced about 10 o'clock on Sunday morning, and a faithful picture of the calamity they declare to be beyond realization. The gale did not abate until Monday morning, and then the rain continued almost without intermission up to the time of their leaving the island (Tuesday), at times the winds rising pretty strong again. The number of victims they estimate at over 200, at least 182 having been already counted. The island was swept by 2 o'clock on Sunday, having been overflowed between noon and that hour. The wind blew first from the north, and the northern part of the island was overflowed. Next the wind came from the east, which beat the waters off from the north side of the island; afterwards the wind shifted to due south, and then the island became overwhelmed by the waters of the Gulf. Horses, cattle, and even fish, lay strewn about the island among the human victims of the storm. It is believed that many bodies were washed out into the Gulf. *Daily Picayune,* 17 Aug 1856.

So much for the contemporary newspaper accounts of the disaster. Of the more recent retellings of the story the most complete and authentic was presented in July 1937 by *The Louisiana Historical Quarterly* with the first publication of a manuscript written "in the year 1856" by Michael Schlatre, Jr., a survivor. Schlatre alone of his family outlived the break-up of his cottage, a broken leg, a perilous trip across the bay on wreckage which served as a raft, and five days in the marshes without food or drink until his rescue by a search party. His manuscript has been carefully edited by Walter Prichard, then editor of the *Quarterly,* and presented in full along with many contemporary press clippings which set the scene for the reader. The following excerpts from the Schlatre account include all remarks of meteorological significance:

. . . Thursday was beautiful day, on Friday a breeze sprang up from the N.E. but nothing extraordinary, on Saturday 9th. August the breeze had increased to a strong gale, and the sea was white with foam and roared like a waterfall, the wind which blew hard but steady, began to swell the tide, and we had a full tide . . . Thus passed Saturday, and during Saturday night, the wind blew with much the same force, strong and regular . . . at noon the weather darkened somewhat, and it blew stronger; wind still N.E. that is blowing from the bay, side across into the Gulf—I had not the least apprehension of anything like a tornado . . . and I made light of the weather, telling them I had seen it blow as hard before this and so I had. We dined at 1 o'clock and already the rain began to come through the roof on the upper floor . . . it was now about 2 P.M. From time to time I looked for the *Star* at anchor, and she still stood steadily; the rain descended in torrents, and it all came down in the rooms . . . At three o'clock the servants came in and told me that the servants house had been blown down, as this was a weak structure, I thought nothing of it—in half an hour more I saw my kitchen nearly doubled, blown close up to our dwelling, for the first time I began to apprehend danger . . . the storm was now at its height, and it was about 5 P.M. . . . Our house was now in ruins and all of us on the floor, amidst bedding, trunks, armoirs etc. At this critical moment the wind shifted E.S.E. My dear, said I, our time has come the Gulf waters will soon cover the Island & drown us all . . . I now asked my wife if the gulf waters were washing over, and looking she said they were just commencing to cover the ground . . .[5]

The story of the tragedy at Last Island soon became interwoven with the legends of the bayou country of the Louisiana Gulf Coast, and like most verbal folklore grew with the years until much fancy became intermingled with fact. It was from the romanticist pen of Lafcadio Hearn that many of the imaginary embellishments to the legend sprang, and these soon became crystallized as fact through the publication of his "The Legend of L'Ile Derniere," first in 1884 in a New Orleans paper,[6] then in *Harper's New Monthly Magazine* in 1888,[7] and finally in book form in 1889 under the title: *Chita: A Memory of Last Island.*[8]

In the most recent retelling of the story that has come to hand, Edward Rowe Snow in *Great Gales and Dire Disasters* (1952) uncritically repeats the Hearn additives and leaves a very misleading impression as to what happened to the people of Last Island that Sunday afternoon and evening of 10 August 1856.[9] Perhaps it would be best at this point to set down in a few short paragraphs the salient meteorological facts as they can be culled from contemporary accounts.

Last Island, or Isle Derniere, as it has been interchangeably called over the years, formed a sandy spit of land whose dimensions were subject to the vagaries of wind and tide. It was described as follows in the 1863 edition of the *American Coast Pilot*: "Isle Derniere is twenty-two miles in length from E. to W.; and at some places more, and at others less than a mile in breadth. It is entirely level and low, with the exception of a small sand ridge, five or six feet high, which runs parallel to the beach. For some miles it is covered by thick chapparal; but the W. end is barren."[10] With its slightly convex configuration, the island formed a barrier beach about five miles offshore protecting Caillou Bay and the Terrebonne bayous. For many years it was visited only by an occasional fisherman, and from time to time had served as a haven for pirates preying on the coastal trade.

In the early 1850's Last Island became a favorite resort place for the residents of southern and western Louisiana, especially among the planters of the Atchafalaya River country nearby. In 1856 there were some twenty unpretentious cottages and one two-story building that served as a hotel. This hostelry was known locally by the unromantic designation, Muggah's Hotel or The Muggah Billiard-House, after its proprietor, and not as The Trade Winds Hotel, as later writers would have us believe. The resort was not the "Newport of the South" nor could it compare with the much more fashionable watering spots along the Gulf Coast east of New Orleans such as Biloxi and Bay St. Louis.[11]

The building of the Opelousas Railroad westward from New Orleans opened a way to reach Last Island easily by the cars to St. Mary (Berwick and Morgan City) and then by small steamer down the Bay to Last Island about 25 miles distant. Though there were plans afoot to expand the resort and to build a new hotel for the next season, at the time the hurricane struck Last Island was essentially a small community where families, many of them prominent in Louisiana life, went primarily to enjoy the cooling breezes of the Gulf, isolation from normal pursuits, and the simple pleasures afforded by the primitive surroundings.

Let us now reconstruct the meteorological story.[12] After a stretch of pleasant weather, the wind picked up from the northeast on Friday, August 8th, and on Saturday increased to a gale: "the sea was white with foam and roared like a water fall, the wind which blew hard but steady, began to swell the tide." On Sunday morning the gale howling from fresh to strong commenced to pile up the waters of the Bay on the north side of the island and by noon the pounding surf had crept higher and higher on the shore. The steamer *Star* entered the bayou near the hotel and came to anchor with difficulty.

As the gale continued to mount during the early afternoon, some of the lightly constructed outbuildings began to give way before the blasts, and the rising waters of the Bay along with the pelting rain undermined some of the blocks which served as foundations for the unsubstantial cottages. The crisis of the storm for Last Island came shortly after mid-afternoon when a rather abrupt shift of the gale to southeast occurred as the storm center drove in from the south. The exact time of the fateful veer of the wind cannot be definitely ascertained at this late date—witnesses gave the hour from 1400 to 1700; perhaps 1600 would be reasonably close to the actual time. With the shift to southeast, winds mounted to full hurricane force and a storm tide, which had been built up and carried along by the whirl at sea, quickly engulfed the shorelines and practically the entire island was overflowed. Witnesses remarked with surprise at the suddenness of the rise of the Gulf waters once the wind went into the southeast: "at the rate of a foot a minute" and "a rise of five feet took place in two minutes." These figures are entirely credible in view of the experiences of others in tropical storm inundations. The waters were said to have risen five feet vertically. Whether every bit of land was completely submerged is open to doubt since a horse, cow, and several sheep survived the flood waters, and many persons were able to cling to small bushes or debris and save themselves. Most of the cottages were lifted from their foundation blocks by the rising waters and floated away. Though early reports said every building on the island had been demolished, this was contradicted by the statement of a steamboat captain, who passed the island two days after the storm and reported five out of the twenty houses on the island still visible, all of these being at the eastern end where the low dunes had some protective vegetation.

Loss of life on the island was finally estimated at about 140 persons. The earlier published lists carrying about 180 names were scaled down as survivors, such as Michael Schlatre who had drifted for five days on a raft, were sought out and rescued by search parties. There were two principal marine disasters involving passenger-carrying vessels. The *Manila* went down with the loss of ten and the highly-regarded *Nautilus* on the Galveston run sank with all except two perishing.

The path of the center of the hurricane certainly passed west of Last Island as the mid-afternoon shift to southeast clearly showed. Our scanty reports indicate a rather erratic behavior of the storm center as it approached the immediate coast, with a very slow movement and perhaps a hesitation when close to shore. At St. Mary Parish on the Atchafalaya the peak of the storm came between 2100 and 2200 on Sunday night, with northerly gales prevailing; then a veer to east followed about 0200; and a final shift to south by morning completed the clockwise gyration. This would indicate a passage of the storm center rather close, but to the west of Atchafalaya Bay and east of the Calcasieu and Cameron area. Heavy rains continued over most of Louisiana all day Monday and the wind remained high from the south, but not with the excessive speeds and disastrous tides of the previous afternoon and evening. The dying hurricane circulation, still packing a punch, probably drifted slowly northward, roughly parallel and to the west of the Mississippi River.[13]

There was one interesting and significant ship report. The *C. D. Mervin* under Captain Mervin out of Cardiff, England, passed directly through the eye of the hurricane when off South West Pass. As the gale commenced out of the northeast on Saturday, the 9th, Capt. Mervin shortened his sail at 1400 and four hours later took in all sail. These precautions proved insufficient, however, as his ship moved toward the center of the whirl. His foremast gave way at 0500 on Sunday morning, to be followed shortly later by the main and mizzen. He checked his barometer at 0800 and noticed a reading of 28.20″, a drop of 1.70″ in 24 hours. At 0900 the *Mervin* experienced a calm which lasted for five minutes. "The sun shone brilliant, and there was every appearance of clearing off, but the wind suddenly struck the ship from the opposite direction." For two hours more a southerly hurricane swept the crippled ship and then gradually abated. We do not know the exact position of the *Mervin* at the height of the hurricane, but when it had abated, South West Pass lay only 60 miles to the east-northeast.[14]

The lateral extent of the storm on land can be traced in contemporary press reports. Pensacola on the east had northeast and east winds of only moderate force (4) on the 10th and a rainshower about 1600 which dropped only 0.15 of an inch.[15] Thus, western Florida must be considered beyond the hurricane's reaches. At Mobile a gale sprang up at 0930 on Sunday morning, but subsided in the harbor area after a brief time. During the evening there was more wind and some very heavy rainshowers. Though again not violent in the harbor, the storm raged outside in sufficient strength to detain the regular mail boats from their scheduled departures.[16]

Westward at Mandeville on Lake Pontchartrain the gale increased in strength:

> Our town was visited on Sunday and Monday by one of the most disastrous storms ever witnessed here. The wind began to blow from the east on Sunday noon, and kept increasing until Monday morning, when it shifted to south by east, and then it was truly awful to witness the progress of destruction. Rain, whirlwinds, waves—all went dashing over our town, bringing with it our bathing houses, wharves, and boats.
>
> About dark on Monday it was thought that the storm had subsided, but at 10 'clock P.M., it assumed its former strength, and continued until today (Tuesday, 12 Aug.).
>
> The storm *raged* on the river above New Orleans on Monday. Boats were hindered from making landings and quite a number of flatboats and small craft were blown ashore. The gale at New Carthage and Baton Rouge subsided on Tuesday, though the heavy rains continued.[17]

South of the city the hurricane struck with even greater fury. At the Parish of Plaquemines the storm "has proved the most destructive which has ever passed over this section of the country." Along the left bank from the Courthouse down to Fort St. Philip the sea rose and covered the fields to a depth of four to five feet, damaging the rice crop severely. The cane suffered less from wind and wave.[18] At Grand Terre, though the storm struck a heavy blow and the flood rose very high, no lives were lost but property suffered:

> We are glad to learn that the damage done on their (L. E. & F. J. Forstall) plantation, which is about three miles from Fort Livingston, was very trivial. It appears that the gale commenced there about 2 o'clock on Sunday morning, the 10th, and at 10 o'clock the water had risen to the height of about one foot at the back of their sugar house. This was the extreme height of the flood at any spot; and in front of their buildings there was still less water. No damage was done to the sugar-house, or any of the houses and out-buildings, except a boat house and an old cooper shop, which were blown down. A few Pride of China trees were also blown down. This was the extent of the damage. None of the cattle were lost.[19]

At the mouth of the Mississippi River the sweep of hurricane winds in the eastern sector severely damaged the installations at South West Pass, which faced Last Island some 80 miles distant to the west. The lighthouse was careened out of position, the pilothouse demolished, and the telegraphic signal house twisted around by the force of the wind.[20] At Caillou Island another disaster was first feared, but a later report stated that the flood rose only five feet which left the higher parts free from water:

> We have had a terrible storm on this our favorite isle, which commenced early on Saturday morning last, and continued until Monday evening. The maximum of it was on Sunday, from 12 to 6 o'clock . . .
>
> There were five dwelling houses destroyed . . . besides a number of outhouses swept away. Also several boats have been lost.
>
> From tradition, and the information of the oldest persons and sailors, this is the severest storm we have had for sixty years.[21]

To the west of New Orleans the rich plantation country received the full brunt of the hurricane as it roared in from the open Gulf. Destruction was particularly severe in St. Mary and Vermillion parishes:

> We are informed by a correspondent at Iberville, Parish of Vermillion, that the storm raged there on the 10th and 11th with terrific violence. Every house in the village was leveled to the ground; trees were torn up by the roots and blown to great distances, even the tombstones in the grave yard were thrown down and broken up. Several persons were injured in the village, but none killed. In the vicinity, sugar houses and dwellings, negro houses and barns, were all blown down and strewn about the fields, and we are pained to learn that five persons were killed by the falling of houses on them.[22]

In western Louisiana press reports from Lake Charles stated that no serious damage to property or crops occurred, though the Mermentou River, which drains the central area east of Calcasieu Lake, flooded and destroyed crops along bottom lands.[23] The western limits of the rainstorm may be judged by the continuance of a severe drought at both Mansfield and Shreveport in western Louisiana and at Galveston and Indianola in northeastern Texas.

[1] SG.
[2] *Daily Crescent,* 11 Aug 1856.
[3] SG.
[4] SG.

[5] The Last Island Disaster of August 10, 1856: Personal Narrative of His Experiences by one of the Survivors. (Introduction by Walter Prichard.) *The Louisiana Historical Quarterly*. 20–3, July 1937. 690–737.
[6] *Times-Democrat* (New Orleans), 14 Sept 1884.
[7] *Harper's New Monthly Magazine* (New York). 76–5, April 1888. 733–67.
[8] Lafcadio Hearn. *Chita: A Memory of Last Island.* New York, 1889.
[9] Edward Rowe Snow. *Great Gales and Dire Disasters.* New York, 1952. 221–29.
[10] Edmund M. Blunt. *The American Coast Pilot.* 19th ed. New York, 1863. 407.
[11] *Daily Picayune,* 17 Aug 1856.
[12] The columns of the *Daily Picayune, Daily Crescent,* and *New Orleans Commercial Bulletin* have supplied most of the data and statements employed here.
[13] Lorin Blodget. *Climatology of the United States.* Philadelphia, 1857.
[14] *New Orleans Commercial Bulletin,* 23 Aug 1856.
[15] SG.
[16] *Mobile Register,* 12 Aug, in *Daily Picayune,* 13 Aug 1856.
[17] *Daily Picayune,* 14 Aug 1856.
[18] *Daily Picayune,* 17 Aug 1856.
[19] *Ibid.,* 19 Aug 1856.
[20] *Mobile Daily Adv.,* 17 Aug 1856.
[21] *Thibodaux Minerva* in *Daily Picayune,* 19 Aug 1856.
[22] *Daily Crescent,* 25 Aug 1856.
[23] *Lake Charles Press,* 23 Aug, in *Daily Picayune,* 2 Sept 1856.

THE SOUTHEASTERN STATES HURRICANE OF 1856

A large, powerful hurricane sustaining great power swept into West Florida from the Gulf of Mexico late on Saturday, 30 August 1856, between Apalachicola and Pensacola. After crossing the sparsely settled coastal area, it roared north-northeastward causing more destruction in interior Alabama and Georgia than any other storm of our period.

First notices of hurricane activity came from east of Cuba on the 27th. On that day Ft. Dallas on Biscayne Bay near present-day Miami had "very strong winds and occasional drops of rain." The wind blew from the northeast at force 6 during all three observations, according to the local Army surgeon's weather report.[1] Key West also felt the first lash of the approaching disturbance very early on the 27th—at daylight a real gale commenced. The storm raged through that day and most of the 28th with the wind on the Keys veering from northeast to southeast as the hurricane swept through the Florida Straits. The barometer reached its lowest point of 29.77″ at 2230 on the 27th and then commenced to rise rapidly, putting the eye abreast, but westward, of Key West at that hour.[2]

The steamship *Daniel Webster* met the gale on the 29th and 30th when in the eastern Gulf of Mexico about halfway from New Orleans to Key West as the hurricane was in the process of recurving close to 85W. Capt. Churchill logged a very full meteorological account of his encounter:

> At 10 A.M. (29th) the wind freshened and a heavy swell rising from the southeast, Capt Churchill sent down the yards and housed the topmast, secured the boats and made all the usual preparations for a gale . . . at 4 P.M. heavy clouds, barometer 29.80″; gale fast increasing; furled all fore and aft sail, sea rising, ship making good weather.
>
> At 8 P.M. the barometer stood at 29.40″; blowing a heavy gale; brought the ship to the wind, North East, with fifteen pounds steam on. At midnight a heavy sea took away part of the starboard paddle-box. The ship's head then fell off four to eight points; had to work the engine by hand; gale increasing; constant lightning all around; no thunder, but a constant roar or rumbling sound as of a tornado.
>
> On the 30th, at 4 A.M., the barometer had gone down to 29.10″; discovered that water was gaining on the pumps, the engine working so slowly; connected the bilge pumps with the engine, but this proved of very little use; formed a line from the store-room hatch and commenced bailing; the ship was now laboring very heavily, lying nearly in the trough of the sea. At 5 A.M. the smoke chimney went over the side, and the engine stopped, the water increasing in spite of all efforts in pumping and bailing. At this time the wind had reached a perfect hurricane . . .
>
> At 10 A.M. the barometer was down to 28.90″, the wind blowing a perfect tornado, taking the tops of the seas horizontally with terrific force. Anything ten yards distant could not be seen from the ship. There were now three feet of water in the hold, the pump and buckets keeping it nearly the same level. At noon the barometer reached its lowest point 28.60″. At 2 P.M. the glass rose to 28.80″, and the gale broke as if it were a passing squall. In two hours it moderated to a common gale, with the encouraging fact of considerable gain being made on the water in the hold. At 4 P.M. the ship lay quite easy at her dredge. At midnight the weather cleared, the sea running down fast (At noon position was 26°31′N, 87°30′W).[3]

The gale on the 28th and 29th proved very disastrous to shipping in the eastern Gulf of Mexico. Many vessels as far westward as 85W reported hurricane force winds. One ship 40 miles southeast of Dry Tortugas had full hurricane blasts on the 28th as the storm center was curving northwestward into the Gulf. On the 30th the *S. S. Florida* was blown ashore in St. Joseph Bay, a complete loss.[4]

The port of Apalachicola lay open to the southeasterly sweep of wind and waves and took its usual battering:

> About one o'clock on the morning of the 30th inst. the wind began to blow violently from the south-east and by three o'clock the water was up to the top of the wharves and was rising rapidly. At noon the water had reached the sidewalks. The wind gradually increased during the afternoon, and by dark blew with the greatest violence. At 7 P.M. the waters were rushing into the stores at Water street. The water was at its maximum height about three or four o'clock in the morning of the 31st, and from that time up to ten, was receding rapidly, with the wind from the south-west and west. The wharves have received comparatively but little damage, though the wooden wharves are destroyed. The water was driven back into the city nearly to the Mansion House side-walk. Commerce street was submerged to the depth of three and a half to four feet. Market street was partially covered, the water reaching about two-thirds the way across the street . . . As regards the violence of the gale, it approximates nearer to that of 1851 than any other—the gale was more protracted than that of 1851, though the water was not quite so high.[5]

The center of this dangerous hurricane struck the West Florida coastline west of Cape San Blas, probably very close to Panama City. The best account of its devastating trail through interior West Florida appeared in the *Marianna Patriot*. The center passed just to the west of Marianna, located on Route 90 about 60 miles northwest of Tallahassee. The wind at this inland point close to the Alabama border picked up from the northeast during Saturday afternoon, the 30th, as rain commenced. Its strength increased gradually until 0100 on Sunday morning when a full gale raged and continued until daylight. It subsided for a short time, then veered around to southeast, and soon regained its original strength. The center must have passed a short distance to the west about daybreak. Very heavy rains accompanied the disturbance. After 0900 on Sunday the winds began to subside. The town was described: "a wreck." [6] The village of Milton farther west on Route 90 also took a severe beating with trees uprooted and lightning rods bent double. A press report had the wind "coming from all directions." [7]

Westward along the coast Pensacola experienced gales reaching force 8 during the night of the 30–31st. But, the direction being from north and northwest, there was no inundation and consequently little harbor damage. Rainfall measured 2.10" there.[8] Mobile had northwest gales late on the 30th with very rough conditions outside the harbor, but no local damage near the city. Inland at Mt. Vernon Barracks, about 30 miles north of Mobile, the storm period lasted from 1600/30th to 0500/31st with winds at northwest, force 7, late on the 30th.[9] Similar conditions with northerly flow existed at New Orleans during this period.

This hurricane differed from many of its predecessors by maintaining great energy after passing inland. Many points in interior Alabama and Georgia reported their heaviest storm in many years. The Montgomery area had high winds and heavy rains which did great damage to the maturing cotton crops.[10] At Eufaula in east-central Alabama the *Spirit of the South* indicated the storm at its height from midnight until noon on Sunday, "raging with unremittent fury." The town was strewn with prostrate trees and fences, with scarcely a lot escaping damage. Some structural injury resulted in the vicinity.[11] Nearby Auburn felt the lash of easterly gales shifting to northwesterly on Sunday morning: fences were down, large limbs off trees, and the countryside deluged.[12]

The storm center cut a track across the State of Georgia, roughly from the southwest corner to the northeast near Augusta. The *Columbus Enquirer* described the scene:

> August went out with storm and tempest, and September opens on many a wreck of nature and art, the work of the angry elements. On Saturday we had a brisk breeze from the east, which gradually veered (sic) around to the northeast and the north, increasing in violence, until on Sunday morning it blew a perfect gale from the north; driving before it in oblique showers a continuous and soaking rain. Throughout the day the tempest and the rain continued with great violence; shade trees were prostrated to such an extent that our streets and sidewalks were partially barricaded in every direction; many fences were prostrated, and sheds and awnings riddled. A portion of the unfinished roof of the Alabama Warehouse was blown off, and several minor injuries to buildings have been reported.[13]

Farther north, though the winds decreased, the rain intensified. The *Augusta Constitutionalist* reported:

> . . . On the 27th and 30th August, we had fine rains, but on Sunday, the 31st, about 9 o'clock A.M. the wind being northeast, a very heavy rain commenced, and for about twelve hours it rained steadily, accompanied with high winds. Such a rain has not descended in this locality, within the memory of any of our residents, as far as we could learn.
>
> The damage by wind and rain in this city has been very light. Several trees and fences were blown down, and awnings torn loose, but the loss is trifling.
>
> The loss to the cotton and corn crop will, no doubt, be immense.
>
> In every portion of the country we have heard from, the storm has been very disastrous. Every milldam on Butler's, Spirit, and Rocky creeks, have given way before the angry floods—the bridges on the common thoroughfares throughout the country, even over small streams, have either been damaged or carried away. The canal at Belville Cotton Factory broke, and carried off the machine and black-

smith shops; the dam at Richmond Cotton Factory was broke, and the bridge on the plank road, below the factory, was floated off, and also several bridges on the line of the plankroad.

The Georgia Railroad embankment in two or three places was injured by the swollen streams, and detained the trains several hours, but the damage was soon repaired, and the trains are now (Tuesday) running regularly.[14]

Along coastal Georgia the storm raged from the southeast. The *Savannah Republican* noted: "We have seldom seen a more blustering day than yesterday. There was little rain, but a heavy gale from southeast prevailed throughout the day and until a late hour of the night. Considerable damage has been done to the trees and roofing in the city, though we have heard of nothing of a very serious nature." [15] At Whitemarsh Island on the immediate coast the wind gradually increased all Sunday forenoon, going from south-southeast to south by 1300 when it became a gale. This continued until 2100 blowing "with great violence and in some of the puffs with very great violence. At 2100 a little rain fell and the wind commenced to abate though it blew strong all night; by morning it was in the south-southwest." [16]

The storm center passed between Columbia and Charleston. The former was "visited with a strong northeaster, accompanied with much heavy rain. The wind at times was blowing strong but not sufficiently so to cause damage." At Charleston: "During Sunday afternoon, the wind being at south-southeast, it began to blow very heavily, and during the night increased to quite a gale, blowing at times with great violence. Fortunately for shipping in port, the wind was just sufficiently far to the southward to prevent the sea acting on them with effect, and the great bulk of the vessels in port escaped without injury." Though the streets were strewn with wrecks of trees and some roofing was removed, none of the instances involved any great losses.[17]

The lowest barometer at Charleston at observation time was 29.88" at 2100/31st. Farther up the coast at Georgetown, the wind on the 31st veered from northeast to southeast in the afternoon and blew at force 8 from the southeast until 0245/1st when it began to move to south and reached southwest at 0315. At that hour the gale commenced to moderate and the clouds soon broke in the west.[18]

At Norfolk the gale was thought to have been unequaled since 1846. It commenced in the east-northeast about 0400 on the 1st and raged all day. The spire of the Baptist Church was blown off, and huge trees of 20 years growth were downed at Portsmouth. There was much damage at the Navy Yard also. The storm passed out to sea in the vicinity of Cape Hatteras and did not affect areas north of Chesapeake Bay. Cape Henlopen had a heavy northeast gale on the 1st. The New England shore experienced a moderate northeast flow on the 1st and 2d, but no storm and little cloudiness developed.[19]

[1] SG.
[2] SG.
[3] *Evening Picayune* (New Orleans), 3 Sept 1856.
[4] *New Orleans Commercial Bulletin*, 13 Sept 1856.
[5] Apalachicola dispatch in *Daily Crescent* (New Orleans), 11 Sept 1856.
[6] *Marianna Patriot*, 2 Sept, in *Daily Picayune*, 11 Sept 1856.
[7] *Milton Phoenix*, 3 Sept, in *Daily Picayune*, 9 Sept 1856.
[8] SG.
[9] SG.
[10] *Montgomery Journal*, 1 Sept, in *Daily Picayune*, 5 Sept 1856.
[11] *Spirit of the South* (Eufaula) in *idem.*
[12] *Auburn Gazette*, 5 Sept, in *Mobile Daily Adv.*, 9 Sept 1856.
[13] *Columbus* (Ga.) *Enquirer* in *Daily Picayune*, 7 Sept 1856.
[14] *Augusta Constitutionalist* in *idem.*
[15] *Savannah Republican* in *idem.*
[16] SI.
[17] *Courier* (Charleston), 1 Sept 1856.
[18] Ms. Met. Reg. Board of Health (National Archives).
[19] *N. Y. Ship List*, 6 & 10 Sept 1856.

1860—HURRICANE I

The first of three severe hurricanes to strike the Middle Gulf Coast in late summer and early autumn of 1860 commenced on the fourth anniversary of the Last Island disaster. It pursued a course some fifty miles to the eastward of its fateful predecessor. From the scanty meteorological evidence available, it would appear that the center moved over the lower Delta area of Louisiana, coming in from the western Gulf at an oblique angle between Last Island and South West Pass at modern Burrwood. Observers at Balize (now Pilottown) reported the gale veering from east to southeast and south.[1] At Pass a l'Outre the gale commenced about 0200 on the morning of the 11th, raging from east-by-north until 1400 in the afternoon when a shift through south to south-southwest occurred. The tide there rose four feet above normal.[2]

Up the river at New Orleans, on the other hand, the winds gradually backed from northeast to north and northwest, from whence they blew with their greatest fury.[3] A report from Plaquemines in the upper Delta also told of a backing wind from east into north and a severe inundation.[4] From this evidence, the center probably moved from southwest to northeast across the lower Delta very close to Fort St. Philip at the English Turn and then across the open water near the Chandeleur Islands to the mainland in the Gulfport-Biloxi-Pascagoula area.

The military observer at New Orleans on the 11th noted: "Very stormy day. High wind in the evening sometimes blowing with force 8 and 9 and doing a great deal of damage in the neighborhood." [5] The *Daily Picayune* on Monday summarized the impact of the storm:

THE GREAT GALE OF SATURDAY

Saturday last was the anniversary of the Last Island disaster, four years ago, and from accounts that are pouring in upon us from all quarters, from river, lake, Gulf and coast, we judge the effects resulting from the gale of last Saturday night will prove, in the aggregate, nearly, if not quite, as disastrous.

During the whole of the 11th, there was a good deal of wind and rain in the city. Portions of the ruins of buildings lately destroyed by fires in Royal and Tehoupitoulas streets, were blown down, but no damage done.

THE LAKE

There was a lively time, during the entire day, at the Lake end of the Pontchartrain and Jefferson Railroads. At the former, the railroad wharf was a good deal broken up, the heavy timbers being washed up over the track, by the surging waves. We have heard the damage to the wharf estimated at over $3,000.

Milneburg (at the lake end of the Pontchartrain railroad) was perfectly flooded, and occupants escaped by being carried off in boats. The pretty gardens, attached to Boudro's and to the Washington Hotel, were sadly cut up and laid waste; and some of the fine trees in front of the Washington were blown down. The shell road, constructed from Milneburg to the bayou St. John, was washed away and destroyed.

LAKE BORGNE—LOSS OF LIFE

The gale and flood were terribly severe at and about Proctorville. We have read in this morning's *Crescent* that the Mexican Gulf cars, on Saturday evening, could not get within four miles of the town for the water on the track; the water being four feet deep on the track near the town. The shell banks and many of the houses were washed away, as many as possible of the people escaping in boats.

We hear this morning, that there was a sad loss of life in that neighborhood. Officer Brooks and others report to us that the loss of life cannot fall short of thirty, and that very likely it will even exceed that number. There is hardly a house remaining at Proctorville.

THE RIVER

There was a rise of some three feet in the Mississippi. Over the river, and the whole line of the levee, the gale swept wildly. The vessels of all kinds lying at their docks, piers and wharves, tossed about and chafed their sides, but we hear of none of them breaking from their moorings except the *Belle Gates,* who took French leave of "the right bank," and came over to this side as if she had all steam up.[6]

PARISH OF PLAQUEMINES

As I informed you yesterday, the hurricane of the 11th inst. proved the most destructive we have had here ever since 1812. From Pointe-a-la-Hache to Fort St. Philip, left bank of the river, an extent of country measuring forty miles, was totally submerged by the sea—the waters rising, on the highest spots of dry ground, to fourteen feet at Fort St. Philip, about ten at Grand Prairie, five or six at Pointe-a-la-Hache. . . . .

* * * * *

The wind blew from the east during ten or twelve hours, and it was during the day time that the waters of the sea rose.

About twenty persons are said to have been drowned and I should not wonder if their number was greater, because all the fishermen encamped on the banks of the bayous and bays on the eastern side of the river, amounting to one or two hundred at this season of the year, must have escaped, if at all, with the greatest difficulty, for there the sea must have risen to twenty feet at least, and, no place of shelter being at hand, the greater part of them must have fallen victims to the fury of the waves. Twelve persons are supposed to have lost life at Pointe-a-la-Hache, and about five or six in the rest of the submerged district.

The cane fields throughout the parish suffered immensely. The cane having been, and still being, laid flat on the ground, almost everywhere. The knowing ones say that two thousand hogshead of sugar less will be made here. The losses, as I said yesterday, amount to above a quarter of a million dollars.

* * * * *

Since writing the above, I have learned that several houses were blown down in the lower section of this parish. Everywhere, trees were uprooted, fences blown down, and other injuries sustained on our plantations and farms.[7]

THE BALIZE

Balize, Aug. 12, 1860

Eds. Pic.—I give you, as interesting to your readers, a brief description of the effects of the storm at this place.

For some time past the weather had been extremely warm, the atmosphere sultry and smoky, and the luminaries of the day and night covered with a lurid mist—certain indications of a coming storm—when, on the evening of the 9th, the very time at which four years ago Last Island was destroyed, a dark, heavy cloud commenced rising in the north-northeast, illuminated by fierce and constant flashes of lightning; but, strange to say, it passed, accompanied by neither thunder, wind, nor rain. During the night the sky was covered with dark clouds, and although a few stars appeared now and then, the more experienced and knowing ones of our people predicted "a bad time coming," and this was too true.

On the morning of the 10th the light breeze which had been prevailing grew stronger, shifting during the day gradually to the east; the sun was hid by the thickly gathering clouds, and the aspect was such that even the spirit of the more strong-hearted grew fainter as the storm grew fiercer. . . .

The wind had now grown into a violent storm, which was driving the waters of the sea towards us; and to add to the scene of horror, rain came pouring down upon us, which being driven by the violence of the wind underneath the shingles wet and soaked every article in the houses, leaving not a dry spot to lay our heads. It was a terrible night. The morning of the 11th found us, without exception, in the most pitiful condition imaginable, for in many houses the sea water had risen to a considerable height—besides there was not the least prospect of a change in the weather. . . .

But the danger of being swept away was approaching: the wind shifted, first to southeast and then to south, when the waters receded in the very direction they came from. Towards noon the weather cleared up, and our poor depressed hearts were relieved.

The devastation done to property, however, is considerable. Quite a number of out-houses, chimnies, and fences are blown down, not to speak of the many shade trees that were rooted up; wharves are destroyed, and two schooners and two smaller craft were blown on the land. One of the dwelling houses stands half unroofed; another is twisted, and still another is level with the ground. The damage done can at present hardly be estimated; but we should judge it to be from $8000 to $10,000. Fortunately no lives were lost.[8]

At Proctorville at the eastern end of Lake Pontchartrain thirteen persons were drowned when the north wind blew the waters of the Lake over the low-lying settlement.[9] To the east at Biloxi the gale, commencing on the morning of the 11th, was thought not to have been rivaled since 1855. The water rose ten feet above normal as the wind, originally from the east and southeast, backed during the night to the north from whence it blew "a perfect gale" until 0300 on the 12th. Its center must have come inshore to the east of Biloxi close to the Alabama border.

In Alabama the wind behavior at Citronelle, close to the Mississippi border, showed a veering pattern into the south and "blowing a tremendous gale" about midnight Saturday. Mt. Vernon Barracks to the north-northeast of Mobile also had its heaviest winds around midnight—4.55 inches of rain fell there during the storm period.[12]

THE GALE AT MOBILE

Finally, about the middle of the afternoon, the gale commenced blowing in fitful gusts, from the east first, which is invariably the case with this sort of storm here, and continued in violence, coming with gusts harder and harder, up to the moment it began to abate. That is to say it raged with increasing violence until about daylight Sunday morning, when it began to subside, gradually. By the time the storm had fully established its character, the wind went around and blew from south by southeast.

With the wind came a slanting, searching rain, that was driven through cracks and crevices not known before to exist, drenching the interior of apartments, always water-proof to ordinary showers. The night was pitchy dark, and the wakeful observer of the howling storm (and many such there were for houses shook and rattled at every blast to a degree to disturb the slumbers of even those who were not scary) who peered through the gloom, saw lights burning during the whole night in the habitations around him.

At 2 A.M. the fire bell rang, to warn the people that the water was rising. Before the alarm, though, the owners of stores had turned out, and were hard at work, moving and putting up things from the floors all along Commerce and Water streets. Much was done, and a great quantity of goods put out of harm's way. These efforts were continued up to about 6 or 7 o'clock, when the wind began to veer around to the west, and all knew then that the storm was abating, and that the water would presently begin to recede, which it did.

The water rose just about high enough to flood Water street from Government to the north end. The depth on the floors of the stores on Commerce street was from four to not more than ten inches at the deepest.[13]

The influence of the storm reached far eastward to Pensacola where on the 11th "a heavy storm of wind and rain set in at 2 P. M. which continued with increasing violence during the night accompanied with much thunder & lightning." With the mercury at 80°, a southeast gale, force 6, was raging at the last observation at 2100. By morning, however, the wind was in the west at force 8 as the storm passed to the north. A downpour of 3.03″ collected in the post rain gage.[14]

[1] *The Eve. Picayune,* 15 Aug 1860.
[2] *Daily Picayune,* 15 Aug 1860.
[3] SI.
[4] *Daily Picayune,* 15 Aug 1860.
[5] SI.
[6] *Daily Picayune,* 15 Aug 1860.
[7] *Idem.*
[8] *Idem.*
[9] *Idem.*
[10] *Daily Delta,* 15 Aug 1860.
[11] *Mobile Daily Adv.,* 14 Aug 1860.
[12] SG.
[13] *Mobile Mercury* in *Daily Picayune,* 15 Aug 1860.
[14] SG.

1860—HURRICANE II

A second severe hurricane swept the Middle Gulf Coast on the night of Friday, September 14th. It continued to pound the shore from the Louisiana Delta to Pensacola until late Saturday afternoon. The storm center again made a landfall close to the mouth of the Mississippi River, striking a severe blow at Pilottown, and then drove northward to the Mississippi Coast where the small villages of Biloxi and Pascagoula again lay directly in the path. The center probably raced over or very close to both places. To the westward, Lake Pontchartrain in the left sector of the advancing whirl received a hard lashing from winds backing into north and northwest—to the east, Mobile lay in the right semicircle with veering winds causing great damage when they reached the southeast.

In the Mississippi River Delta area, the Balize and the Passes, already battered by the mid-August storm, sustained the greatest damage in many years. The blow, commencing from the northeast on Friday afternoon, continued until 1000 Saturday morning, the 15th. It was accompanied by large hailstones and heavy rain, according to a local observer. Nearly every building at the Balize was either blown down by the gale or washed away by the accompanying inundation.

Nine houses along with three lookout buildings, their boats, and sheds were destroyed by the rising waters which reach six feet above normal high water mark. All personal property of the residents was lost.[1] The death toll at Pass-a-l'Outre, alone, was estimated at ten persons. Those who escaped had to wade through water up to their necks and then cling to floating logs. One man carried his baby over his head to safety while up to his shoulders in water. Four tow boats were blown ashore; two large ships grounded on the bar; and two were blown into the marshes. Only the customs house seemed to have survived the fury of the wind and waves.[2]

On Lake Pontchartrain the winds, after rising to hurricane blasts from the northeast, backed around to north, from which direction they drove the piled-up waters of the lake southward through the narrow passes and over low beaches into the Gulf. The small resort and fishing communities along the south shore of the lake suffered severely in the inundation. The long wharf at Milneburg serving the resort steamships was washed away completely. All wharves, bathhouses, and sheds along the lower lake shore were destroyed, according to early press reports. The storm there was judged as severe as the August blow, but did not continue as long.[3]

At New Orleans there was "a heavy storm of wind and rain," but no flooding was mentioned in the press. The barometer at the Crescent City dropped to 29.86" at 1500 on the 15th when the storm was well to the east.[4] The main effect of Hurricane No. 2 was felt along the Gulf Coast east of Bay St. Louis. Biloxi, where the blow reached its height about 1100 on Saturday morning, appeared to have suffered the worst damage. The village was reported "a wreck" after the winds subsided. The bluffs were undermined as far as 20 to 30 feet back from the shore, and the lighthouse was completely carried away.[5] The long wharf used by the steamboats was covered with nine feet of water. The town was strewn with loose timber and furniture. One life, at least, was lost when Brown's Hotel collapsed. The gale at Biloxi was considered "far worse than the fearful storm of 1855." [6]

East Pascagoula witnessed the tide rising to a high pitch by midnight of Friday and continuing its destructive course until 2200 on Saturday night. Water rose three feet higher than in any storm since the historic inundation of July 1819. The wharf at Pascagoula was completely destroyed. Damage in the area was put at $40,000 in an early estimate.[7]

Mobile also lay within the sweep of dangerous winds and surging tides. The storm there commenced at 2200, Friday night, with the wind from north working around to southeast by midnight. For the next thirteen hours the continued southeasterly flow raised the water level of the harbor to almost record height—at the peak of the storm tide the waters were only 19 inches below the high mark established on 26 August 1852. Losses were estimated as high as $1 million with three-quarters of this consisting of cotton stored on the flooded wharves. By 1330 on Saturday the wind veered to south-southwest and both waters and wind abated. "In some respects," concluded the *Daily Advertiser,* "it has proved more disastrous than the memorable one of 1852." [8]

[1] *Daily Picayune,* 18 Sept 1860.
[2] *Idem.*
[3] *Ibid.,* 16 Sept 1860.
[4] SG.
[5] *Mobile Daily Adv.,* 20 Sept 1860.
[6] *Daily Picayune,* 18 Sept 1860.
[7] *Ibid.,* 19 Sept 1860.
[8] *Mobile Daily Adv.,* 16–19 Sept 1860.

1860—HURRICANE III

With the headline, "Another Terrific Storm," the New Orleans *Picayune* announced the third hurricane in seven weeks to descend on Louisiana. This one was considered much "blacker" than either the August or September visitations.[1] Damage in and around New Orleans, itself, as well as up-river, was much greater in the October blow since the richest and most densely populated areas of Louisiana and Mississippi this time lay in the dangerous eastern semicircle of the vast storm system. The center moved inland from the Gulf of Mexico well west of the Mississippi Delta, while the previous two had crossed the mouth of the great river, putting the Mississippi, Alabama, and West Florida coasts in the chief danger zone. From the future course of the hurricane, a landfall over Atchafalaya Bay appeared logical and then a course north-northeastward, carrying the center well west of New Orleans, close to Baton Rouge, and into central Mississippi east of Natchez and Vicksburg.

The cyclone struck inland with great power about noon on October 2d. At Plaquemines, down-river from New Orleans, residents considered it the greatest *wind* storm ever experienced, although the usual dreaded inundation failed to materialize.[2] A rise of only 12 to 15 inches at Port-a-la-Hache proved relatively unharmful in contrast to the flood devastation meted out by the two earlier hurricanes.[3] The rich sugar growing areas south and southwest of New Orleans were close to the central fury of the storm and suffered maximum losses since the hurricane struck just at the commencement of the sugar producing season. Not only was the cane laid flat in the wet fields where further sprouting might destroy its saccharine qualities, but many of the flimsy sheds housing the machinery employed to grind the grain and produce the sugar were flattened by the terrific blasts out of the southeast.[4] At Thibodaux the "wind carried everything with it"—there was "unparalleled destruction." The large Presbyterian Church was severely damaged as were many other buildings and houses.[5] In St. James Parish the gales raged at their height from 1200 to 1600 on the 2d and caused major structural damage.[6] Throughout south-central Louisiana it was considered the greatest hurricane since the memorable storm of August 1812. Similar reports came from Vermillion, Feliciana, Albemarle, Bayou Lafourche, Pointe Coupee, St. Bernards, and Terrebone.[7]

At New Orleans the winds began to rise early Tuesday morning soon after midnight and continued for more than 24 hours. The Medical Corps weather observer described: "A very stormy day blowing at times almost a hurricane doing much damage to shipping and also houses—causing great loss of life in this neighborhood." The wind flow at the Crescent City, which had been southeast on Monday afternoon and evening at moderate speeds (force 3), came out of the northeast at force 6 by 0700/2d. During that morning the direction worked around to east rising to force 8 at the 1400 observation. It continued from the east at least until 2100 that evening. Next morning the wind vane had moved to southeast, constantly veering, until it reached south by early afternoon, with speed down to force 3 however. Rainfall totaled 5.04 inches on the 1st and 2d.[8] The press reported the barometer had taken a precipitous fall on Tuesday, but the lowest pressure recorded by the local observer, 29.62", would indicate that the center stayed well west of the city, though wind blasts of great force swept the area for many hours.[9]

A detailed account of the storm's havoc in the New Orleans area appeared next day in the *Daily Picayune*:

ANOTHER TERRIFIC STORM

BUILDINGS BLOWN DOWN

SUPPOSED LOSS OF LIFE

As we surmised, the wind increased in violence yesterday, until it blew a perfect hurricane, and we have to report much damage both in the city and river; and fear that our record of disasters will yet have another chapter added to it, blacker perhaps than those relating to the fearful storms of the 8th (sic) of August and 11th (sic) of September last.

* * * * *

In the City

During the whole day there was a continual blowing down of chimneys, fences, and signs, to the great danger of passers by, and many trees in our public squares and avenues were uprooted and broken.

Towards half past 2 o'clock, the wind increased in violence and blew with such force that it was hardly possible to walk through the streets. As to carrying an open umbrella, that was out of the question.

About 4 o'clock a large brick building in the course of construction, five stories high, came down with a terrific crash. It crushed two other buildings along side. Two believed lost. Stables blown down. Tin roof of beef market ripped off. Glass in clock of old cathedral shattered. Shrubbery and fine trees on Jackson Square suffered some damage. Telegraph wires down, also police wires severed.

At Milneburg the damage not as great as feared. Water rose very high, inundating the village, but did not cause much damage except washing away some of the timbers of the pier and obstructing the track with drift wood.[10]

Driven by the easterly hurricane, the waters of Lake Pontchartrain invaded the eastern and northern portions of the city within three-quarters of a mile of the river, an "unprecedented event." The flood was channeled between the New Canal and Carondelet Canal, back of Clairborne Street as far as Rampart Street. The waters ceased rising at 0300 on the 4th and then receded very slowly.

Along the westerly and northwesterly rim of the Lake, the tracks of the Jackson Railroad were under water as much as four to five feet deep from Bayou La Branche and Bayou Desert, near the stations of Frenier and Manchac. Eleven miles of track were washed out, and seventeen families living along the right-of-way were forced to flee as their homes were inundated. All railroad communication to the northeast was severed for a number of days.[11]

To the north, the Baton Rouge area suffered its first serious hurricane damage in years. Twenty-one coal boats were swamped and four steamboats sunk at their moorings.[12] At Natchez the "terrible gale" reached its peak after dark on the night of the 2d–3d. "Since the tornado of 1840, we have never experienced the wind so high in this city," commented the *Natchez Courier*.[13] Across the river at Concordia a powerful north wind prostrated trees, and at Vicksburg farther up the river trees were pulled out by the roots and laid low.[14] Westward on the Red River at Angola Plantation, the gale raged out of the east until 2300, Tuesday night, when it went into the north still increasing. The height of the gale there came from 0500 to 1100 on Wednesday morning. Never had there been such destruction to crops and trees in the vicinity, according to a dispatch to the *Picayune*.[15]

The subsequent course of the storm has not been traced. To the east along the Gulf shoreline little damage occurred at Bay St. Louis or the resort communities of Mississippi. At Mobile a hurricane scare developed Tuesday afternoon when wind and water commenced to rise about 1600. A two-foot tidal increase was noted by 1800, and the water rose at the rate of six inches per hour. The winds, however, failed to mount to hurricane proportions so the Alabama port was spared a third inundation in a single season.[16] At Fort Barrancas (Pensacola) the wind mounted to force 6 from the east on the evening of the 2d and a heavy fall of 5.85 inches of rain descended, the main portion of a total storm catch of 7.38 inches.[17]

[1] *Daily Picayune*, 3 Oct 1860.
[2] *Ibid.*, 6 Oct 1860.
[3] *Ibid.*, 7 Oct 1860.
[4] *Ibid.*, 4 Oct 1860.
[5] *Idem.*
[6] *Thibodaux Gazette*, 6 Oct, in *Picayune*, 9 Oct 1860.
[7] *New Orleans Price Current*, 6 Oct 1860.
[8] SG.
[9] *Picayune*, 7 Oct 1860.
[10] *Picayune*, 3 Oct 1860.
[11] *Delta* in *N. Y. Trib.*, 10 Oct 1860.
[12] *N. Y. Trib.*, 6 Oct 1860.
[13] *Natchez Courier* in *Price Current*, 10 Oct 1860.
[14] *Idem.*
[15] *Picayune*, 5 Oct 1860.
[16] *Mobile Tribune* in *Evening Picayune*, 4 Oct 1860.
[17] SG.

THE SABINE RIVER-LAKE CALCASIEU STORM OF 1865

A Gulf hurricane of the Audrey 1957 type struck in the same area on September 13th in 1865. No dependable meteorological data are available. The principal damage seemed concentrated in the lower Sabine River Valley near Orange, Texas, and around Lake Calcasieu near Cameron and Lake Charles, Louisiana.

A dispatch to the *Vermillion Advertiser* from Niblet's Bluff on the Sabine supplied our principal information:

> Niblet's Bluff was utterly destroyed by a terrific storm. He describes the scene as terrible indeed. After a stormy day, about sunset the wind began to increase in violence and continued till 11 o'clock. All the buildings in the place were blown down or badly injured. The inhabitants had to flee from their houses, and prostrate themselves on the ground to keep from being blown into the river. Fortunately no lives were lost, though many received severe bruises from falling timber. The forest for four miles on this side of the Bluff was completely prostrated.[1]

The above account reads more like a tornado than a hurricane except for the long period of rising winds. Another report, however, from the Sabine River country mentioned strong northeast gales with heavy rain; the wind later veered to north, northwest, and west—to confirm that this was a hurricane and not a local tornado. Severe damage occurred at the town of Orange where only three houses survived intact. At Johnson's Bayou many houses were blown down and one person killed.[2]

The area around Lake Calcasieu was inundated by a storm tide. Grand Cheniere was put under water and many houses washed away. Several persons perished there.[3] Eastward along the Gulf Coast as far as the Mississippi River tides ran very high and inundated fields, but there are no reports of high winds except in extreme western Louisiana.[4]

[1] *Vermillion Advertiser* in *Morning Star* (New Orleans), 1 Oct 1865.
[2] *Galveston Civilian* in *Daily Picayune* (New Orleans), 26 Sept 1865.
[3] *Opelousas Courier*, 23 Sept, in *New Orleans Bee*, 26 Sept 1865.
[4] *Daily Picayune*, 23 Sept 1865.

THE GALVESTON HURRICANE OF 1867

Almost thirty years to the day after Racer's Storm had struck a devastating blow at Galveston Island, another great hurricane approached from the south. As in 1837, the storm-driven waters of the Gulf before the city and of the Bay behind joined forces over the submerged land to put the inhabitants through the terrors of another inundation. One will recall that similar storms, though of probably somewhat lesser intensity, hit the same region—on 12–14 September 1818 and again on 15–17 September 1854. And Galveston would again receive even greater onslaughts from tropical intruders—on 19–20 August 1886, 8 September 1900, 16–17 August 1915, and 11–12 September 1961.

The *Galveston News* described the scene of damage and desolation on the day following the storm of 3 October 1867:

> All Wednesday night the strong winds from the east prevailed, doing, however, very little damage. About 5 o'clock Thursday morning it increased, accompanied by rain. From that time the water began to rise and overflow the island, creeping up from one street to another, until, at noon, it had, on Tremont street, reached as high as Church street.
>
> The lower floors of the stores on the Strand were from two to four feet under water by noon, and goods and property damaged to the amount of near a quarter of a million dollars.
>
> The wharves were submerged by the waves at noon, and before, and vessels tied up to them were chafed and damaged, the bay being very rough. Lumber was floating all through the streets, and signs were more plentiful on the waters than on the houses. Hacks and drays were covered with passengers, hunting high ground. Merchants were hunting laborers to remove their goods from the lower floors to high places, and paying them from $4 to $6 per hour. Vast piles of salt along the sidewalks of the Strand melted away and returned to mother sea, leaving the sacks where they belonged. Sugar boxes in the warehouses, under cover of the flood, gave out their sweetness to the deluge. The great top-knot on the front of Sauter's building, about 10 o'clock in the morning, came down with a crash that startled the whole town. The third story of the new brick hotel, commenced by M. A. Thompson, was blown down on the Odd Fellows' Hall, crushing it to a shapeless mass of splintered boards and timbers, and later in the day the next lower story came down.
>
> The most of the slate roof of the Masonic Hall was torn off. Freedman's Hall was reported to have been blown down, and one man killed. The water submerged Mechanicsville, and did great damage to the property in that vicinity, which is very low ground, and covered with water in ordinary high tides. Messers. A. Sessums & Co., T. B. Stubbs & Co., Spooner & Co., and all the old storehouses in their vicinity were invaded, and to each one more or less damage done. The bayous were all out of bank, spreading water in places entirely across the island. The city railroad track was submerged, and the trains stopped running at an early hour.
>
> There was no mail or passenger train on the G. H. & H. railroad, and it was reported early in the day that the bay bridge had been washed down. Sheds and galleries and china trees were twisted off by the wind, and fences blown down in every part of town. A drenching rain accompanied the storm all day, preventing the people from coming out to tell their mishaps and report the damages. The yellow fever patients, the doctors say, will feel the change in the air in spite of every precaution, and advise us to look for increased mortality in consequence.
>
> The water on the Gulf beach was belly deep to a horse. The City railroad depot had its foundation so washed by the Gulf water that the blocks on one side gave way, and the building settled to the ground. The wind was much more violent on the Gulf shore than in the city. A good many of the buildings put up by the United States officers were blown down. The City Cemetery was covered with water a foot deep, and all the ground around the cemetery and to the westward, as far as could be seen, was a sea of water.
>
> (The summary of many ships damaged and sunk is omitted.)
>
> All the wharves are badly damaged . . . All the buildings on the block east of Hendley's wharf are blown down or washed away from their foundations.
>
> Several of our citizens, who have been all over the city, estimate the total damage at not less than a half a million dollars. Mr. Hawley, the wharf builder, estimates the damage to the wharves fully one fourth their value . . . Many of the tin roofs of the city have been blown off, and some few slate roofs damaged . . . The wharves at 4 P.M. yesterday presented the most completely wrecked and desolate appearance we ever witnessed . . . The water at midday had so inundated the gas works that it became impossible to build fires, and the city was without gaslight last night.[1]

The Smithsonian weather observer was on the spot and briefly noted the storm in his monthly report to Washington: "Terrible gale on the 3rd—water enters yard at 6 A.M.—wind from SE & changes NE at 1100 and north at 1600. Water over whole of the orchard. Trees and houses blown down & washed away. People drowned."[2]

Fortunately for our study, Prof. C. G. Forshey was also on the spot and has left a careful eyewitness account of the main meteorological events. Prof. Forshey seemed to have been a johnny-on-the-spot in weather matters as 27 years before he had been at Natchez to examine the track of the Great Tornado of 8 May 1840 and had written a valuable study of its nature and movement. His hurricane account appeared in full in the *Galveston News*:

STORM OF OCTOBER 3, 1867

BY PROF C. G. FORSHEY

Violent storms are common along the coast of the Mexican Gulf; and occasionally, at long intervals, very destructive. Hurricanes visit the Islands and coast of Texas, and Louisiana.

The great storms of —, 1837, of Sept. 17 and 18, 1854, and again the Last Island gale of 1856: and this final storm of Oct. 3, 1867, have been so terrific in their fury, as to be worthy a careful record of their phenomena.

I propose to collate the history of the storm that has just swept our city and Island, while the information is fresh; and I respectfully ask communications from all sources, in answer to the following questions:

1. Where did the storm begin?
2. Where did it abate?
3. What was the course of the wind at all hours during the storm?
4. The movements of the Thermometer?
5. The movements of the Barometer?
6. The heights of the water?
7. The violence or force of wind?

I herewith submit the observations and minutes I have made of the storm at this place.

APPROACH.

During the afternoon of Wednesday the heat of the sun became sickening and fierce; the perspiration flowed, without effort, and the air was so humid as to prevent the cooling effect of evaporation. Frequent weeping showers swept over the city, from the fleecy clouds; but this lasted only a minute. The tide was very high in the evening, and as I learn since, the Barometer fell quite low. But no registry was kept.

During the night the same continued, the thermometer standing at 80°.

THURSDAY, 3D.

ARRIVAL.—Before dawn the windstorm was upon us, blowing with vigorous blasts from the east, the sea beating with tremendous force upon the east end of the Island. At 7 A. M., wind 15° N. of East. This direction is slightly oblique to the coast line, and will drive water into the Bay. Temperature 80°.

Barometer down to 30 inches. The whole heavens are portentous, and the fury of the gale increases as the rain sets in.

PROGRESS.

8 A. M.—Wind, rain and temperature unchanged. Trees and fences giving way, and slender buildings falling.

9 A. M.—Drifting rain renders objects invisible at a distance. Think we can see ship masts (at times) out of their place. Wind increases, and water stands without draining off.

10¼ A. M.—No abatement of violence. The wind has prodigious force. The sea appears nearly all over town.

Looking west and north—cannot see far.

(What a morning for funerals!)

My neighbor, a young maiden, the mirth of her family, has expired in the fury of this storm. To bury now is impossible, except in a sea of water. The pestilence cannot pause for the tempest! (Have they combined for the utter desolation of our fair city?)

There is no human form on these streets. Thermometer stands 81°.

Wind—full 30° N. of East.

11—The fury has not abated, though the rain has subsided in a great measure. Wind sensibly veering.

11:15—Gusts increase in violence, with longer intervals. Wind N. 45°. Clouds thin and drifting with immense velocity. Trees and fences falling—trees cannot rise, the ground softening with water.

12:15—Fury increases—rain still drifting in spray, and water rising as wind passes round towards N.

Thermometer 80°.

Barometer 29 in.

Wind N. 60° E.

News comes of great destruction downtown—Sauter's Store, Odd Fellows' Hall, Freedmen's Hall, Thompson's Hotel. Water in all the stores on the Strand.

1 P.M.—Wind force unabated. Rain increases. A man can scarce keep his feet in the wind. Must try to measure velocity of gusts. Have no watch. Clock ticks twice a second. My pulse is 65 per minute by clock. In a little over two pulsations the gust runs 60 feet, the length of the fence opposite. Count repeatedly, and find it a near approximation to 8–10 mile per minute, or 70 feet per second. This is prodigious velocity.

MAXIMUM VIOLENCE OF HURRICANE.

At 1H. 35M., P.M.—Wind due NE.

2 P.M.—Rain slackens, sky lightens up, but wind gusts are terrific; house shakes and trembles. In fact there seems a deep roar beneath the dashing, splashing and crackling noises round us.

Dogs come in cowering; horses seek human protection; whole storm ominous; Heavens filled with portent; the earth seems to tremble; air filled with fragments and trash; temperature stubbornly 80°. This is a genuine tropical hurricane, but the duration of the N.E. wind is inexplicable.

3 P.M.—The water passes over the ridge on 28th street, and the brackish bay water covers all west and north of us. Land appears still along the ridge east on Broadway. Rain and wind N. N. E., slightly moderated.

3½ P.M.—Wind due north. No rain, but dark, and wind furious. Temperature still 80°.

4 P.M.—Wind E. of N. Scud clouds flying very low; looks like a blustering norther, but fiercer and quite too warm.

Water begins to recede here, after falling near two hours on the Strand. This cannot last long, for the wind has at no time been in a direction to fill the bay. I believe it has run out nearly as fast as in, and that West Bay will feel it most. The head of this bay is scarcely affected, I predict.

5 P.M.—North wind, still violent and hot; never below 80 deg.; must explore if my pony can breast the storm.

Found the water fallen one foot on Strand, but it, and all the streets near it, covered with wreck; nearly all the shipping destroyed or damaged; wharves all gone, but Central; many vessels away up streets and out on the prairies, wrecked or damaged.

Stores on Strand all under water, and the damage done to goods and buildings incalculable.

It was difficult, and somewhat dangerous to reach the points I attempted, for the salt spray cut my face, and blinded me, and nearly capsized me and horse in the flood. The bay and harbor looked mad with turbulence and destruction.

* * * * * * * *

Darkness closed upon me in the flood, and the storm raved on.

8 P.M.—Storm slightly abating; wind still N. N. E., and warm as ever—80 deg. Clouds run from N.E., and have the same great velocity.

Wind nothing like a "Norther"; it seems to veer round further east again.

At 9 and 10 wind subsiding a little, but not changing course much.

Oct. 4.—The morning wind is gentle and balmy, the air clear, the waters entirely subsided, and all pretence of violence is ignored by the Heavens.

* * * * *

During the whole day the wind is N. and N. NW.; gentle but *very* warm.

At 7 P.M., the thermometer is again at 80°; perfectly clear; and at 11 P.M. the wind continues North, but has heated up to 85 deg.!

This is nearly unparalleled, and gives some threat of further storm. Wind hurried; high dew point. But more of this anon.

Highest water on Strand 9½ inches above mark of 1854.

WATER HEIGHTS.—In 1854, I made a high-water mark on Hendley's building, Menard & Co., and by information then furnished me by Michael Menard this mark was near 1 foot above the high water of 1837.

PERMANENT REGISTER.—I have driven two nails in the west wall of Kahn's building. Inside, beneath the stairs, 13 inches above the floor, and with marking-ink inscribed along a black line "H. W. Oct. 3, 1867." Let this be remembered.

The floor of Hendley's block was three inches above the flood.

The floor of Osterman & Wood's iron-front buildings, on Strand, between 22d and Tremont, was two inches under water.

The water was at no time level; so many were the obstructions to its flow, and so short was the duration of the highest water. This was true all over the city. Hence, marks are valuable as levels only near the Bay—say, on the Strand.

As anticipated, the water was not very high at the head of the Bay, the wind not favoring it as in 1854. Then there was S.E. wind nearly two days, and the wind passed round by South to West and N.W. because the center of violence passed north of the city. [Forshey in error here.]

No storm was felt in Houston or at Lynchburg—but blustering and showery wind from N.E.

At Indianola no violent storm, but high wind from North and Northwest, yet with sudden and violent gusts from N.E.

INFERENCE—It is already manifest from these facts that the center of violence was South of us, and that we are to hear of this storm along the Gulf and Atlantic coast—at Savannah and Charleston 24 to 30 hours later than here.

These hurricanes all gyrate, and the wind blows a tangent to the circle. They all travel, at this latitude, from S.W. to N.E., sometimes expiring in the Gulf, but oftener extending to the coast of Labrador; sometimes to Norway and Iceland, keeping parallel to, or along, the Gulf Stream.

P.S.—Without knowing anything of the result, I am prepared to hear that the inner bar entering the harbor has been deepened and widened materially; so long did the wind drive in a direction to favor that result, and so violent were the water currents into the harbor.

For the same reason the channels of the west Bay ought to be benefited; as the winds bore down directly on the Bay nearly all day. Hence the higher waters in that direction, and the destruction of the railroad bridge.[8]

The extent of this hurricane has not been traced adequately. W. Armstrong Price in his study of hurricane and flooding of the Texas coast stated that both the old towns at the mouth of the Rio Grande river—Bagdad in Mexico and Clarksville in Texas—were destroyed by the 1867 hurricane.[4] To the northeast the storm center must have passed close to South West Pass at the mouth of the Mississippi River. The lighthouse there was almost completely destroyed by the wind, though fortunately the keeper was rescued by a passing steamer.[5] Six houses at Pilottown were demolished.[6]

New Orleans felt the gale on the 4th: "the wind continued to increase until 1630 when it blew a heavy gale accompanied by torrents of rain. Towards sunrise 5th the gale abated." The highest winds came from the east-northeast as the storm passed to the southward, probably very close to the coast. Lake Pontchartrain, as usual with such a wind, rose very high in the rear parts of the city.[7] Down at Port à la Hache the same conditions prevailed with the Gulf waters driven over the plantations. The quarters of the troops at Fort St. Philip were blown down and the parade ground flooded.[8]

[1] *Galveston News,* 4 Oct 1867.
[2] SI.
[3] *Galveston News,* 5 Oct 1867.
[4] W. Armstrong Price. *Hurricanes Affecting the Coast of Texas from Galveston to Rio Grande.* Washington?, 1956. A-3.
[5] *Price Current* (New Orleans), 16 Oct 1867.
[6] *Daily Crescent* (New Orleans), 8 Oct 1867.
[7] *Idem.*
[8] *Idem.*

THE LOWER TEXAS COAST HURRICANE OF 1869

REFUGIO

I have been through all the storms which struck this county in my time. The worst of them were those of 1869, 1875, 1886, 1887, 1916, and 1919. The 1869 storm came on August 16. It was one of the worst. It did considerable damage along the coast at Rockport and St. Mary's. There was considerable damage at Refugio. Several houses were blown off their foundations. One was Sheriff Billingsley's. The roof of the Catholic Church was blown off, and the town was pretty well messed up.[1]

Mrs. L. B. Randolph had opened a private school in the Joe Dugat house and lived there. The young Russell children, including myself, attended this school. The roof of this house was blown away and all our school books got wet.

There was some, but not serious, injury to the wharf installations.

Our family was at St. Mary's during the 1869 storm. My father was at Refugio attending court at the time the storm came up and my mother and the younger children were alone in our big house. The storm appeared about dark and continued through the night, abating about daybreak. The wind was so strong and loud that you couldn't hear one another speak without shouting. The rain poured down all night.

The people were afraid their homes would blow down and crush them beneath the wreckage. When roofs began to blow off and chimneys began to crumble, many people quit their houses and took to the open, seeking shelter from the wind in gullies, which did not have too much water running in them. I remember the Howard family as having spent the night in the storm. These people remained in the storm all night and got thoroughly drenched, but the banks of the gullies protected them from the howling wind.

Our family went out into the storm and stayed long enough to get drenched, but, after seeing that our roof did not blow off, we decided the house to be safest, and went back into it. No harm came to us. Our house had been built differently from most, inasmuch as the gable faced the bay instead of the length of the roof.

The roofs of practically all houses along the beach were blown off. The two-story Presbyterian church and school was blown flat. The front porch and shed room of the Dr. Carpenter house were torn off and blown onto the prairie. Emil A. Perrenot was lodging with Mrs. Carpenter and had the shed room. He was in it when the storm blew it away. Aside from sticking a nail in his foot he escaped injury. The chimney of this house crashed in and crushed the cradle in which young Eugene Low had been just a moment before. His mother happened to take him out just before the chimney fell.[2]

INDIANOLA

Last Monday night our city was visited by a severe gale of wind which increased toward 8 p.m. to almost hurricane in its fury, blowing in fitful gusts at times that was almost terrific. Several houses were blown from their foundations, but without serious damage. The Episcopal Church was almost demolished, being crushed in by the force of the wind tumbling over the tower and otherwise receiving serious damage. The large building of Mr. Cassimer Villeneuve on the corner of Main and Kaufman streets was considerably injured. The tide was very high, and the water on some of the back streets was over a foot deep.

The damage done to the shipping was not as serious as first apprehended. . . .

The numerous coasting vessels in the bayou met with little or no loss, being well sheltered from the storm. A few rubs or bruises, easily repaired, were the only consequence.

The total loss was trifling when compared with the violence of the gale, amounting in the aggregate to more than eight or ten thousand dollars.

No lives were lost, nor was anyone injured. We may consider ourselves fortunate in coming out of it with so trifling a loss.

We learn that the storm did some damage at Saluria, blowing down houses and scattering furniture in every direction. The gale was the most violent that we ever experienced on this coast during a twenty years' residence. The damage done to shipping and property is comparatively small.[3]

PORT LAVACA

On Monday last about 10 o'clock A.M. the tide was observed to increase in volume notwithstanding a prevalent north wind; the heavens were overcast—the atmosphere was humid and variable in temperature. About dusk the wind rapidly increased in volume from the North, and ripened into a gale. The tide was observed to increase with great rapidity; about 11 o'clock the wind chopped around to the east and increased its fury. This brought in one of the highest tides witnessed in this bay for some years past. Soon after midnight, the wind veered somewhat to the South, and appeared to gyrate in resistless gusts. The storm reached its climacteric about 2 o'clock A.M., but did not fully abate before daylight.

The casualties, to our marine were inconsiderable, in view of the magnitude of the storm.[4]

ROCKPORT AND CORPUS CHRISTI

From Rockport and Corpus Christi we learn that the storm was severely felt, though damage to landed property and shipping was inconsiderable—being confined to the destruction of a few small bay craft and the unroofing and removing from their foundations of a few small buildings.[5]

[1] William Louis Rea. Ms. Memoirs. Hobart Huson. *A Comprehensive History of Refugio County from Aboriginal Times to 1955.* Woodsboro, Texas, 1955. II, 246.

[2] Sallie J. Burmeister. Reminiscences in Refugio. *Ibid.*, 143–144.

[3] *Indianola Bulletin,* 19 Aug, in *The Galveston Tri-Weekly News,* 23 Aug 1869.

[4] *Lavaca Commercial,* 18 Aug, *idem.*

[5] *Idem.*

THE MOBILE STORM OF JULY 1870

The wind which had hung for some time in the east, got around to the southerly quarter about 12 o'clock today and began to blow a full gale with rain. The water, already nearly up to the wharves rose rapidly, soon covering them, and the boats which were fast to them immediately felt the influence of the wind and storm.

At half past one, the wind, which slackened, is increasing from the southeast, and the alarm is ringing again. The water is now fairly over Water Street, and it is difficult to learn any particulars from "the front."

Fifteen minutes to two—The dry dock has just drifted up past St. Michael St.

The Girard house is partially unroofed.[1]

The storm was brief in duration, commencing in all its reality just about 12, and being over before 2 o'clock. All direct damage occurred almost at the commencement, and the whole, including that from collision of drifting craft, was probably accomplished in the space of two hours. By half past two or sooner, the water began to subside, and was soon below the wharves. At no time was it many inches deep in any of the Front Street stores, barely reaching the floor of the Trade Company's and Mobile and Montgomery railroad office.[2]

The *Daily Register* in another column added some facts. The wind set in from east by south, then shifted to south. The barometer was reported to have fallen two inches in two hours to a low of 27.50 inches (? ? ?). The water rose higher than any time since 1860. Along the shore of the bay considerable damage occurred from high water, especially to the Shell Road which was cut through in several places.[3]

[1] *Evening Register,* 30 July, in *Daily Register* (Mobile), 31 July 1870.

[2] *Daily Register,* 31 July 1870.

[3] *Daily Register,* 31 July and 2 Aug 1870.

THE TWIN KEY WEST HURRICANES OF OCTOBER 1870

Two major hurricanes crossed the Isle of Cuba in October 1870 and passed close enough to Key West to give all the line of keys a severe lashing, though no major structural damage occurred and there were no unusual inundations. The first passed about 30 miles east of Havana on 7 October, moved on a northerly track, and passed up the East Coast of Florida.[1] Key West experienced hurricane force winds for four successive days from northeast to north as the storm moved very slowly during recurvature.[2]

The second storm moved about 80 miles west of Havana on 19 October. The center passed over Pinar del Rio where a 15 minute calm was observed and the barometer sank to 29.32 inches.[3] It moved west of Key West where hurricane force winds from the southeast were experienced for a short period during the morning of the 20th.[4]

KEY WEST

A heavy gale was raging here (commencing the 8th inst. and ending the 11th) which caused considerable damage to the shipping. The wind was from the northeast and upon moderating it turned to NW.[5]

	0700	1400	2100
Oct 8	NE 8	NE 8	NE 10
9	NE 9	NE 9	NE 10
10	NE 9	N 9	N 10
11	N 9	N 10	N 8
12	NW 4	NW 3	NW 5

In the night from the 19th to 20th the wind commenced to blow very heavy from the SE and turned into a gale by about 1100 on the 20th which lasted all day, commenced to moderate during night when gale was heaviest the wind blew from the SE. This last gale also did a great deal of damage to Buildings of the Island and shipping.[6]

	0700	1400	2100
Oct 19	E 1	SE 1	SE 6
20	SE 10	SW 9	N 7
21	N 4	N 2	NE 2

[1] Vines, B. Huracanos del 7 y 19 Octubre 1870. Habana Observatorio magnetico y meteorologico. *Observaciones 1869–70*. Havana, 1870. 6.
[2] SI.
[3] Vines, 14.
[4] SI.
[5] SI.
[6] SI.

A BIBLIOGRAPHY OF EARLY AMERICAN HURRICANES TO 1870

The following list of 125 references is limited to: (1) published accounts of hurricanes which have crossed or closely approached the coastline of the United States, (2) books or articles on the subject of storm dynamics by Americans or by others if the material was discussed in periodicals and affected the thinking of American meteorologists, and (3) any works on the subject published by an American press. Only material published prior to 1870 is included.

1610 (1625)

Strachey, William. *A true reportory of the wracke, and redemption of Sir Thomas Gates, Knight; upon, and from the Ilands of the Bermudas: his coming to Virginia, and the estate of that Colonie then, and after, under the government of Lord La Warre, July 15, 1610.* London, 1625.

1635 (1684)

Thatcher, Anthony. Narrative of his shipwreck. First published in Increase Mather. *Remarkable Providences.* Boston, Joseph Browning, 1684.

1638

Taylor, John. *Newes and strange newes from St. Christophers of a tempestuous spirit which is called by the Indians Hurry-Cano or whirlewind.* (5 August 1638.)

1649 (1790)

Winthrop, John. *Winthrop's Journal. "History of New England."* James K. Hosmer, ed. New York, Charles Scribner, 1908. 2 vols. First edition appeared in 1790.

1667

Strange news from Virginia, being a true relation of a great tempest in Virginia, by which many people lost their lives, great number of cattle destroyed, houses, and in many places whole plantations overturned, and whole woods torn up by the roots. As a further addition to this calamity, the sea exceeded its usual height above twelve foot, overflowing all the plain country, carrying away much corn and tobacco, with many cattle, forcing the inhabitants into the mountains for the security of their lives. London, W. Thackeray, 1667. 7 p.

1684

Mather, Increase. *An essay for the recording of illustrious providences: wherein an account is given of many remarkable and very memorable events, which have hapned this last age, especially in New England.* Boston, Joseph Browning, 1684.

1694 (1697)

Scarburgh, —. Extract of a letter from Mr. Scarburgh, dated Acomack, July 14, 1694, giving a relation of the effects of a violent storm of the 19th October, on the rivers of that country. (Communicated by the Honorable Sir Robert Southwell, F.R.S.) Royal Society, *Philosophical Transactions.* 19–231, 1697, 659.

1698

Langford, Capt. Observations of his own experiences upon hurricanes and their prognosticks. Royal Society, *Philosophical Transactions.* 20, 1698, 407.

1703

Dampier, William. *A new voyage round the world.* 2 v. London, J. Knapton, 1703–05.

1728

Laval, Antoine Francois. *Voyage de la Louisiane fait par ordre du roi en l'année mil sept vingt: dans lequel sont traites diverses matieries de physique, astronomie, geographie et marine.* Paris, J. Mariette, 1728. 304 p.

1731

Catesby, Mark. *Natural History of Carolina, Florida, and the Bahama Islands.* London, 1731–43.

1733

Robertson, Rev. —. *A short account of the hurricane, that pass'd thro' the English Leeward Caribee Islands, on Saturday the 30th of June 1733.* London, 1733. 20 p.

1747 (1822)

Franklin, Benjamin. Letter to Jared Eliot. July 16, 1747. *The Papers of Benjamin Franklin.* L. W. Larabee, ed. New Haven, Yale, 1961. 3, 149. First publication: *American Journal of Science and the Arts* (New Haven). 4–2, 1822, 361–362.

1749

Evans, Lewis. *Map of Pensilvania, New-Jersey, New-York, and the Three Delaware Counties.* Phila., 1749. First published account of Franklin's northeast storm hypothesis.

1750 (1822)

Franklin, Benjamin. Letter to Jared Eliot. Philadelphia, Feb. 13, 1750. *The Papers of Benjamin Franklin.* L. W. Larabee, ed. New Haven, Yale Univ. Press, 1961. 3, 463. First publication: *American Jour. Sci.* 4–2, 1822, 364–365.

1760 (1904)

Franklin, Benjamin. Letter to Alexander Small of London. 12 May 1760. *The Works of Benjamin Franklin.*

John Bigelow, ed. New York, G. P. Putnams, 1904. 3, 260–263.

1763 (1770)

Milligen-Johnston, George. *A short description of the province of South Carolina, with an account of the air, weather and diseases at Charleston.* (Written in the year 1763.) London, John Hinton, 1770. 96 p.

1765

Franklin, Benjamin. Physical and meteorological observations, conjectures and suppositions. Royal Society, *Philosophical Transactions*. 55, 1765, 182–192.

1766 (1942)

Bartram, John. Diary of a Journey through the Carolinas, Georgia, and Florida from July 1, 1765 to April 10, 1766. Francis Harper, ed. *Trans. Amer. Phil. Soc.* n.s. 33–1, 1942, 20.

1772

Knox, Hugh. *Discourse, delivered on the 6th of September, 1772, in the Dutch Church of St. Croix. On occasion of the hurricane which happened on the 31st day of August; being the most dreadful known among these Islands, since their first settlement.* St. Croix, Daniel Thibou, 1772.

1773 (1786)

Perkins, John. Conjectures concerning wind and waterspouts, tornadoes and hurricanes. *Trans. Amer. Phil. Soc.* 2, 1786, 335–347. Submitted in 1773.

1776

Chalmers, Lionel. *An account of the weather and diseases of South-Carolina.* London, Dilly, 1776.

Romans, Bernard. *A concise natural history of East and West Florida.* New York, 1776. 4–8.

1781

Fowler, John. *A general account of the calamities occasioned by the late tremendous hurricanes and earthquakes in the West-India Islands, foreign as well as domestic: with petitions to and resolutions of, the House of Commons, in behalf of the sufferers at Jamaica and Barbados. Also a list of the committee appointed to manage the subscriptions of the benevolent public, towards their further relief. Carefully collated from authentic papers, by Mr. Fowler.* London, J. Stockdale, 1781.

1784

Touch, P. *Thanksgiving sermon, preached at St. Lucia, the Sunday after the hurricane in October 1780 on board His Majesty's Ship Vengeance, Capt. Holloway; and before Commordore Hotham.* London, 1784. 35 p.

1801

Capper, James. *Observations on the winds and monsoons; illustrated with a chart, and accompanied with notes, geographical and meteorological.* London, J. Debrett, 1801. 234 p.

1801 (1804)

Dunbar, William. Meteorological Observations, made by William Dunbar, Esq., at the Forest, four miles east of the River Mississippi . . . for the year 1800; with remarks on the state of winds, weather, vegetation, &c. calculated to give some idea of the climate of that country. (Natchez, Aug. 22 1801. Read Dec. 18, 1801.) *Transactions of the American Philosophical Society.* Philadelphia, 1804. 6–1, 43–55.

1804

Violent Storm. Boston, Oct 1804. Broadside. Also contains: A Poem on the Late Hurricane.

1805

Carpenter, S. C. Domestic occurrences: Particular account of the storm which happened on the coast of South Carolina, in September 1804. *Monthly Register, Magazine, and Review of the United States* (Charleston). 1–1, Jan 1805, 33–38.

Carpenter, S. C. Domestic occurrences: Retrospective account of the tremendous storm in October 1804. *Monthly Register, Magazine, and Review of the United States* (Charleston). 1–2, Feb 1805, 75–79.

Harris, Tucker. Some account of the great hurricane of 1804. *Journal of Medical and Physical Science* (Philadelphia). 1805, 53–58.

Mitchill, Samuel Latham. Some particulars of a terrible hurricane. *Medical Repository* (New York). 8–4, 1804–05, 354–365.

Ramsay, David. Facts concerning the yellow fever which prevailed in Charleston (South-Carolina), during the hot season of 1804; with observations on the damage done to the crops by the hurricane and insects. *Medical Repository* (New York). 8–4, 1804–05, 365–367.

1809

Burt, —. *Narrative of proceedings on board His Majesty's Ship Theseus, Sept. 4–15, 1804; being an account of a hurricane which she encountered in the Atlantic Ocean.*

Ramsay, David. The natural history of South Carolina. *History of South Carolina.* Newberry, S. C., E. J. Duffie, 1858. 2d ed. 175–184.

1817

Mitchill, Samuel Latham. Intelligence: The storm of September 23, 1815. *Medical Repository* (New York). 18–2, 1817, 190–192.

1819

Beck, John B. Observations on salt storms. *American Journal of Science and Arts* (New Haven). 1, 1819, 389.

1821

Farrar, John. An account of the violent and destructive storm of the 23d of September 1815. *Memoirs of the American Academy of Arts and Sciences.* Boston, 1821. 4–1, 92–97.

Remarkable storm of wind and rain on Oct. 9, 1804 that resembled that of Sept. 23, 1815, but was not so severe. *Memoirs of the American Academy of Arts and Sciences.* Boston, 1821. 4–1, 96.

1822

Moreau de Jonnes, M. *Histoire physique des Antilles Francaises.* Paris, 1822.

1823

Goodwin, Ezra S. Notices, by Rev. Ezra S. Goodwin of Sandwich, of the effects in that vicinity of the Great Storm of 23 September, 1815. *Collections of the Massachusetts Historical Society.* 2d ser., 10, 45–54; *Boston Journal.* 1, 1823, 370–377.

1828

Dove, Heinrich Wilhelm. Ueber barometrische minima. *Annalen der Physik und Chemie* (Leipzig). 13, 1828, 596–613.

1831

Mitchell, Elisha. On the proximate causes of certain wind and storms. *Amer. Jour. Sci.* 19, April 1831, 248–292.

Redfield, William C. Remarks on the prevailing storms of the Atlantic Coast, of the North American States. *Amer. Jour. Sci.* 20, July 1831, 17–51.

Redfield, W. C. Note on the hurricane of August, 1831. *N. Y. Journal of Commerce,* 28 Sept 1831; *Amer. Jour. Sci.* 21, Oct 1831, 191.

1833

Espy, James Pollard. Meteorological remarks—aurora—theory of waterspouts. *Journal of the Franklin Institute* (Philadelphia). n.s. 12, Nov 1833, 292–296.

Redfield, W. C. Observations on the hurricanes and storms of the West Indies and the coast of the United States. *Blunt's American Coast Pilot.* 12th ed., 1833, 626.

1834

Redfield, W. C. Observations on the hurricanes and storms of the West Indies and the coast of the United States. *Amer. Jour. Sci.* 25, Jan 1834, 114–121. (Revised from *Blunt's American Coast Pilot,* 12th ed.)

1835

Redfield, W. C. On the evidence of certain phenomena in tides and meteorology, in reply to an Observer [Mr. Espy]. *Jour. Franklin Institute.* 25, June 1835, 372–379; *Amer. Jour. Sci.* 28, Oct 1835, 310–318.

1836

Espy, J. P. Essays on meteorology. Examination of Hutton's, Redfield's and Olmsted's theories. *Jour. Franklin Institute.* n.s. 18, Aug 1836, 100–108.

Espy, J. P. Essays on meteorology, No. IV: North east storms, volcanoes, and columnar clouds. *Jour. Franklin Institute.* n.s. 18, Oct 1836, 239–246.

Redfield, W. C. On the gales and hurricanes of the Western Atlantic. *London Nautical Magazine.* 5, April 1836, 199; *U. S. Naval Magazine.* 1, July 1836, 301; *Amer. Jour. Sci.* 31, Jan 1837, 115–130.

1837

Hare, Robert. On the causes of tornadoes or waterspouts. *Trans. Amer. Phil. Soc.* 5, 1837, 375.

Redfield, W. C. Mr. Redfield in reply to Mr. Espy on the whirlwind character of certain storms. *Jour. Franklin Institute.* n.s. 19, Feb 1837, 112–127.

Redfield, W. C. Remarks on the supposed connection of the Gulf Stream with opposite currents, on the coast of the United States. *U. S. Naval Magazine.* 2, May 1837, 243; *Jour. Franklin Institute.* 19, May 1837, 384–387; *Blunt's American Coast Pilot,* 13th ed. June 1837, 666; *Amer. Jour. Sci.* 32, Oct 1837, 349–354.

White, Capt. Carleton. *Narrative of the loss of the Steam-Packet Home.* New York, 1837.

1838

Arago, Francois. Opinion de M. Arago sur la theorie d'Espy, Bache, Reif, et Redfield. *Compte Rendue de L'Academie des Sciences* (Paris). 7, 1838, 708.

Reid, Lt. Col. William. *An attempt to develop The Law of Storms by means of facts, arranged according to place and time; and hence to point out a cause for the variable winds, with a view to practical use in navigation.* Illustrated by charts and woodcuts. 1st ed. London, John Weale, 1838. 431 p; 2d ed. 1841; 3d ed. 1850.

1839

Reid, Lt. Col. William. Redfield's law of storms: notice of Col. Reid's work on hurricanes. *Amer. Jour. Sci.* 34, Jan 1839, 182–184.

Espy, J. P. Examination of Col. Reid's work on the Law of Storms. *Jour. Franklin Institute.* n.s. 23, Jan 1839, 38–50, 149–158, 217–231, 289–298.

Espy, J. P. Facts collected by Mr. Espy, taken from newspapers of the time. (Storm of Sept 3, 1821.) *Jour. Franklin Institute.* n.s. 23, Mar 1839, 156–158.

Redfield, W. C. On the courses of hurricanes, with notices of the typhoons of the China Sea and other storms. *Amer. Jour. Sci.* 35, Apr 1839, 201–223; *London Nautical Magazine.* Jan 1839, 1.

Redfield, W. C. Remarks on Mr. Espy's theory of centripetal storms, especially his positions relative to the storm of Sept 3d, 1821, with some notice of the fallacies which appear in his examination of other storms. *Jour. Franklin Institute.* n.s. 23, May, June 1839, 323–336, 363–378.

Redfield, W. C. Additional facts relating to Raleigh's Typhoon of August 5th and 6th, 1835, in the China Sea. *Amer. Jour. Sci.* 36, July 1839, 59–60; *London Nautical Magazine.* 1839, 461.

Redfield, W. C. Further notice of the fallacies of Mr. Espy's examination of storms. *Jour. Franklin Institute.* n.s. 24, July 1839, 1–4.

1840

Redfield, W. C. Account of the circular storm of Dec. 1839. *N. Y. Journal of Commerce,* 6 Jan 1840; *London Nautical Magazine.* 1840, 424.

1841

Babinet, Jacques. Expose de la theorie de M. Espy sur les ouragans. *Annales des Chemie et Physique.* 3d ser., 1, 372; *Comptes Rendue.* 12, 1841.

Espy, J. P. Examination of Reid's, Piddington's, and Loomis's storms. *The philosophy of storms.* Boston, Little and Brown, 1841. 188–293.

Hare, Robert. Objections to Mr. Redfield's theory of storms, with some strictures on his reasoning. *Amer. Jour. Sci.* 41, Oct 1841, 140–147.

Redfield, W. C. Observations on the storm of Dec. 15th, 1839, with a map showing the direction of the wind at noon of that day, as observed at various places. (Read before the Amer. Phil. Soc. Jan. 15, 1841.) *London, Edinburgh and Dublin Philosophical Magazine.* 17, July 1841, 17; *Amer. Jour. Sci.* 42, Jan 1842, 112–119; *Trans. Amer. Phil. Soc.* n.s. 8, May 1843, 77–82.

Reid, William. *The progress of the development of the law of storms and of variable winds, with the practical application of the subject to navigation.* London, J. Weale, 1841. 424 p.

1842

Darling, Noyes. Notice of a hurricane that passed over New England in September, 1815. *Amer. Jour. Sci.* 42, Jan 1842, 243–252.

Hare, Robert. Strictures on Prof. Dove's essay "On the Law of Storms." *Amer. Jour. Sci.* 44, Oct 1842, 137–146.

Redfield, W. C. Reply to Dr. Hare's objections to the whirlwind theory of storms. *Amer. Jour. Sci.* 42, April 1842, 299–316; *Lond., Ed., and Dubl. Phil. Mag.* 20, May 1842, 353.

Redfield, W. C. Reply to Dr. Hare's further objections to whirlwind storms, with some evidence of the whirling action of the Providence Tornado of August, 1838. *Amer. Jour. Sci.* 43, July 1842, 250–278; latter portion of article in *Amer. Reportory of Arts and Sciences.* 3, March 1841, 81; *Jour. Franklin Institute.* n.s. 4, Oct 1842, 252; first portion in *Lond., Ed., and Dubl. Phil. Mag.* 21, Dec 1842, Suppl. 481.

1843

Dove, H. W. On the Law of Storms (from Poggendorff's *Annalen der Physick und Chemie,* 1841). *Amer. Jour. Sci.* 44, Jan 1843, 315–339.

Hare, Robert. Additional objections to Redfield's theory of storms. *London Phil. Magazine.* 23, Aug 1843, 92.

Redfield, W. C. Notice of Dr. Hare's strictures of Prof. Dove's Essay on the Law of Storms. *Amer. Jour. Sci.* 44, April 1843, 384–392; *Jour. Franklin Institute.* 3d ser., 1842, 252–264.

Redfield, W. C. Remarks on the tides and prevailing currents of the ocean and atmosphere. (Read before the Amer. Phil. Soc. at centennial meeting, May 27th, 1843.) *Proc. Amer. Phil. Soc.* 3, May 1843, 86–89; *Amer. Jour. Sci.* 45, July 1843, 293–309; *London Nautical Magazine.* 1843, 655.

Tracy, Charles. On the rotary action of storms. *Amer. Jour. Sci.* 45, Apr 1843, 65–72; reprint: *Smithsonian Miscellaneous Collections.* Washington, D. C. 51–4, 1910, 16–22.

1844

Resumen de los desastres ocurrides en la puerto de la Habana y sus jurisdicciones immediates, del Departamento Occidental de la isla Cuba. Dias 4 y 5 de Octubre de 1844. Habana, Impr. de Vidal y compania, 1844? 20 p.

1846

Herrera y Cabrera, Desiderio. Memoria sobre los huracanes en la isla de Cuba. *Memorias de la Societa Economica de la Habana.* 1, Jan 1846, 56–101; also issued as a separate: Habana, Impr. de Barcona, 1847. 72 p.

Redfield, W. C. Effects of the earth's rotation upon falling bodies, and upon the atmosphere. *Amer. Jour. Sci.* 2d ser., 3, March 1847, 283–284.

Redfield, W. C. On three several hurricanes of the American seas, and their relation to Northers, so-called, of the Gulf of Mexico and the Bay of Honduras, with charts illustrating the same. *Amer. Jour. Sci.* 2d ser., 1, 1846, 1–16, 153–169, 333–367, and 2, 162–187, 311–334. Reprinted as a separate: New Haven, B. L. Hamlen, 1846. 117 p.

1847

Peirce, Charles. *A meteorological account of the weather in Philadelphia from Jan. 1, 1790 to Jan. 1, 1847.* Philadelphia, Lindsay & Blakiston, 1847. 208 p.

1848

Evans, Lt. "Stormy Jack." A chronological list of hurricanes which had occurred in the West Indies since the year 1493; with interesting descriptions. *London Nautical Magazine.* 1848, 397, 453, 524.

Piddington, Henry. *The sailor's horn-book for the law of storms.* New York, 1848.

Schomburgk, Robert H. *The history of Barbados.* London, 1848.

1849

Redfield, W. C. Remarks on a letter from R. N. Shufeldt of the U. S. Ship Marion to E. and G. Blunt, relative to a hurricane encountered by Ship *Marion,* Sept. 1848. *London Nautical Magazine.* 1849, 39.

1850

Hare, Robert. On the whirlwind theory of storms. *Proc. Amer. Assoc. Adv. Sci., 4th meeting, New Haven, 1850.* 231–242.

Johnston, Alexander Keith. Chronological table of the principal hurricanes which have occurred in the West Indies, within 150 years. *The Physical Atlas of Natural Phenomena.* Phila., 1850.

Redfield, W. C. The Law of Storms and its penalties for neglects. *N. Y. Journal of Commerce,* 19 June 1850; *London Nautical Magazine.* 1850, 470. *Bermuda Royal Gazette,* 16 July 1850; also *N. Y. Courier and Enquirer,* n.d.

Redfield, W. C. On the apparent necessity revising the received systems of dynamical meteorology. *Proc. Amer. Assoc. for the Advancement of Science, 4th meeting, New Haven.* 1850, 366–369.

Reid, William. *An attempt to develop the law of storms by means of facts, arranged according to place and time; and hence to point out a cause for the variable winds, with a view to practical use in navigation.* 3d ed. London, John Weale, 1850. 530 p.

Webster, Lt. J. D. Survey of the coast at the mouth of the Rio Grande. *Senate Executive Document No. 65,* 31st Congress, 1st Session, 14, July 17, 1850. 12 p.

1851

Maury, Matthew F. Circulation of the atmosphere. *Explanations and sailing directions.* 3d ed. Washington, C. Alexander, 1851. 42–57.

Maury, Matthew F. *Gales, typhoons, and tornadoes.* Washington, 1851.

1852

Hare, Robert. Questions relating to the theory of storms. *The Merchants' Magazine and Commercial Review* (New York). 27, July 1852, 191.

Piddington, Henry. *Conversations about hurricanes for the use of plain sailors.* London, Smith, Elder & Co., 1852. 109 p.

1854

Coffin, James H. On the winds of the northern hemisphere. *Smithsonian Contributions to Knowledge.* Washington, 1854. 6, 5–197.

Redfield, W. C. On the first hurricane of September, 1853, in the Atlantic, with a chart and notices of other storms (Cape Verde and Hatteras hurricane). *Amer. Jour. Sci.* n.s. 18, July, Sept 1854, 1, 176; *London Nautical Magazine.* Sept, Oct, Nov 1854, 466, 526, 597; reprinted as a separate: *Cape Verde and Hatteras Hurricane of Aug.–Sept. 1853, with a hurricane chart, and notices of various storms in the Atlantic and Pacific oceans north of the equator.* New Haven, B. L. Hamlen, 1854. 32 p.

1855

Maury, Matthew F. *The physical geography of the sea.* New York, Harper & Brothers, 1855. 287 p.

Poey y Aguirre, Andres. A chronological table comprising 400 cyclonic hurricanes which have occurred in the West Indies and in the North Atlantic within 362 years, from 1493 to 1855, with a bibliographical list of 450 authors, books, etc. and periodicals where some interesting accounts may be found, especially on the West and East Indian hurricanes. *The Journal of the Royal Geographical Society* (London). 25, 1855, 291–328.

Redfield, W. C. On the storm of October 7th, 1854, near the coast of Japan, and the conformity of its progression with other cyclones. *Proc. Amer. Assoc. Adv. Sci., 9th meeting, Providence, 1855.* 183.

1856

Ferrel, William. Essay on the winds and currents of the ocean. *Nashville Journal of Medicine and Surgery.* 12–4, 5, Oct, Nov 1856. Reprinted in *Professional Papers of the Signal Service, No. 12.* Washington, 1882.

Redfield, W. C. On the spirality of motion in whirlwinds and tornadoes. *Proc. Amer. Assoc. Adv. Sci., 10th meeting, Albany, 1856.* 212–214; *Amer. Jour. Sci.* 2d ser., 23, Jan 1857, 23–24.

Redfield, W. C. Observations in relation to the cyclones of the western Pacific, embraced in a communication to Commodore Perry. Report of United States Expedition to Japan. *House of Representatives Executive Document, No. 97. 2d Session, 33d Congress.* Washington, 1856. 337–359.

Sanchez-Layas, E. Huracan en las Antilles en Agosto, 1856. *Cron. Naval Esp.,* Madrid, 4, 1856, 129–155, 257–280, 396–415, 532–546.

History of Louisiana Hurricanes. *Daily Crescent* (New Orleans), 16 August 1856.

1857

Blodget, Lorin. List of Hurricanes on the Coast of the South Atlantic States, and on the North Coast of the Gulf of Mexico. *Climatology of the United States, and the Temperate Latitudes of the North American Continent.* Philadelphia, J. B. Lippincott, 1857. 397–403.

Dove, Heinrich Wilhelm. *Über das gesetz der stürme.* (On the law of storms.) Berlin, Dietrich Reimer, 1857. 115 p. Also *Das gesetz der stürme.* 3d ed. enl. Berlin, Dietrich Reimer, 1864. 346 p. An English translation of 2d edition appeared in London in 1862.

Redfield, W. C. Notes attached to a communication "On the avoidance of the violent portions of cyclones, with notices of a typhoon at the Bonin Islands," by John Rodgers, USN, and Anton Schonborn. *Amer. Jour. Sci.* 2d ser., 23, March 1857, 205–211.

Redfield, W. C. On the cyclones and typhoons of the North Pacific Ocean; with a chart showing their courses and progression. *Amer. Jour. Sci.* 2d ser., 24, July 1857, 21–38.

1861

Bell, John H. *A practical illustration of the movements of hurricanes with plain directions how they may be avoided.* Baltimore, 1861. 8 p.

1862

Poey y Aguirre, Andres. *Table chronologique de quatre cents cyclones que ont sevi dans les Indes Occidentales and dans l'Ocean Atlantique Nord pendant un intervalle de 362 annees, depuis 1493 jusqu'en 1855.* Paris, Paul Dupont, 1862. 49 p. A translation of the 1855 work.

1863

History of Louisiana Hurricanes. *Daily Picayune* (New Orleans). 13 & 27 September 1865.

1865

"Gleaner." History of Hurricanes. Letter from Plaquemines. *Daily Picayune* (New Orleans). 16 July 1865.

1866

Blunt, George W. *The way to avoid the violent center of our gales.* New York, 1866. 31 p.

Poey y Aguirre, Andres. Bibliographie cyclonique: catalogue comprenant 1,008 ouvrages, brochures et ecrits qui ont paru jusqu'a ce jour sur les ouragans et les tempetes cycloniques. (Cyclone bibliography: a catalog of 1,008 works, brochures, and writings which have appeared up to now on hurricanes and cyclonic storms.) 2d ed. corrected and enlarged. Paris, Paul Dupont, 1866. 96 p. An extract from *Annales Hydrographiques* (Paris), 1865.

1867

Forshey, C. G. Storm of October 3, 1867. *The Galveston News.* 4 Oct 1867.

1868

Eastman, J. R. *Discussion of the West Indian Cyclone of October 29 and 30, 1867.* Washington, 1868.

Loomis, Elias. *A treatise on meteorology.* New York, Harper & Brothers, 1868. 305 p.

Moret, J. C. Storms on the sea coast of Mississippi. *De Bow's Review* (New Orleans). 38–9 (n.s. 5–9), Sept. 1868, 791–796.

1869

The Great September Gale of 1869, in Providence and Vicinity. Providence, 1869.

1870

Vines, Benito. Huracanos del 7 y 19 Octubre 1870. *Habana Observatorio magnetico y meteorologico. Observationes 1869–70.* Habana, 1870.

1870 (1871)

Observatorio Magnetico y Meteorologico del Real Colegio de Belem. *Observaciones correspondientes al mes de Octubre de 1870, con las descripcion de los huracanes ocurridos en las isla endico mes.* Habana, Impr. Liberia Religiosa, 1871. 18 p.

BIBLIOGRAPHIES OF HURRICANE MATERIAL

1866

Andres Poey y Aguirre. *Bibliographie cyclonique: catalogue comprenant 1,008 ouvrages, brochures et ecrits qui ont paru jusqu'a ce jour sur les ouragans et les tempetes cycloniques.* (Cyclone bibliography: a catalogue of 1,008 works, brochures, and writings which have appeared up to now on hurricanes and cyclonic storms.) 2nd ed. corrected and enlarged. Paris, Paul Dupont, 1866. 96p. An extract of the *Annales Hydrographiques 1865.*

1891

Signal Corps, U. S. Army. *Bibliography of meteorology. Part 4. Storms.* Washington, 1891.

1935

Isaac Cline. *A century of progress in the study of cyclones.* New Orleans, 1935. 24–29.

1955

Ivan Ray Tannehill. *Hurricanes.* 9th ed. Princeton, 1955. 295–303.

1956

American Meteorological Society. Bibliography on Tropical Storms, Hurricanes and Typhoons. *Meteorological Abstracts and Bibliography.* 7–9, Sept 1956, 1115–1163.

——. Bibliography on Tropical Cyclone Effects. *Ibid.* 7–10, Oct 1956, 1270–1309.

——. Bibliography on Tropical Cyclone Theory. *Ibid.* 7–11, Nov 1956, 1405–1443.

1960

Gordon Dunn and Banner Miller. *Atlantic hurricanes.* Baton Rouge, La., 1960, 313–318.

A CHRONOLOGICAL INDEX OF EARLY AMERICAN HURRICANES: 1528-1870

THE FIRST TWO CENTURIES: 1501–1700

1528	Oct 2	Middle Florida	Apalachee Bay: W. Lowery. *The Spanish Settlements.* I, 198.
1559	Aug 29	Pensacola-Mobile	*The Luna Papers.* 2, 245.
1565	Sept 23	St. Augustine	Lowery. II, 167–68.
1566	Sept 26	Florida East Coast	*Ibid.*, 190.
1586	June 23–26	Roanoke Island	Hakluyt Society. *The Roanoke Voyages.* 2–104, 1, 302.
1587	Aug 31	Roanoke Island	*Ibid.*, 2, 532.
1591	Aug 26	Roanoke Island	*Ibid.*, 2, 608.
1635	Aug 25	S.E. New England	The Great Colonial Hurricane.
1638	Aug 13	Eastern New England	The Triple Storms of 1638.
1638	Oct 5	Eastern New England	The Triple Storms of 1638.
1667	Sept 6	North Carolina, Virginia, New York	The Dreadful Harry Cane of 1667.
1675	Sept 7–8	Southern New England	The New England Hurricane of 1675
1683	Aug 23	Virginia and southern New England	Hurricane and Flood of 1683.
1686	Sept 4–5	South Carolina	The Spanish Repulse Hurricane of 1686.
1693	Oct 29	Virginia and Delaware	The Great Storm of 1693 at Acomack.
1696	Oct 3–4	Florida East Coast	Jupiter Inlet: *Jonathan Dickinson's Journal*, 27.
1700	Sept 16	South Carolina	The Rising Sun Hurricane of 1700.

HATTERAS NORTH: 1701–1814

1703	Oct 18–19	Virginia to New England	The Hurricane of 1703.
1706	Oct 14–15	E. New York and W. New England	The Stormy Season of 1706.
1716	Oct 24–25	Eastern Mass.	The 1716 Hurricane.
1723	Aug 9	New York	Stokes: *Iconography Manhattan Island*, 584.
1723	Nov 10	Rhode Island	*Boston News Letter.* 7 Nov 1723.
1724	Aug 23	Virginia	The Great Gust of 1724 in Virginia.
1724	Aug 28–30	Virginia and Penna.	*Amer. Mercury* (Phila.). 20 Aug 1724.
1727	Sept 27	E. New England	A Great Rain and Horrible Wind.
1743	Nov 1–2	Entire Coast	Benjamin Franklin's Eclipse Hurricane.
1749	Oct 18–19	Entire Coast	The Coastal Hurricane of 1749.
1757	Sept 23–24	E. New England	*Boston Gaz.* 17 Oct 1757.
1761	Oct 23	S.E. New England	The Southeastern New England Hurricane of 1761.
1766	Sept 11	Virginia	Redfield Mss.
1769	Sept 7–8	Entire Coast	The September Hurricane of 1769.
1770	Oct 20	S.E. New England	The Late Season Storm of 1770.
1773	Aug 26	Virginia	*Va. Gaz.* (PD). 2 Sept 1773.
1775	Sept 2–3	Entire Coast	The Independence Hurricane of 1775.
1778	Aug 12–13	S.E. New England	The Ordering of Providence Hurricane of 1778.
1783	Oct 7–8	Entire Coast	The Stormy October of 1783.
1785	Sept 23–25	Entire Coast	The Equinoctial Storm.
1788	July 22–24	Virginia	George Washington's Hurricane.

1788	Aug 19	New Jersey to New Hampshire	Hurricane, Tornado, or Line Squall?
1795	Aug 2–3	North Carolina and Virginia	The Twin North Carolina Hurricanes of 1795.
1795	Aug 12–13	North Carolina and Virginia	The Twin North Carolina Hurricanes of 1795.
1803	Oct 2–3	Norfolk	The Norfolk Storm of 1803.
1804	Sept 11–12	S.E. New England	The Gale of 1804.
1804	Oct 9–10	New York and S.E. New England	The Snow Hurricane of October 1804.
1805	Oct 2–3	Maine	W. Bentley. *Diary.* III, 196.
1806	Aug 23–24	Entire Coast	The Great Coastal Hurricane of 1806.

HATTERAS SOUTH: 1701–1814

1713	Sept 16–17	South Carolina	The Carolina Hurricane of 1713.
1715	Aug ?	Florida East Coast	Douglas. *Hurricane.* 138.
1722	Sept 19–21 ?	South Carolina	The Great Hurricane in Louisiana.
1724	Aug 28	South Carolina	*Cal. State Papers.* 29 (1724–25), 214.
1728	Aug 13–14	South Carolina	The Carolina Hurricane of 1728.
1730	n.d.	South Carolina	Bartram. *Diary of a Journey.* 20
1744	Sept 8	Offshore	*S.C. Gaz.* 17 Sept 1744.
1749	Oct 17–18 ?	Offshore	*S.C. Gaz.* 30 Oct 1749.
1750	Aug 29	Cape Hatteras	*Col. Records North Carolina.* 4, 1031.
1752	Sept 15	South Carolina	The Great Hurricane of 1752.
1752	Sept 30– Oct 1	North Carolina	The Second Hurricane in 1752.
1757	Sept 22–24	Entire Coast	D. Warden. *Account of the U.S.* 1, 155.
1766	Sept 11–12	Offshore	*S.C. Gaz.* 20 Oct 1766.
1767	Oct 16–17	North Carolina	*S.C. Gaz.* 19 Oct 1767.
1769	Sept 7–8	North Carolina	The Great September Hurricane of 1769.
1769	Sept 28–29	Charleston	*S.C. Gaz.* 5 Oct 1769.
1769	Oct 29	Florida East Coast	Romans. *A Concise Nat. Hist.* 2.
1772	Sept 1	North Carolina	*Va. Gaz.* 25 Nov 1772.
1775	Sept 2–3	North Carolina	The Independence Hurricane of 1775.
1781	Aug 10	Charleston	The Occupation Storm at Charleston in 1781.
1783	Oct 7–8	Charleston	The Charleston Hurricane of 1783.
1786	Aug 29	North Carolina	*Hist. Rec. North Carolina, County.* 3, 62.
1787	Sept 19	Charleston	Redfield Mss.
1792	Oct 30–31	Charleston	The End of October Storm of 1792.
1795	Aug 2–3	North Carolina	The Twin North Carolina Hurricanes of 1795.
1795	Aug 12–13	North Carolina	The Twin North Carolina Hurricanes of 1795.
1797	Oct 19–20	Charleston	The Storm of October 1797.
1800	Oct 4–5	Charleston	The October Storm of 1800.
1803	Aug 31	North Carolina	The Carolina Hurricane of 1803.
1804	Sept 7–9	Georgia and Carolinas	The Great Gale of 1804.
1806	Aug 23–24	North Carolina	The Great Coastal Hurricane of 1806.
1806	Sept 2	North Carolina	N. Webster. *Diary.* 1, 567.
1806	Sept 15	Florida East Coast	*Courier* (Charleston), 2 Oct 1806.
1806	Sept 28	Cape Hatteras	D. Stick. *The Outer Banks of North Carolina.* 81.
1806	Oct 8–9	Charleston	*Courier*, 10 Oct 1806.
1810	Aug 12	North Carolina	J. Sprunt. *Early Hist. Cape Fear.*
1811	Sept 10	Charleston	The Hurricane and Tornado of 1811.
1811	Oct 6	N.E. Florida	*Public Ledger* (Norfolk), 16 Oct 1811.

1813	Aug 27–28	Charleston	The Dreadful Storm of 27–28 August 1813.
1813	Sept 16	Georgia	Savannah, 23 Sept, in *Richmond Enquirer*, 5 Oct 1813.

THE GULF COASTS: 1701–1814

1715	n.d.	Dauphin Island (near Mobile Bay)	D'Artaguiette (1722) quoted by Mereness, 25.
1722	Sept 22–23	Ala.-Miss.-La.	The Great Hurricane in Louisiana.
1733	n.d.	Mobile	C. Gayarre. *Hist. of La.* 457.
1736	n.d.	Pensacola	Tannehill, 244. No other ref. located.
1740	Sept 22	Mobile	The Twin Mobile Hurricanes of 1740.
1740	Sept 29	Mobile	The Twin Mobile Hurricanes of 1740.
1746	n.d.	Ala.-Miss.-La.	C. Gayarre. *Hist. of La.* 35
1752	Nov 3	Pensacola	*Fla. Hist. Quar.* 20–2, 163.
1758	n.d.	N.W. Florida	Dunn & Miller. *Atlantic Hurricanes.* 297.
1759	Sept ?	S.W. Florida	Douglas. *Hurricane.* 149.
1760	Aug 12	Pensacola.	*Fla. Hist. Quar.* 20–2, 168.
1764	Nov 16	Apalachee	Fla. Hist. Quar. 19– , 195.
1766	Sept 4	Galveston Bay ??	*Mon. Wea. Rev.* 49–8, Aug 1921, 455
1772	Aug 30–Sept 3	Fla.–La.	Bernard Romans' Gulf Coast Hurricane of 1772.
1778	Oct 7–10	Fla.–La.	The October 1778 Storm.
1779	Aug 18	Louisiana	Dunbar's Hurricane at New Orleans.
1780	Aug 24	Louisiana	The Louisiana Hurricane of 1780.
1780	Oct 18–21	Gulf of Mexico	Solano's Hurricane.
1793	Aug ?	Louisiana	Mss. Cabildo Records.
1794	Aug 31 ?	Louisiana	The Storm Tide of 1794 in Louisiana.
1806	Sept 17	New Orleans	*La. Gaz.* 19 Sept 1806.
1812	June 11–12	Louisiana	The June Storm of 1812.
1812	Aug 19	Louisiana	The Great Louisiana Hurricane of 1812.

HATTERAS NORTH: 1815–1870

1815	Sept 4–5	Norfolk	The North Carolina Hurricane of 1815.
1815	Sept 23	New England	The Great September Gale of 1815.
1817	Aug 8–9	Middle Atlantic States	*Nat. Int.* (Wash., D. C.), 14 Aug 1817
1821	Sept 3	Virginia to New England	The Norfolk and Long Island Hurricane of 1821.
1825	June 4–5	Entire Coast	The Early June Hurricane of 1825.
1827	Aug 26–27	Virginia to New England	The Great North Carolina Hurricane of 1827.
1830	Aug 17	Del-Mar-Va and Nantucket	Twin Hurricanes Feature the Season of 1830—II.
1830	Aug 26	S.E. New England	Twin Hurricanes Feature the Season of 1830—II.
1833	Oct 13	Entire Coast	*N.Y. Ship List*, 16 Oct 1833.
1834	Sept 4–5	Entire Coast	South Carolina Hurricane of 1834.
1839	Aug 29–30	Entire Coast	The Atlantic Coast Hurricane of Late August 1839.
1841	Oct 3–4	S.E. New England	The Memorable October Gale of 1841.
1846	Sept 9	New Jersey and Long Island	The Hatteras Inlets Hurricane of 1846.
1846	Oct 13–14	Entire Coast	The Great Hurricane of 1846—II.
1849	Oct 6–7	Entire Coast	The October Hurricane of 1849.
1850	July 18–19	Middle Atlantic and W. New England	The Triple Storms of 1850.
1850	Aug 24–25	Middle Atlantic and S. New England	The Triple Storms of 1850.

1850	Sept 7–8	Nantucket	The Triple Storms of 1850.
1853	Sept 7–8	Norfolk	Cape Verde and Cape Hatteras Hurricane (offshore).
1854	Sept 9–10	Entire Coast	The Coastal Hurricane of 1854.
1856	Aug 19–20	S. New England	The Charter Oak Storm of August 1856.
1856	Sept 1	Norfolk	The Southeastern States Hurricane of 1856.
1858	Sept 16	New England	The New England Tropical Storm of 1858.
1861	Sept 27	Entire Coast	"Equinoctial Storm." *N.Y. Trib.*, 1 Oct 1861.
1861	Nov 2–3	Entire Coast	The "Expedition" Hurricane of 1861.
1866	Oct 30	S.E. New England	"Violent SE winds, high tides" SI.
1867	Aug 2–3	S.E. New England	The Early August Offshore Hurricane of 1867.
1869	Sept 8	East New England	The September Gale of 1869 in Eastern New England.
1869	Oct 3–4	Entire Area	Saxby's Gale and the Great Northeastern Rainstorm and Flood of October 1869.

HATTERAS SOUTH: 1815–1870

1815	Sept 3	Eastern N. Carolina	The North Carolina Hurricane of 1815.
1815	Sept 22	Cape Hatteras	The Great September Gale in New England.
1816	Sept 23	Eastern North Carolina	Nine vessels wrecked at Ocracoke Bar. *Nat. Int.* 10 Oct 1816.
1817	Aug 7–8	Georgia-Carolinas	Severe at St. Marys, Amelia Island: *Savannah Republican*. 9 Sept 1817.
1820	Sept 10	Charleston to Cape Lookout	The Winyaw Hurricane of 1820.
1821	Sept 2–3	Eastern N. Carolina	The Norfolk and Long Island Hurricane.
1822	Sept 27–28	N. and S. Carolina	The Great Carolina Hurricane of 1822.
1824	Sept 14	St. Augustine to Beaufort, S.C.	The Georgia Coastal Hurricane of 1824.
1825	June 3–4	Entire Coast	The Early June Hurricane of 1825.
1827	Aug 23–25	Eastern N. Carolina	The Great North Carolina Coastal Hurriance of 1827.
1830	Aug 15–17	Entire Coast	The Twin Atlantic Coast Hurricanes of 1830.
1830	Aug 24–25	Cape Hatteras	The Twin Atlantic Coast Hurricanes of 1830.
1831	June 10	East Florida	St. Augustine: "a violent gale" SG.
1831	Aug 14–15	South Florida	The Barbados-Louisiana Hurricane of 1831.
1833	Aug 21	Eastern N. Carolina	Cape Hatteras: *N.Y. Ship List*. 28 Aug 1833.
1834	Sept 3–4	Georgetown and Wilmington	The Carolina Hurricane of 1834.
1835	Sept 12–15	N.W. Fla. to N. Carolina	Couvier (Charleston), 21 Sept 1835.
1835	Sept 15–18	South Florida	The South Florida Hurricane of 1835.
1836	Oct 10–11	Eastern N. Carolina	New Bern: "gale did great damage." *N.Y. Ship List*. 2 Nov 1836.
1837	Aug 1	N.E. Florida and Georgia	1837—No. 2—The Barbados-Florida Hurricane.
1837	Aug 6	North Florida and Georgia	1837–No. 3—The Antigua-Florida Hurricane.
1837	Aug 16–20	N.E. Florida to N. Carolina	1837—No. 4—Calypso.
1837	Aug 31–Sept 1	N.W. Florida to N. Carolina	1837—No. 6—The Apalachee Bay Storm.
1837	Sept 12–15	Cape Hatteras	1837—No. 7—The Bahamas Hurricane.
1837	Sept 25–26	East Florida	1837—No. 8—The East Florida Hurricane.
1837	Oct 8–10	Interior Georgia, S. Carolina, coastal N. Carolina	1837—No. 10—Racer's Hurricane.

1837	Oct 29	Off Cape Hatteras	1837—No. 11—The Cuba-Hatteras Hurricane.
1839	Aug 28–29	Charleston and north	The Atlantic Coast Hurricane of 1839.
1841	Aug 23–24	Entire Coast	Gales Miami to Carolinas: SG.
1841	Sept 25	Off Hatteras	Very severe: *N.Y. Ship List.* 6 Oct 1841.
1841	Oct 3	Off Hatteras	The Memorable October Gale of 1841.
1842	July 13–14	Eastern N. Carolina	The Destructive North Carolina Hurricane of 1842.
1842	Sept 4	South Florida	Antje's Hurricane.
1842	Oct 4–6	N.W. Florida to N. Carolina	The Gulf to Bermuda Hurricane.
1843	Sept 14	N.W. Florida to N. Carolina	Great Storm—Port Leon—1843.
1844	Sept 9	N.W. Florida to S. Carolina	Apalachicola—1844.
1844	Oct 4–7	South Florida	The Cuba to Bermuda Hurricane of 1844.
1846	Sept 7	Eastern N. Carolina	The Hatteras Inlets Hurricane of 1846.
1846	Oct 10–12	Entire area	The Great Hurricane of 1846.
1848	Sept 17–18	N.E. Florida	"Hardest blow for many years on St. Johns River," *Sav. Republican.* 20 Sept 1848.
1848	Oct 12–15	W. Florida to Cape Hatteras	The Tampa Bay Hurricane of 1848.
1850	Aug 24	Georgia-Carolinas	Severe Storm at Apalachicola in August 1850.
1851	Aug 24	Georgia-Carolinas	The Great Middle Florida Hurricane of August 1851.
1852	Aug 27	Georgia-Carolinas	The Great Mobile Hurricane of 1852.
1852	Oct 9–10	Georgia	The Middle Florida Storm of October 1852.
1853	Sept 7–8	Cape Hatteras	The Cape Verde-Cape Hatteras Hurricane of 1853.
1854	Sept 7–9	Jacksonville and north	The Great Carolina Hurricane of 1854.
1856	Aug 31–Sept 1	Georgia-Carolinas	The Southeast States Hurricane of 1856.
1857	Sept 12	Carolina Coasts	The *S.S. Central America* Disaster in 1857.
1861	Aug 14–16	South Florida	The Key West Hurricane of 1861.
1861	Nov 1	Cape Hatteras	The "Expedition" Hurricane of 1861.
1863	Sept 17–18	Charleston and Wilmington	Hurricane offshore: NA.
1865	Oct 22	South Florida	"Severe hurricane at Key West." SG.
1866	Oct 1–4	Carolina Coasts	The Great Nassau Hurricane of 1866.
1870	Oct 8–11	South Florida	Key West: SG.
1870	Oct 19–20	South Florida	Key West: SG.

THE GULF COASTS: 1815–1870

1818	Sept 12–14	N.E. Texas	The Lafitte Hurricane.
1819	July 27–28	La.–Ala.	The Bay St. Louis Hurricane of 1819.
1821	Sept 15–17	La.-Ala.-W. Fla.	The September Hurricane of 1821.
1822	July 7–8	Ala.-W. Fla.	The Early Tropical Storm of 1822.
1823	Sept 12–14	La.-Ala.	*La. Courier* (New Orleans). 15 Sept 1823.
1829	July 9	Rio Grande	*N.Y. Ship List.* 15 Aug, 16 Sept 1829
1831	Aug 14–18	Key West–La.	The Barbados to Louisiana Hurricane of 1831.
1831	Aug 28–29	W. La.	*La. Adv.* 31 Aug 1831.
1833	Sept 4–5	W. La.	Redfield: Record of Storms.
1834	Sept 6	W. La.	*La. Courier.* 18 Sept 1834.
1834	Sept 29	W. La.	*La. Sentinel.* 25 Oct 1834.
1835	Aug 15–18	Key West–Rio Grande	The Antigua-Gulf of Mexico-Rio Grande Hurricane.
1835	Sept 12–15	Apalachicola	*N.Y. Ship List,* 30 Sept 1835.
1837	Aug 3	Fla. West Coast	1837—No. 2—Barbados-Florida.
1837	Aug 6–7	St. Marks–Pensacola	1837—No. 3A—Middle and West Florida.
1837	Aug 30–31	Middle Florida	1837—No. 6—The Apalachee Bay Storm.

1837	Oct 3–7	Texas–La.–Fla.	1837—No. 10—Racer's Storm.
1841	Sept 14–15	Middle Fla.	The Late Gale at St. Joseph
1841	Oct 19	Key West	*N.Y. Ship List.* 17 Nov 1841.
1842	Sept 4–8	Key West–Rio Grande	Antje's Hurricane.
1842	Sept 18–19	Galveston	*Pensacola Gaz.* 1 Oct 1842.
1842	Oct 4–6	Middle Fla.	The Gulf-to-Berumda Hurricane of 1842.
1843	Sept 13–14	Middle Fla.	Great Storm—Port Leon—1843.
1844	Aug 4–6	Rio Grande	Matamoros—1844.
1844	Sept 8	Middle Fla.	Apalachicola—1844.
1844	Oct 4–5	Key West	The Cuban and Florida Straits Hurricane of 1844.
1845	Sept 20–21	Tampa Bay	Ft. Brooke: SG.
1846	Oct 11–12	Key West and Fla. Peninsula	The Great Hurricane of 1846.
1848	Sept 25	Fla. West Coast	The Tampa Bay Hurricane of 1848.
1849	Sept 13–14	Rio Grande	Gale at Brazos Santiago in 1849.
1850	Aug 23	Middle Fla.	Severe Storm at Apalachicola—August 1850.
1851	Aug 23–24	Middle Fla.	The Great Middle Florida Hurricane of August 1851.
1852	Aug 25	Ala.–W. Fla.	The Great Mobile Hurricane of 1852.
1852	Sept 11	Fla. West Coast	Ft. Brooke: SG.
1852	Oct 9	Apalachee Bay	The Middle Florida Storm of October 1852.
1854	Sept 17–18	Central Texas	The Matagorda Hurricane of 1854.
1855	Sept 15–16	Ala.-Miss.-La.	The Middle Gulf Shore Hurricane of 1855.
1856	Aug 10–11	Louisiana	The Last Island Disaster.
1856	Aug 30	Ala.-W. Fla.	The Southeastern States Hurricane of 1856.
1859	Sept 15	Mobile	*Mercury* (Mobile), 16 Sept 1859.
1860	Aug 11	Ala.-Miss.-La.	1860—Hurricane I.
1860	Sept 14–15	Ala.-Miss.-La.	1860—Hurricane II.
1860	Oct 2–3	Miss.-La.	1860—Hurricane III.
1861	Aug 14–16	Key West	The Key West Hurricane of 1861.
1865	Sept 13	West La.–N.E. Texas	The Sabine River–Lake Calcasieu Storm of 1865.
1865	Oct 22	Key West	"Severe hurricane," SG.
1867	Oct 3	Texas	The Galveston Hurricane of 1867.
1869	Aug 16	S. Texas	The Corpus Christi Area Storm of 1869.
1870	July 30	Mobile	The Mobile Storm of July 1870.
1870	Oct 8–11	Key West	The Twin Key West Hurricanes in 1870.
1870	Oct 19–20	Key West	The Twin Key West Hurricanes in 1870.

A GEOGRAPHICAL INDEX BY STATES OF EARLY AMERICAN HURRICANES: 1528-1870